T0358314

WAVEFRONTS AND RAYS AS CHARACTERISTICS AND ASYMPTOTICS

Third Edition

Other Related Titles from World Scientific

Waves and Rays in Elastic Continua
Fourth Edition
by Michael A. Slawinski
ISBN: 978-981-122-640-3

Waves and Rays in Seismology: Answers to Unasked Questions
Third Edition
by Michael A. Slawinski
ISBN: 978-981-122-643-4

On Foundations of Seismology: Bringing Idealizations Down to Earth
by James Robert Brown and Michael A. Slawinski
ISBN: 978-981-4329-49-1

WAVEFRONTS AND RAYS AS CHARACTERISTICS AND ASYMPTOTICS

Third Edition

Andrej Bóna
Michael A. Slawinski

World Scientific

NEW JERSEY • LONDON • SINGAPORE • BEIJING • SHANGHAI • HONG KONG • TAIPEI • CHENNAI • TOKYO

Published by

World Scientific Publishing Co. Pte. Ltd.
5 Toh Tuck Link, Singapore 596224
USA office: 27 Warren Street, Suite 401-402, Hackensack, NJ 07601
UK office: 57 Shelton Street, Covent Garden, London WC2H 9HE

Library of Congress Control Number: 2020945157

British Library Cataloguing-in-Publication Data
A catalogue record for this book is available from the British Library.

WAVEFRONTS AND RAYS AS CHARACTERISTICS AND ASYMPTOTICS
Third Edition

ISBN 978-981-122-646-5 (hardcover)
ISBN 978-981-122-647-2 (ebook for institutions)
ISBN 978-981-122-648-9 (ebook for individuals)

For any available supplementary material, please visit
https://www.worldscientific.com/worldscibooks/10.1142/11996#t=suppl

Desk Editor: Ng Kah Fee

Typeset by Stallion Press
Email: enquiries@stallionpress.com

Printed in Singapore

We dedicate this book to those who engage in the mathematical study of wave phenomena.

Preface to First Edition

[L]a physique ne nous donne pas seulement l'occasion de résoudre des problèmes; elle nous aide à en trouver les moyens, et cela de deux manières. Elles nous fait pressentir la solution; elle nous suggère des raisonnements.[1]

Henri Poincaré (1999)

Rays and wavefronts are intrinsic entities contained in hyperbolic differential equations. In particular, they are solutions of the characteristic equations associated with these differential equations. Quantitative descriptions of physics rely on differential equations. Hence, any method for examining wave phenomena must acknowledge the results of ray theory in relation to its description. A thorough understanding of characteristic equations in the context of ray theory illustrates why such equations can be utilized to describe the trajectory and speed of the propagation of disturbances in a medium. In addition, such an understanding can be useful in delineating limitations, thereby helping us to assess the accuracy of the description of wave phenomena that is obtained from the solutions of differential equations. In this book, we investigate the mathematics used to describe wavefronts and rays, which we exemplify in the context of elasticity and electromagnetism.

Several key aspects of characteristics of partial differential equations were introduced in the second half of the eighteenth century by Paul Charpit de Villecourt and Joseph-Louis Lagrange, and further elaborated at the beginning of the nineteenth century by Gaspard Monge, Augustin-Louis

[1] *[P]hysics not only provides us with the opportunity to solve problems, but also helps us to find the methods to get these solutions, this being achieved in two ways. Physics forecasts the solution; and it suggests the path of reasoning.*

Feynman (1967) states a similar view: *The physicist, who knows more or less how the answer is going to come out, can sort of guess part way, and so go along rather rapidly.*

Cauchy and William Rowan Hamilton.[2] In the first half of the twentieth century, the concept of characteristics was elegantly described in the context of wave phenomena by Jacques Hadamard (1903), (1932).

Asymptotic methods, commonly associated with ray theory, date back to the eighteenth century with works by James Stirling, Colin Maclaurin, Leonhard Euler and Adrien-Marie Legendre. Henri Poincaré (1886) introduced more rigour to the treatment of asymptotic series. Asymptotic methods had a renaissance in the middle of the twentieth century and many mathematical books on the subject emerged, among them[3] Courant and Hilbert (1989, Vol. II), Luneberg (1944), Erdélyi (1956), De Bruijn (1981), Jeffreys (1962), Copson (1965), followed by Olver (1974), Bleistein and Handelsman (1986). Asymptotic theory has been mathematically deepened and formulated in geometrical language, as shown in the book by Guillemin and Sternberg (1990).

Our book is intended for graduate students, researchers and teachers of physical sciences and engineering. We assume that the reader is familiar with linear algebra, differential and integral calculus, vector calculus and tensor analysis. This book consists of five chapters. Each chapter begins with *Preliminary remarks*, where we provide the motivation for concepts discussed therein, links to other chapters, and an outline of the chapter. *Closing remarks* emphasize the importance of the concepts discussed and show their relevance to other chapters. Each chapter ends with *Exercises* and their solutions, which provide steps omitted in the main text, and useful examples.

In Chapter 1, we introduce the concept of characteristics of the first-order linear partial differential equations. Also, we discuss systems of first-order equations, as illustrated by the Maxwell equations.

In Chapter 2, we study the second-order equations, and systems of such equations. We discuss the classification of hyperbolic, parabolic and elliptic equations in the context of characteristics. Using the fact that the characteristic equations are determined by the highest-order derivatives of differential equations, we consider equations whose lower-order terms are nonlinear.

[2]Readers interested in a mathematical description of the history of development of the method of characteristics might refer to Hamilton (1828–1837) and Kline (1974, Vol. II, pp. 531–538).

[3]Books are listed according to the date of their first publication; numbers in parentheses refer to the edition we used.

The characteristic equations of the second-order equations are first-order nonlinear equations, namely the Hamilton-Jacobi equations, also referred to as the eikonal equations. Consequently, in Chapter 3, we study the characteristics of first-order nonlinear equations.

In Chapter 4, we relate results obtained in Chapters 1, 2 and 3 to the concepts of asymptotic solutions and high-frequency approximations closely related to characteristics. These are the paths along which discontinuities of solutions can propagate; such discontinuities are exemplified by wavefronts. We complete the book with a chapter on caustics, which are a common occurrence in studies of wave propagation in the context of ray theory.

The five chapters are followed by appendices, which provide background to subjects covered in the main text. In Appendix A, we derive the integral theorems that allow us to formulate the elastodynamic equations derived in Appendix B and the Maxwell equations derived in Appendix C. In Appendix D we discuss Fourier analysis, which allows us to consider the frequency dependence of equations. Distributional solutions of differential equations, which allow us to consider discontinuities of solutions, are discussed in Appendix E.

Ray theory allows us to solve many problems associated with the propagation of disturbances. The physical meaning of characteristics as rays and wavefronts furnishes us with both a tractable approach to, and an intuitive understanding of, the solution. The link between physics and mathematics in the context of characteristics and asymptotics provides the underlying thread of our understanding of ray and wavefront description of the wave phenomena.

Changes from First Edition

Pure scientific research is either basic or based. Basic research is concerned with source theories, such as classical mechanics, and with basic empirical procedures, such as time measurements. The conceptual and empirical tools wrought by basic research are then worked out in based research, e.g., solid state physics, which in turn poses problems requiring further basic research, and in addition keeps applied research alive.

Mario Bunge (1967)

Changes in the second edition are a result of teaching from this text, readers's enquiries, and research into concepts discussed in the first edition. A summary of these changes, and the reason for their inclusion, are given below.

In Chapter 1—to familiarize the reader with various formulations of the Maxwell equations—Example 1.7 is refined and Exercises 1.13 and 1.14 are added.

The remainder of additions and changes is associated with the appendices, which can serve—on their own—as a course on mathematical methods. The presentation of all appendices is refined to enhance explanations therein and to ensure references among them and with the main body of the book. Also, Appendix F is added.

Appendix A.1.3 is added to broaden the scope of Appendix A and to provide a connection with Appendix F. Also in Appendix A, Example A.2 is added, which provides an insight into a relationship between the classical and distributional interpretations of differential equations; the latter is discussed in Appendix E, where in Example E.4 we revisit the issue raised in Example A.2, but invoking a more general formulation.

In Appendix B, Example B.1 is added. Together with other refinements in that appendix, it allows the reader to appreciate concepts underlying the elastodynamic equations, in particular, properties of the material and spatial descriptions in continuum mechanics, and such operators as the material time derivative.

In Appendix D, addition of Examples D.1 and D.2 emphasizes the Fourier series as an expansion in terms of orthonormal functions. Addition of Example D.3 and Figure D.1 allows the reader to gain an insight into the Fourier series of functions that are periodic but not differentiable. In Example D.4, we illustrate the Fourier transform as a method for solving partial differential equations.

In Appendix E, Figure E.1 is refined to illustrate—in a quantitative manner—the effect of higher-order terms used to model the Dirac delta. Three examples are added: the aforementioned Example E.4, and Examples E.2 and E.5, in which, respectively, we illustrate the concept of a distributional derivative of a nondifferentiable function and the Fourier transform of a distribution.

The addition of Appendix F, which is devoted to Green's functions, is motivated by their importance in studying differential equations, in general, and the wave and elastodynamic equations, in particular. The material within this appendix allows the reader to gain an insight into issues addressed in other appendices and in the main text. Also, it increases the scope of the appendices as the text for a course on mathematical methods.

Bibliographic references are added; several among them refer to books and articles published since the first edition. Furthermore, in Chapter 1, expressions of Example 1.3 are corrected; in Chapter 4, expressions of Section 4.1 are corrected; in Chapter 5, several expressions of Section 5.3 are corrected by including Fresnel's modification of Huygens's principle required for an examination of caustics. Also, to enhance the coherency of the book, index entries and *List of symbols* are updated.

Changes from Second Edition

[...] we can see that asymptotic reasoning plays a major role in our understanding of how various theories "fit" together to describe and explain the workings of the world. [...] One can learn much more about the nature of various theories by studying these asymptotic limits than by investigating reductive relations according to standard philosophical models.

Robert W. Batterman (2002)

Changes in the third edition continue to be a result of teaching from this text, readers's enquiries, and research into concepts discussed in the second edition. A summary of these changes, and the reason for their inclusion, are given below.

A clarifying comment, which relates the directional and partial derivatives, is given in Remark 1.1. An insightful comment is introduced by Remark 1.2, wherein we distinguish—among the Maxwell equations—between those referring to intrinsic field properties and those referring to evolutions of these fields.

Most additions of the third edition appear in *Appendices*, which contain the material that is a prerequisite for the main part of the book. These appendices constitute about a third of the entire book, and can serve—on their own—as a course on mathematical methods.

Significant additions appear in Appendix A. Therein, Examples A.1, A.2 and A.3 are expanded, including a new figure, Figure A.3, as well as new remarks, Remarks A.2 and A.3. Also, there is a new section, namely, Appendix A.2.3, where we formulate Corollary A.2, in terms of the Curl Theorem; this corollary is commonly referred to as Green's Theorem. Appendix A.2.3 also contains Example A.4 and Remark A.4. *Closing remarks* are extended by comments on the Gradient Theorem, stated in expression (A.20).

Appendix C.1 is expanded by several explanatory descriptions, including explicit relations between the integral and differential equations, as stated in Appendix C.1.5, in the context of Ampère's law,

Another significant addition appears in Appendix E.2, where the section on *Operations on distributions* is now split into two subsections: *Derivative of distributions* and *Fourier transform of distributions.* Apart from clarifying several details and correcting a few typographical mistakes of x and t, this addition strengthens the derivation of formulæ for operations on distributions.

The book contains many footnotes, which play an important—but secondary—part; herein, *secondary* means that the continuity of initial reading need not be interrupted by referring to a footnote, which can be consulted upon deeper study of the material. Since this book is the second volume of the series that contains Slawinski (2015), Slawinski (2016) and Brown and Slawinski (2017), the footnotes marked by *see also* direct the reader to specific locations with further details in these three volumes. Other footnotes refer to pertinent work whose reference might broaden the reader's perspective and facilitate the understanding of material. Continuous work with, and lecturing from, this book results in new index entries, which facilitate its use as a reference for a specific topic.

Adjustments of the LaTeX template enhance the coherency of references within the document. The figure in *Afterword* exhibits its proper reference number, and so do figures, theorems, remarks, examples, definitions and footnotes in *Appendices.*

Contents

List of Figures

Acknowledgments

This book originated as graduate-course notes, whose quality benefitted from collaborations with Len Bos, Nelu Bucataru and Yves Rogister.

The first edition benefitted from discussions with Vassily Babich, Norm Bleistein and Michael Rochester, from editorial revisions of Cathy Beveridge, graphic support of Elena Patarini, and proofreading of David Dalton.

The second edition benefited from discussions with Misha Kochetov and Michael Rochester, from proofreading and scientific editing of David Dalton, as well as from the graphic support of Elena Patarini.

The third edition benefited from proofreading and scientific editing of Md Abu Sayed, Filip Adamus, David Dalton and Theodore Stanoev, as well as from the graphic support of Elena Patarini. Kah-Fee Ng ensured the adjustments of final details prior to printing.

Chapter 1

Characteristic equations of first-order linear partial differential equations

To construct solutions of partial differential equations by means of power series, a very restrictive assumption is required—the data must be analytic. This excludes many relevant problems. However, for partial differential equations of first order a more direct and complete theory of integration can be developed under rather weak assumptions of continuity and differentiability. [...] *The key to the theory is the concept of characteristics, which will play a decisive part also in higher order problems.*

Courant and Hilbert (1989, Vol. II)

Preliminary remarks

In this chapter we introduce the concept of characteristics by studying first-order linear partial differential equations. These equations are common in physics, and their characteristics are endowed with physical meaning. More complicated equations and their characteristics are discussed in subsequent chapters.

The characteristic equations define both the directions of differentiation given by the differential equation and the directions along which we are not free to specify the side conditions. In other words, along these characteristic directions the behaviour of the solution is restricted more by the equation itself than in other directions. This is the defining property of the characteristics for second-order linear partial differential equations.

We begin this chapter by a motivational example in which we introduce the concept of characteristics. Subsequently, we use the directional derivative to find characteristics and define them rigorously. Then, we relate the characteristics to the compatibility between the differential equation and its side condition. We complete the discussions of this chapter by considering systems of first-order linear equations.

Readers might find it useful to study this chapter together with Appendix C, where we formulate the Maxwell equations.

1.1 Motivational example

1.1.1 *General and particular solutions*

There are physical quantities that do not change along a particular direction. An example of such a quantity is a plane wave: a wave whose amplitude is constant for all directions in a plane.

To describe such a quantity, let us consider an equation whose solution remains constant in a particular direction. We write

$$\frac{\partial f(x_1, x_2)}{\partial x_1} = 0\,, \tag{1.1}$$

which implies that f does not change along the x_1-axis. The general solution of equation (1.1) is

$$f(x_1, x_2) = g(x_2)\,; \tag{1.2}$$

in other words, g is an arbitrary function of x_2.

To obtain a specific solution, which is tantamount to finding g explicitly, we impose conditions that the solution must satisfy. We refer to these conditions as side conditions. This terminology avoids a common distinction between the initial and boundary conditions, which can be misleading for cases of differential equations whose variables are not associated with time or position. However, the distinction between the initial and boundary condition is useful for certain types of differential equations, as discussed in Section 2.1.3. We exemplify the use of a side condition in the following way.

We can specify values of f along curve γ that is given by, say, $x_2 = 2x_1$—illustrated in Figure 1.1—by, say,

$$f(x_1, x_2)|_\gamma = x_1^2; \tag{1.3}$$

in other words, by requiring that along this curve the solution exhibits particular values. Writing this curve as $x_1 = x_2/2$, we express f in terms of x_2 alone; we write side condition (1.3) as

$$f\left(\frac{x_2}{2}, x_2\right) = \left(\frac{x_2}{2}\right)^2.$$

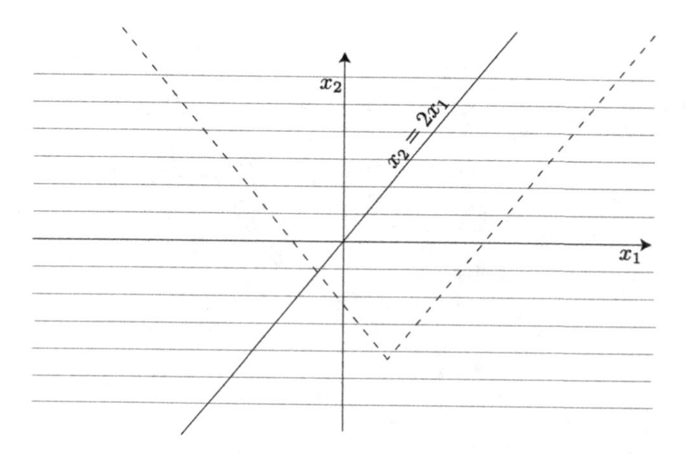

Fig. 1.1 Differential equation (1.1) specifies the rate of change of a solution along the curves that are parallel to the x_1-axis. These curves are called characteristics. Side condition (1.3) is specified along $x_2 = 2x_1$. The dashed line exemplifies a curve that crosses the same characteristics twice, and hence cannot be used to specify such a condition.

Since f does not change along the x_1-axis, the form of solution (1.2) that satisfies both equations (1.1) and (1.3) is $g(x_2) = (x_2/2)^2$. We can verify directly that $\partial (x_2/2)^2 / \partial x_1 = 0$ and, along $x_2 = 2x_1$ the solution is $(x_2/2)^2 = (2x_1/2)^2 = x_1^2$, as required. This solution is shown in Figure 1.2. Its graph consists of a surface in the three-dimensional space spanned by x_1, x_2, y , where the range of f is along the y-coordinate. The level of this surface is constant along the x_1-axis. As we show in Section 1.1.2, the side condition cannot intersect this surface along multiple curves.

1.1.2 *Characteristics*

Using the concept of a side condition along a given line, we can show that there are lines along which we cannot specify $f(x_1, x_2)$ arbitrarily. To this end, let us require that

$$f(x_1, x_2)\big|_\gamma = x_1^2 \,, \tag{1.4}$$

where γ is the line given by $x_2 = C$, with C denoting a constant. Substituting $x_2 = C$ into solution (1.2), we obtain

$$f(x_1, x_2) = g(C) \,,$$

which is a constant; this contradicts expression (1.4), and thus cannot be a solution. We conclude that along γ we cannot set the side conditions

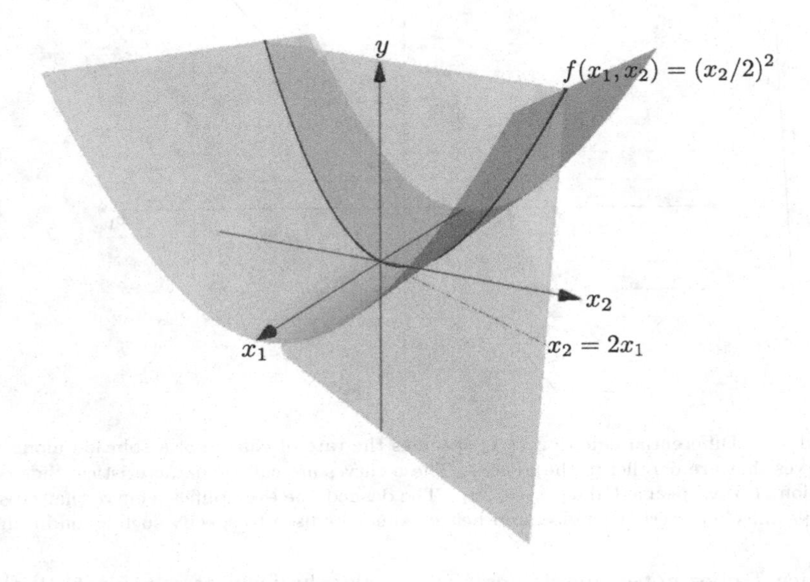

Fig. 1.2 The parabolic surface represents the graph of the solution of equation (1.1) with side condition (1.3) shown by the intersection of the surface and the x_2y-plane. The value of the solution is constant along the characteristics, $x_2 = \text{const}$.

freely since the behaviour of the solution along $x_2 = C$ is constrained by the differential equation itself. These curves are the characteristics of equation (1.1).

Similarly, we cannot choose freely side conditions if the curve along which we specify them intersects the same characteristic twice, as illustrated in Figure 1.1. The values at these intersections are related to one another, since the value at one intersection is propagated to the other intersection along the characteristics.

1.2 Directional derivatives

The approach we use to find the characteristics of a first-order partial differential equation with constant coefficients can be used to investigate equations with variable coefficients. We consider the general form of such an equation in n variables:

$$
\begin{aligned}
A_1(x_1, x_2, \ldots, x_n) \frac{\partial f}{\partial x_1} &+ A_2(x_1, x_2, \ldots, x_n) \frac{\partial f}{\partial x_2} \\
&+ \cdots + A_n(x_1, x_2, \ldots, x_n) \frac{\partial f}{\partial x_n} \\
= B(x_1, x_2, \ldots, x_n) f &+ C(x_1, x_2, \ldots, x_n) \qquad (1.5)
\end{aligned}
$$

This equation states that the solution, f, changes along direction $[A_1(x), A_2(x), \ldots, A_n(x)]$ at the rate of $B(x) f + C(x)$, as discussed in Exercise 1.1. In other words, the behaviour of solutions is prescribed along the curves whose tangent vectors are $A(x)$, and thus satisfy the following system of ordinary differential equations.

$$
\begin{aligned}
\frac{\mathrm{d}x_1}{\mathrm{d}s} &= A_1(x_1(s), x_2(s), \ldots, x_n(s)), \\
\frac{\mathrm{d}x_2}{\mathrm{d}s} &= A_2(x_1(s), x_2(s), \ldots, x_n(s)), \\
&\vdots \\
\frac{\mathrm{d}x_n}{\mathrm{d}s} &= A_n(x_1(s), x_2(s), \ldots, x_n(s)).
\end{aligned}
\qquad (1.6)
$$

Curves satisfying these equations are the characteristic curves of equation (1.5); also, we refer to these curves as characteristics. The equations themselves are the characteristic equations. Thus, the original equation can be expressed as a derivative along the characteristics:

$$
D_A f := A \cdot \nabla f \equiv \frac{\mathrm{d}}{\mathrm{d}s} f(x(s)) = B(x(s)) f(x(s)) + C(x(s)); \qquad (1.7)
$$

symbol x without a subscript stands for $x_1, x_2, \ldots, x_n$. This is a restatement of the fact that equation (1.5) specifies the behaviour of the solutions only along the characteristics. The behaviour of the solutions in directions transverse to the characteristics must be given by the side conditions.

If we combine equations (1.6) and (1.7), we obtain a system of equations describing a vector field in the $x_1 \ldots x_n f$-space—the $(n+1)$-dimensional space spanned by the independent variables, x_i, and the functional values, f. The solutions of this system are curves whose projections onto the $x_1 \ldots x_n$-space are characteristics. For this reason, we refer to these curves as lifted characteristics. If we choose the lifted characteristics that pass through points on the graph of the side condition, we obtain a surface, which is the solution to the differential equation and its side condition.

Expressing a differential equation using a directional derivative is equivalent to a change of variables. In the new variables the differential equation has a simpler form, as can be seen by equation (1.7). This is not the only change of variables that simplifies the equation, as discussed in Exercise 1.5.

Expressing partial differential equations as ordinary differential equations along the characteristics, as exemplified by equation (1.7), is a general property of the first-order partial differential equations. This property plays an important role in our studies, and it results in the Hamilton equations, which are discussed in Chapter 3.

Also, examining equation (1.5) and its characteristic equations (1.6), we see that the characteristics do not depend on the right-hand side; in other words, they depend only on the terms containing derivatives. The right-hand side governs the rate of change of a solution along the characteristics. The fact that characteristics do not depend on the right-hand side is used in Section 1.6.

Characteristic curves are not only useful to discuss the side conditions of a differential equation; they can be used also to find its solution, as illustrated in the following example.

Example 1.1. Herein, we construct the characteristics and use them to find solutions of a first-order linear partial differential equation with variable coefficients,

$$\frac{\partial f\left(x_1, x_2\right)}{\partial x_1}+x_2 \frac{\partial f\left(x_1, x_2\right)}{\partial x_2}=0 . \tag{1.8}$$

Since we can write equation (1.8) as

$$\left[1, x_2\right] \cdot\left[\frac{\partial f}{\partial x_1}, \frac{\partial f}{\partial x_2}\right]=0,$$

we recognize that it is the directional derivative of f in the direction $[1, x_2]$. Following the definition of the directional derivative given by expression (1.65) in Exercise 1.1, we write equation (1.8) as

$$\mathrm{D}_{\left[1, x_2\right]} f\left(x_1, x_2\right)=0, \tag{1.9}$$

which means that f does not change along the curve whose tangent vector is $[1, x_2]$; once we fix the value of f at a single point on this curve, the value of f is determined for all points along the curve. This curve represents a characteristic of equation (1.8).

We can rewrite characteristic equations (1.6) using the slope of the tangent to the characteristic as

$$\frac{\mathrm{d} x_2}{\mathrm{~d} x_1}=\frac{x_2}{1}=x_2 . \tag{1.10}$$

Equation (1.10) is an ordinary differential equation whose solution is the family of curves given by

$$x_2(x_1) = C \exp x_1 , \tag{1.11}$$

with C being a constant, which is the x_2-intercept of a given curve, as shown in Figure 1.3.

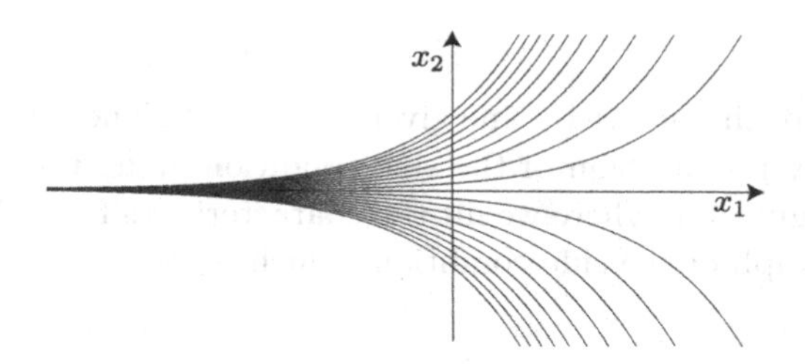

Fig. 1.3 Curves represent characteristics of equation (1.8), which are given by exponentials (1.11). These curves do not intersect, which is a behaviour of characteristics of linear differential equations; they are integral curves of the vector field defining the differential equation.

Solving expression (1.11) for C, we obtain $C = x_2 \exp(-x_1)$. Using the fact that f does not change along the characteristics and each characteristic is specified by the value of C, we write the general solution of equation (1.8) as

$$f(x_1, x_2) = g(x_2 \exp(-x_1)) , \tag{1.12}$$

where g is a differentiable function of a single argument. To obtain a particular solution of equation (1.8), we can specify the value of g at a single point of each characteristic.

Formally, equation (1.8) requires f to be differentiable. However, a non-differentiable function g accommodates solution (1.12), if we interpret the differential equation as the directional derivative (1.9). Let us specify the value of f at $x_1 = 0$ for each characteristic. This means that we specify the value of g along the x_2-axis. For instance, we let $f(0, x_2) = |x_2|$, which is not a differentiable function. Since for each point along the x_2-axis, $x_2 = C$, we set $f = |C|$, for each characteristic. We return to the general solution of equation (1.8). At $x_1 = 0$, solution (1.12) is

$$f(0, x_2) = g(x_2) .$$

Since we let $f = |x_2|$ at $x_1 = 0$, this implies that $g(x_2) = |x_2|$; g stands for a rule according to which we take the absolute value of the argument. Following solution (1.12), we can write this rule for all values of x_1 to get

$$f(x_1, x_2) = |x_2 \exp(-x_1)| = |x_2| \exp(-x_1) .$$

This is a particular solution of equation (1.8) with the side condition given by

$$f(x_1, x_2)|_\gamma = |x_2| , \tag{1.13}$$

where γ is a noncharacteristic line given by $x_1 = 0$. The graph of the solution formed by propagation of the side condition along the characteristics is shown in Figure 1.4, where we lift the characteristics from the x_1x_2-plane to cross the graph of the side condition, which is $|x_2|$.

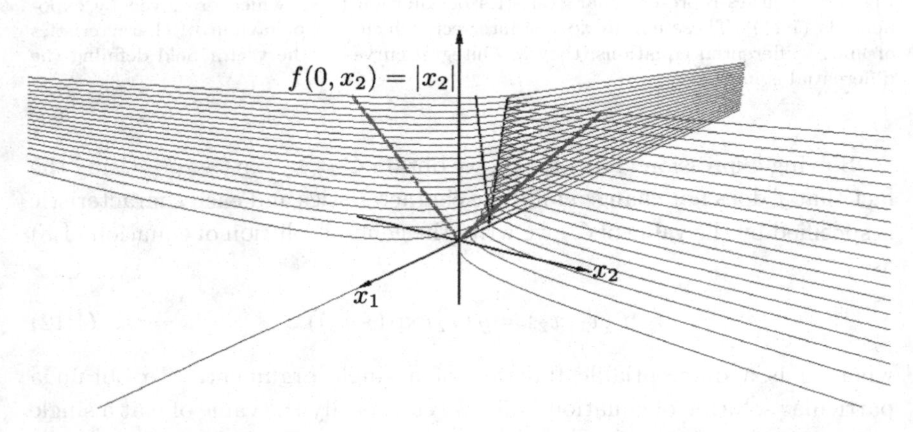

Fig. 1.4 Propagation of the side condition along the characteristics for equation (1.8). The wide v-shape curve at the origin determines the side condition given at $x_1 = 0$ and the side condition propagated along the characteristics is shown as a narrow v-shape curve away from the origin. The horizontal axes are the x_1-axis and the x_2-axis, and the vertical axis gives the value of $f(x_1, x_2)$.

Let us try to specify the value of f at $x_2 = 0$. In other words, let us specify the value of g for each point along the x_1-axis. For instance, we let $f(x_1, 0) = x_1^2$. We return to the general solution of equation (1.8). At $x_2 = 0$, solution (1.12) is

$$f(x_1, 0) = g(0) .$$

This means that once we set the value at $x_2 = 0$, it remains the same for all points along the x_1-axis. Since $g(0)$ is a constant, we cannot set it to be equal to x_1^2, which represents a function whose value changes with x_1.

To understand this result, let us return to the characteristics of equation (1.8) by recalling expression (1.11), namely, $x_2(x_1) = C \exp x_1$. We realize that $x_2 = 0$ is one of the characteristics; it corresponds to $C = 0$. Since equation (1.8) requires f to be constant along the characteristics, we cannot specify f to be x_1^2 along $x_2 = 0$.

We note that the solution prescribed along the characteristics need not be constant, as illustrated in Exercise 1.2.

Remark 1.1. Partial derivatives are directional derivatives along the coordinate axes. For instance, for the Cartesian coordinates in $\mathbb{R}^3$, $\partial/\partial x_1 = [1,0,0] \cdot \nabla$, $\partial/\partial x_2 = [0,1,0] \cdot \nabla$ and $\partial/\partial x_3 = [0,0,1] \cdot \nabla$.

Equation (1.8) together with its side conditions (1.13) is referred to as a Cauchy problem. In general, a Cauchy problem consists of finding the solution of a differential equation that satisfies also the side conditions given along a hypersurface and consisting of the values of the derivatives of the order lower than the order of the differential equation. These side conditions are called the Cauchy data.

1.3 Nonlinear digression: Inviscid Burgers's equation

The method of finding the solutions of linear differential equations described in Section 1.2 can be applied to particular types of nonlinear first-order equations. In this section we look at an example of such equations, the inviscid Burgers's equation,

$$\frac{\partial f}{\partial t} + f \frac{\partial f}{\partial x} = 0, \tag{1.14}$$

which, with t and x standing for time and space, describes shock-wave formation or mass transport. This equation can be solved by finding directional derivatives. We write equation (1.14) as

$$D_{[1, f(x,t)]} f = [1, f(x,t)] \cdot \left[\frac{\partial f}{\partial t}, \frac{\partial f}{\partial x}\right] = 0.$$

Curves $[x(s), t(s)]$ that are tangent to the directions of derivatives satisfy

$$\frac{\mathrm{d}t}{\mathrm{d}s} = 1 \tag{1.15}$$

$$\frac{\mathrm{d}x}{\mathrm{d}s} = f(x(s), t(s)). \tag{1.16}$$

The solution, f, along these curves is fully described by the differential equation; since the right-hand side is zero, it follows that f is constant along these curves:

$$\frac{\mathrm{d}}{\mathrm{d}s} f\left(x\left(s\right), t\left(s\right)\right) = 0.$$

Thus, curves that satisfy equations (1.15) and (1.16) are the characteristics of equation (1.14). Since f is constant along $x(s)$, the slope of the characteristics, $\mathrm{d}x/\mathrm{d}t = \mathrm{d}x/\mathrm{d}s = f$, is constant for each curve. It follows that the characteristics are straight lines whose slope can change depending on the side conditions. The dependence of the characteristics on the side condition is due to the nonlinearity of the equation, as discussed in Chapter 3; for a linear equation, the characteristics depend on the equation only. To illustrate this dependence, we look at the following example.

Example 1.2. Consider equation (1.14) with the following side condition:

$$f(x, 0) = \sin x. \tag{1.17}$$

The characteristics are shown in Figure 1.5; they are straight lines with slopes given by $\mathrm{d}t/\mathrm{d}x = \sin x$.

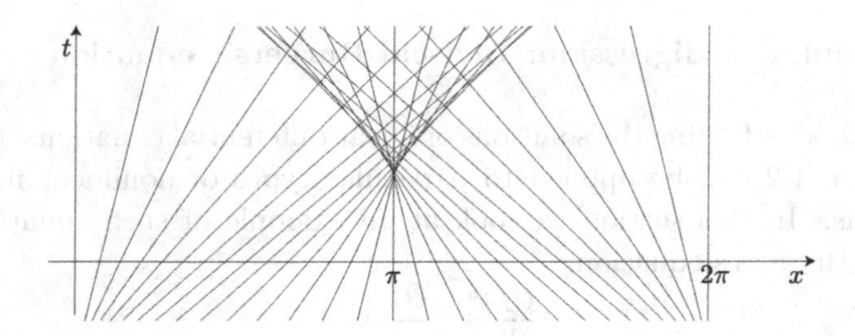

Fig. 1.5 Lines represent the characteristics of the inviscid Burgers's equation corresponding to side condition (1.17). In contrast to linear equations, characteristics depend on side conditions; also in contrast to linear equations, they cross each other, which results in multivalued solutions.

Since the characteristics cross, the corresponding solution is a multivalued function obtained by propagating the side condition along the characteristics, as illustrated in Figure 1.6.

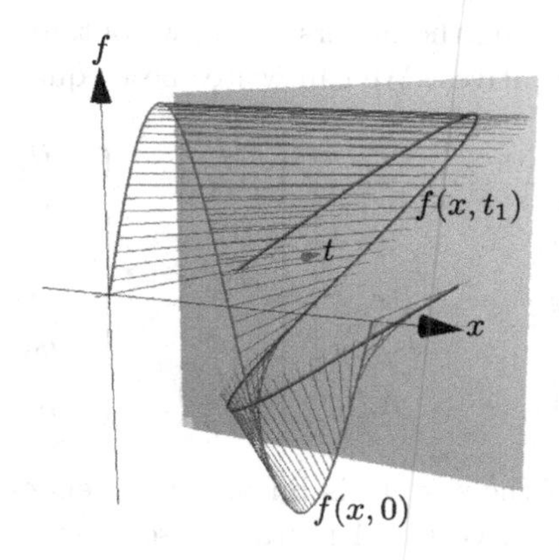

Fig. 1.6 Propagation of the side condition along the characteristics. The sine curve determines the side condition given at $t = 0$. The vertical plane, shown in light gray, contains the propagated solution, shown in black, along the lifted characteristics, shown by straight lines. The graph shows the multivalued solution. The horizontal axes are x and t, and the vertical axis gives the functional value $f(x, t)$.

1.4 Taylor series of solutions

A differential equation can be viewed as a relation among derivatives of its solution. One can use this relation to determine the derivatives of the solution at a point to construct the Taylor series of the solution. In this section, we use the first-order linear equations to explore this approach. An approach analogous to the one presented in this section is used on page 70 to find solutions of higher-order differential equations.

Consider the Cauchy problem for a general first-order linear partial differential equation,

$$\sum_{i=1}^{n} A_i(x) \frac{\partial f}{\partial x_i} = B(x) f + C(x).$$

The Cauchy data along the hypersurface are given by $x_1, \ldots, x_n$ as functions of parameters $s_1, \ldots, s_{n-1}$, which we denote—with a common abuse of notation that does not distinguish between a function and its value—as $x = x(s_1, \ldots, s_{n-1})$:

$$f(x(s_1, \ldots, s_{n-1})) = f_0(s_1, \ldots, s_{n-1}),$$

where x stands for $x_1, \ldots, x_n$.

To find the first derivatives along the hypersurface, we differentiate the side condition with respect to the parameters s to obtain $n-1$ linear equations for the n derivatives, $\partial f/\partial x_i$. By evaluating the original differential equation at a point along the hypersurface, we obtain another linear equation for the first derivatives. We can write these equations as

$$\begin{bmatrix} \frac{\partial x_1}{\partial s_1} & \frac{\partial x_2}{\partial s_1} & \cdots & \frac{\partial x_n}{\partial s_1} \\ \vdots & \vdots & \ddots & \vdots \\ \frac{\partial x_1}{\partial s_{n-1}} & \frac{\partial x_2}{\partial s_{n-1}} & \cdots & \frac{\partial x_n}{\partial s_{n-1}} \\ A_1 & A_2 & \cdots & A_n \end{bmatrix} \begin{bmatrix} \frac{\partial f}{\partial x_1} \\ \frac{\partial f}{\partial x_2} \\ \vdots \\ \frac{\partial f}{\partial x_n} \end{bmatrix} = \begin{bmatrix} \frac{\partial f_0}{\partial s_1} \\ \vdots \\ \frac{\partial f_0}{\partial s_{n-1}} \\ B f_0 + C \end{bmatrix}. \tag{1.18}$$

The last equation of the system is linearly independent of the first $n-1$ equations if and only if vector A is transverse to the hypersurface. If the vector were tangent to the hypersurface, we would not be able to solve the system since the last equation would be a linear combination of the other equations. Since it is transverse, we can find the first derivatives of the solution at any point of the hypersurface. Subsequently, we can differentiate the first derivatives with respect to s to obtain linear expressions for all the second derivatives except those in a transverse direction to the hypersurface. To complete this system of equations, we consider the derivative of the original differential equation in the transverse direction. We can proceed in a similar manner to obtain all derivatives of the solution at x_0. Having found the derivatives, we construct the Taylor series for function f of n variables:

$$\sum_{\alpha_1=0}^{\infty} \cdots \sum_{\alpha_n=0}^{\infty} \frac{1}{\alpha_1!\alpha_2!\cdots\alpha_n!} \frac{\partial^{\alpha_1+\alpha_2+\cdots+\alpha_n} f(x_0)}{\partial x_1^{\alpha_1} \partial x_2^{\alpha_2} \cdots \partial x_n^{\alpha_n}} (x_1 - x_{0_1})^{\alpha_1} \cdot (x_2 - x_{0_2})^{\alpha_2} \cdots (x_n - x_{0_n})^{\alpha_n},$$

which can be written using the multiindex notation as

$$\sum_{|\alpha|\geqslant 0} \frac{1}{\alpha!} \frac{\partial^{|\alpha|} f(x_0)}{\partial x^{\alpha}} (x - x_0)^{\alpha},$$

where α is a multiindex, $\alpha = (\alpha_1, \alpha_2, \ldots, \alpha_n)$, $|\alpha| = \alpha_1 + \alpha_2 + \cdots + \alpha_n$, $\alpha! = \alpha_1!\alpha_2!\cdots\alpha_n!$ and $x^{\alpha} = x_1^{\alpha_1} x_2^{\alpha_2} \cdots x_n^{\alpha_n}$; to become familiar with this notation, the reader may refer to Exercises 4.3, 4.4 and 4.5.

The convergence of this series to the solution of the Cauchy problem is guaranteed if the functions used in the differential equation and the side conditions are analytic in a neighbourhood of x_0, as stated by the Cauchy-

Kovalevskaya Theorem.[1] In the following example we find the solution of an equation using the convergent Taylor series.

Example 1.3. Let us find the solution of the Cauchy problem consisting of equation (1.68), discussed below in Exercises 1.2–1.5, namely,

$$x_2 \frac{\partial f(x_1, x_2)}{\partial x_1} - x_1 \frac{\partial f(x_1, x_2)}{\partial x_2} = x_2 , \tag{1.19}$$

together with the Cauchy data given by

$$f(0, x_2) \equiv f_0 = x_2^2 . \tag{1.20}$$

A straightforward parametrization of the Cauchy data is obtained by choosing s to be x_2. Thus the hypersurface along which the Cauchy data is defined is given by curve $[x_1(s), x_2(s)] = [0, s]$, which allows us to write the Cauchy data as $f_0(s) = s^2$. Let us choose point x_0 along the x_2-axis, where the Cauchy data is defined; say, $x_0 = [0, 1]$. We want to find the derivatives of the solution at this point to be included in the initial terms of the Taylor series at the same point, which are

$$f(0,1) + \frac{\partial f}{\partial x_1}(0,1)\, x_1 + \frac{\partial f}{\partial x_2}(0,1)\,(x_2 - 1) \tag{1.21}$$
$$+ \frac{1}{2}\frac{\partial^2 f}{\partial x_1^2}(0,1)\, x_1^2 + \frac{\partial^2 f}{\partial x_1 \partial x_2}(0,1)\, x_1 (x_2 - 1) + \frac{1}{2}\frac{\partial^2 f}{\partial x_2^2}(0,1)\,(x_2 - 1)^2 + \cdots .$$

Let us proceed to find terms of this series.

From the Cauchy data in expression (1.20), we see that $f(0,1) = 1$. To find the first-order derivatives, we need to use both the differential equation and the Cauchy data, which we can differentiate only along s, namely,

$$\frac{\mathrm{d}f_0}{\mathrm{d}s} = \frac{\partial f}{\partial x_1}\frac{\mathrm{d}x_1}{\mathrm{d}s} + \frac{\partial f}{\partial x_2}\frac{\mathrm{d}x_2}{\mathrm{d}s} .$$

Since the Cauchy data are defined along a one-dimensional hypersurface, the derivative along this hypersurface is a total derivative.

According to system (1.18), the derivative of the Cauchy data along s together with the differential equation can be written as

$$\begin{bmatrix} \frac{\mathrm{d}x_1}{\mathrm{d}s} & \frac{\mathrm{d}x_2}{\mathrm{d}s} \\ A_1 & A_2 \end{bmatrix} \begin{bmatrix} \frac{\partial f}{\partial x_1} \\ \frac{\partial f}{\partial x_2} \end{bmatrix} = \begin{bmatrix} \frac{\mathrm{d}f_0}{\mathrm{d}s} \\ Bf_0 + C \end{bmatrix} . \tag{1.22}$$

[1]The proof of this theorem can be found, for example, in Courant and Hilbert (1989, Volume 2, pp. 48-54). Also, there exists a stronger version of this theorem—Holmgren's Theorem—that does not require the analyticity of the side conditions. For more details, the reader might refer to Courant and Hilbert (1989, Volume 2, pp. 237-239).

In view of the parametrization: $\mathrm{d}x_1/\mathrm{d}s = \mathrm{d}x_1/\mathrm{d}x_2 = 0$ and $\mathrm{d}x_2/\mathrm{d}s = \mathrm{d}x_2/\mathrm{d}x_2 = 1$; in view of side condition (1.20): $\mathrm{d}f_0/\mathrm{d}s = \mathrm{d}x_2^2/\mathrm{d}x_2 = 2x_2$; in view of differential equation (1.19): $A_1 = x_2$, $A_2 = -x_1$, $B = 0$, $f_0 = x_2^2$ and $C = x_2$. Hence, we rewrite expression (1.22) as

$$\begin{bmatrix} 0 & 1 \\ x_2 & -x_1 \end{bmatrix} \begin{bmatrix} \dfrac{\partial f}{\partial x_1} \\ \dfrac{\partial f}{\partial x_2} \end{bmatrix} = \begin{bmatrix} 2x_2 \\ x_2 \end{bmatrix}. \tag{1.23}$$

Solving this system, we get

$$\begin{aligned} \frac{\partial f}{\partial x_1}(0, x_2) &= 1, \\ \frac{\partial f}{\partial x_2}(0, x_2) &= 2x_2. \end{aligned} \tag{1.24}$$

Evaluating at x_0, we obtain

$$\begin{aligned} \frac{\partial f}{\partial x_1}(0, 1) &= 1, \\ \frac{\partial f}{\partial x_2}(0, 1) &= 2. \end{aligned}$$

The second derivatives can be obtained by differentiating expressions (1.24) with respect to x_2 and the original differential equation with respect to x_1.

$$\begin{aligned} \frac{\partial^2 f}{\partial x_2 \partial x_1}(0, x_2) &= 0, \\ \frac{\partial^2 f}{\partial x_2^2}(0, x_2) &= 2, \\ x_2 \frac{\partial^2 f(x_1, x_2)}{\partial x_1^2} - \frac{\partial f(x_1, x_2)}{\partial x_2} - x_1 \frac{\partial^2 f(x_1, x_2)}{\partial x_2 \partial x_1} &= 0. \end{aligned} \tag{1.25}$$

Evaluating at x_0, we obtain

$$\begin{aligned} \frac{\partial^2 f}{\partial x_2 \partial x_1}(0, 1) &= 0, \\ \frac{\partial^2 f}{\partial x_1^2}(0, 1) &= 2, \\ \frac{\partial^2 f}{\partial x_2^2}(0, 1) &= 2, \end{aligned}$$

where in evaluating the last expression we use the values of derivatives obtained above.

Let us consider the third-order derivatives. The third derivatives of expressions (1.25) that contain derivatives with respect to x_2 and are evaluated at x_0 are zero, as can be seen by differentiating and evaluating each of them with respect to this variable. The result is immediate for the first two expressions: $\partial^3 f/\partial x_2^2 x_1 = 0$ and $\partial^3 f/\partial x_2^3 = 0$. From the third one, we get

$$\frac{\partial^2 f(x_1,x_2)}{\partial x_1^2} + x_2 \frac{\partial^3 f(x_1,x_2)}{\partial x_2 \partial x_1^2} - \frac{\partial^2 f(x_1,x_2)}{\partial x_2^2} - x_1 \frac{\partial^3 f(x_1,x_2)}{\partial x_2^2 \partial x_1} = 0\,,$$

which results in $\partial^3 f/\partial x_2 \partial x_1^2 = 0$. If we assume the equality of mixed partial derivatives, the third derivative with respect to x_1, can be obtained by differentiating twice the original differential equation to get

$$x_2 \frac{\partial^3 f(x_1,x_2)}{\partial x_1^3} - 2\frac{\partial^2 f(x_1,x_2)}{\partial x_1 \partial x_2} - x_1 \frac{\partial^3 f(x_1,x_2)}{\partial x_2 \partial x_1^2} = 0\,. \tag{1.26}$$

Solving for $\partial^3 f/\partial x_1^3$ and evaluating at x_0, we get

$$\frac{\partial^3 f}{\partial x_1^3}(0,1) = 0\,.$$

All higher derivatives evaluated at x_0 are zero.

Using the above results in expression (1.21), we write the Taylor series as

$$f(x_1,x_2) = 1 + x_1 + 2(x_2-1) + \frac{1}{2}\left(2x_1^2 + 2(x_2-1)^2\right) = x_1 + x_1^2 + x_2^2\,,$$

which is the particular solution of the Cauchy problem; it satisfies equations (1.19) and (1.20). Also, it agrees with the general solution given in Exercise 1.4: $f = x_1 + g(x_1^2 + x_2^2)$.

An insight into the concept of characteristics can be gained by examining Example 1.3. The solubility of system (1.23) requires that the determinant of the coefficient matrix,

$$\det\begin{bmatrix} 0 & 1 \\ x_2 & -x_1 \end{bmatrix} = -x_2\,,$$

be nonzero, as discussed in Exercise 1.2. As explored in Chapter 2, characteristics of partial differential equations are hypersurfaces along which we cannot set the side conditions to obtain the unique solutions for the nth partial derivatives of the solutions of these equations, where n is the order of the equation.

Another insight into characteristics can be gained by formulating the convergent Taylor series around $x_0 = [0,0]$, which is along the x_2-axis as required by the use of expression (1.20) to find the partial derivatives of f. Differentiating expressions (1.24) with respect to x_2 and evaluating at x_0, we get

$$\frac{\partial^2 f}{\partial x_2 \partial x_1}(0,0) = 0$$

and

$$\frac{\partial^2 f}{\partial x_2^2}(0,0) = 2\,.$$

To obtain $\partial^2 f / \partial x_1^2$, we differentiate equation (1.19) with respect to x_1; we get

$$x_2 \frac{\partial^2 f(x_1, x_2)}{\partial x_1^2} - \frac{\partial f(x_1, x_2)}{\partial x_2} - x_1 \frac{\partial^2 f(x_1, x_2)}{\partial x_2 \partial x_1} = 0\,.$$

Evaluating at x_0 using the values obtained above, we write

$$0\,\frac{\partial^2 f(0,0)}{\partial x_1^2} = 2\,,$$

which is a contradiction. To understand this result, we recall from Exercise 1.2 that the characteristics of equation (1.19) are circles, $x_1^2 + x_2^2 = C$. Hence, x_0 is a characteristic point corresponding to $C = 0$; it cannot be used to set the side conditions, as discussed in Section 1.5, and it cannot be used as a point about which we construct the Taylor series.

1.5 Incompatibility of side conditions

To obtain the Taylor-series solution discussed in Section 1.4, we require that the side conditions not be given along a curve that is parallel to vector A, which is the direction of the differentiation in the original differential equation. This directional derivative is discussed in Section 1.2, and allows us to obtain the characteristics. Since the behaviour of the solution along the characteristics is prescribed by the equation itself, we cannot arbitrarily set the side conditions along the characteristic curves. This conclusion suggests a new way of looking at the characteristics and, consequently, leads us to another method of obtaining them. In this method we look at curves along which arbitrary side conditions lead to an incompatibility with the differential equation. Even though for the linear first-order equation these two methods are equivalent, the new approach is more general, as we see by studying the second-order equations in Chapter 2.

Let us return to the study of the general differential first-order equation in two variables given by expression (1.5), namely,

$$A_1(x_1,x_2)\frac{\partial f}{\partial x_1}+A_2(x_1,x_2)\frac{\partial f}{\partial x_2}=B(x_1,x_2)f+C(x_1,x_2)\,. \tag{1.27}$$

To find curves γ along which we cannot arbitrarily set the side conditions, let us find conditions under which the side conditions given along $\gamma(s)=[x_1(s),x_2(s)]$ are compatible with the differential equation itself. In other words, we determine under which conditions the solutions can satisfy both the differential equation and the side condition.

If f is given along $\gamma(s)=[x_1(s),x_2(s)]$ by a side condition, its derivative $f'(s):=\mathrm{d}f/\mathrm{d}s$ along this curve is known. This derivative can be expressed in terms of the partial derivatives along x_1 and x_2 as

$$f'(x_1(s),x_2(s))=\frac{\partial f}{\partial x_1}x_1'(s)+\frac{\partial f}{\partial x_2}x_2'(s)\,. \tag{1.28}$$

Thus, we have two linear equations for unknowns $\partial f/\partial x_1$ and $\partial f/\partial x_2$: one from the differential equation (1.27) and one from the side condition (1.28), namely,

$$\begin{bmatrix} x_1'(s) & x_2'(s)\\ A_1(x_1,x_2) & A_2(x_1,x_2)\end{bmatrix}\begin{bmatrix}\dfrac{\partial f}{\partial x_1}\\ \dfrac{\partial f}{\partial x_2}\end{bmatrix}=\begin{bmatrix} f'(s)\\ B(x_1,x_2)f+C(x_1,x_2)\end{bmatrix}. \tag{1.29}$$

This system cannot be solved uniquely for the two unknown derivatives if and only if the determinant of the coefficient matrix is zero, namely,

$$A_2(x_1(s),x_2(s))\,x_1'(s)-A_1(x_1(s),x_2(s))\,x_2'(s)=0\,. \tag{1.30}$$

Since $x_i'(s)$ stands for $\mathrm{d}x_i/\mathrm{d}s$, this condition is equivalent to

$$\mathrm{d}x_1A_2(x_1,x_2)=\mathrm{d}x_2A_1(x_1,x_2)\,, \tag{1.31}$$

which is the characteristic equation.

In this case, equations (1.27) and (1.28) are either incompatible with one another or equivalent to one another. If the two equations are equivalent to one another, then the characteristic equation (1.31) is satisfied and the equations are scalar multiples of one another. This equivalence can be written as the compatibility condition along a characteristic.

$$[A_1,A_2,Bf+C]=\zeta\,[x_1',x_2',f']\,, \tag{1.32}$$

where ζ is a proportionality constant. In other words, the compatibility condition tells us whether or not the side condition given along a characteristic is compatible with the differential equation. Equation (1.32) describes the characteristic curves and the rate of change of the solution along them in agreement with the directional derivative (1.7). This can be seen from the fact that equation (1.32) describes characteristics as curves whose tangent is equal to vector $(1/\zeta)[A_1, A_2]$ and the derivative along them is $(1/\zeta)(Bf + C)$. Compatible side conditions along the characteristics do not give any additional information about the solutions.

Example 1.4. To illustrate the above results, let us revisit equation (1.8), namely,

$$\frac{\partial f(x_1, x_2)}{\partial x_1} + x_2 \frac{\partial f(x_1, x_2)}{\partial x_2} = 0, \tag{1.33}$$

and find the family of characteristic curves for this equation. Examining equation (1.33) and taking into account equation (1.27), we see that $A_1(x_1, x_2) = 1$, $A_2(x_1, x_2) = x_2$ and $B(x_1, x_2) = C(x_1, x_2) = 0$. First, we see that equation (1.31) becomes equation (1.10). Hence, the family of characteristic curves is given by equation (1.11), namely,

$$x_2(x_1) = C \exp x_1 . \tag{1.34}$$

Secondly, for equation (1.33) we write the compatibility equation as

$$\frac{1}{x_1'(s)} = \frac{x_2(s)}{x_2'(s)} = \frac{0}{f'(s)}.$$

From this expression we infer that

$$\frac{\mathrm{d}x_2}{\mathrm{d}x_1} = x_2$$

and $f'(s) = 0$, where the latter equation results from the assumption that $|x_1'| < \infty$. In the case of $f'(s) \neq 0$, we see that $x_1'(s) = \pm\infty$ and $x_2(s) = 0$. This results in a curve that coincides with the x_1-axis, which can be obtained by setting $C = 0$ in equation (1.34). The first equation is characteristic equation (1.10). The second equation states that the solution, f, does not change along the characteristic curves, which is consistent with results in Section 1.2. If the side condition is given by a constant function along the characteristic, the solution of the differential equation would not be uniquely determined; any function that is constant along the characteristics and has a proper value along the side condition would be a solution. The compatible side condition along characteristics does not restrict the possible solutions.

In the two-dimensional case, the hypersurfaces are curves. The characteristic hypersurfaces coincide with the characteristic curves, since the side conditions cannot be specified along these curves. In this case, both methods give the same result, as expected and as illustrated in Exercise 1.2.

In the higher-dimensional case, the characteristic hypersurface is a surface that is composed of characteristic curves. This can be illustrated by expression (1.37). These curves are given by the direction discussed in the directional derivatives approach.

1.6 Semilinear equations

The characteristics of linear partial differential equation (1.5) do not depend on its right-hand side. This observation leads to a generalization of the class of differential equation for which we can find characteristics. In this section we turn our attention to semilinear first-order partial differential equations. Such equations are given by

$$A_1(x)\frac{\partial f}{\partial x_1} + A_2(x)\frac{\partial f}{\partial x_2} + \cdots + A_n(x)\frac{\partial f}{\partial x_n} = B(x, f), \tag{1.35}$$

where B is an arbitrary function. In other words, the terms with the derivatives are linear, whereas the remaining terms might not be linear. A typical side condition for this equation would be given as a fixed value of f along a hypersurface. In the n-dimensional case, this hypersurface can be parametrized, at least locally, by $n-1$ parameters, say, $s_1, \ldots, s_{n-1}$, namely,

$$\begin{aligned}
x_1 &= x_1(s_1, \ldots, s_{n-1}), \\
x_2 &= x_2(s_1, \ldots, s_{n-1}), \\
&\vdots \\
x_n &= x_n(s_1, \ldots, s_{n-1}).
\end{aligned}$$

Hence, we can write the side condition as

$$f(x(s_1, \ldots, s_{n-1})) = f_0(s_1, \ldots, s_{n-1}).$$

Having stated the side condition, we can ask if it is compatible with the differential equation. More precisely, we can write a system of equations for $\partial f/\partial x_1, \partial f/\partial x_2, \ldots, \partial f/\partial x_n$, namely

$$\underbrace{\begin{bmatrix} \frac{\partial x_1}{\partial s_1} & \frac{\partial x_2}{\partial s_1} & \cdots & \frac{\partial x_n}{\partial s_1} \\ \frac{\partial x_1}{\partial s_2} & \frac{\partial x_2}{\partial s_2} & \cdots & \frac{\partial x_n}{\partial s_2} \\ \vdots & \vdots & \ddots & \vdots \\ A_1 & A_2 & \cdots & A_n \end{bmatrix}}_{M} \begin{bmatrix} \frac{\partial f}{\partial x_1} \\ \frac{\partial f}{\partial x_2} \\ \vdots \\ \frac{\partial f}{\partial x_n} \end{bmatrix} = \begin{bmatrix} \frac{\partial f_0}{\partial s_1} \\ \frac{\partial f_0}{\partial s_2} \\ \vdots \\ B \end{bmatrix}. \tag{1.36}$$

This system has a unique solution only if the determinant of M is nonzero. If the determinant is zero, there are either no solutions or infinitely many solutions.

The determinant of M is zero if and only if the rows of matrix M are linearly dependent vectors. Since we consider a hypersurface, the first $n-1$ rows are linearly independent of each other. The only possible linear dependence of the rows can be expressed as

$$\begin{aligned} [A_1, A_2, \ldots, A_n] = \zeta_1 \left[\frac{\partial x_1}{\partial s_1}, \ldots, \frac{\partial x_n}{\partial s_1}\right] + \zeta_2 \left[\frac{\partial x_1}{\partial s_2}, \ldots, \frac{\partial x_n}{\partial s_2}\right] \\ + \cdots + \zeta_{n-1} \left[\frac{\partial x_1}{\partial s_{n-1}}, \ldots, \frac{\partial x_n}{\partial s_{n-1}}\right], \end{aligned} \tag{1.37}$$

which states that vector A must be tangent to the characteristic surface; a characteristic surface is composed of characteristic curves whose tangents are parallel to A. This result is consistent with the fact that differential equation (1.35) determines the rate of change of the solution along direction A.

Example 1.5. In this example we illustrate the fact that for the first-order equations the characteristics depend only on the terms with derivatives and the solutions depend on all terms. We modify equation (1.8) by adding to its right-hand side a nonlinear term in f. We consider

$$\frac{\partial f(x_1, x_2)}{\partial x_1} + x_2 \frac{\partial f(x_1, x_2)}{\partial x_2} = \frac{1}{f(x_1, x_2)}, \tag{1.38}$$

which is a semilinear equation. The characteristic curves are the same as for equation (1.8), and are given by expression (1.11), namely,

$$x_2(x_1) = C \exp x_1 .$$

They are shown in Figure 1.3. However, the side condition propagates differently along these curves. To see this, we choose the same side condition, namely,

$$f(0, x_2) = |x_2| \, . \tag{1.39}$$

If we parametrize the characteristics by s, the solution along a characteristic is governed by an ordinary differential equation,

$$\frac{\mathrm{d}f(x_1(s), x_2(s))}{\mathrm{d}s} = \frac{1}{f(x_1(s), x_2(s))} \, . \tag{1.40}$$

Integrating, we get

$$\frac{1}{2}f^2 = s + D \, ,$$

where D is an integration constant, which is constant along a given characteristic but can vary for different characteristics. The solution of equation (1.40) can be written implicitly as

$$f^2(x_1(s), x_2(s)) = 2s + 2D \, . \tag{1.41}$$

In view of the double-valued solution along the characteristics, it is important to choose the parametrization of the characteristics to allow propagation of the side condition to both values of the double-valued solution. A suitable reparametrization can be obtained by changing the parameter in equation (1.40),

$$\frac{1}{f} = \frac{\mathrm{d}f}{\mathrm{d}s} = \frac{\mathrm{d}f}{\mathrm{d}t}\frac{\mathrm{d}t}{\mathrm{d}s} \, ,$$

using

$$\frac{\mathrm{d}t}{\mathrm{d}s} = \frac{1}{t} \, . \tag{1.42}$$

The equation along the characteristic becomes

$$\frac{1}{f} = \frac{\mathrm{d}f}{\mathrm{d}s} = \frac{\mathrm{d}f}{\mathrm{d}t}\frac{1}{t} \, ,$$

and its solution is

$$f^2 = t^2 + K \, , \tag{1.43}$$

where K is an integration constant. Change of parametrization (1.42) implies that

$$t^2 = 2s + L \, .$$

In this new parametrization, the characteristics are

$$x_1(t) = \frac{t^2 - L}{2} \tag{1.44}$$

and

$$x_2(t) = C \exp \frac{t^2 - L}{2} . \tag{1.45}$$

Equation (1.43) describes the rate of change of the solution along the characteristics parametrized by t. We want to choose the value of L such that the characteristics can cover the entire graph of the solution. A possible choice is $L = C^2$. To find the value for K, we invoke the side condition along $x_1 = 0$,

$$f^2(0, x_2) = x_2^2 ,$$

which is equivalent to

$$K = 0 ,$$

since, for $x_1 = 0$, $t^2 = C^2$ and $x_2 = C$. Thus, along the characteristics,

$$f^2 = t^2.$$

To express this solution in terms of x_1 and x_2, we use expressions (1.44) and (1.45) to write

$$C = x_2 \exp(-x_1)$$

and

$$t^2 = 2x_1 + C^2 = 2x_1 + x_2^2 \exp(-2x_1) ,$$

which implies that

$$f^2(x_1, x_2) = 2x_1 + x_2^2 \exp(-2x_1) . \tag{1.46}$$

Implicit differentiation allows one to verify that it is indeed the required solution. The solution formed by the propagation of the side condition along the characteristics is illustrated in Figure 1.7.

Let us examine the linear equation (1.8) and the semilinear equation (1.38) with their side condition (1.39), and their solutions shown in Figures 1.4 and 1.7, respectively. Orthogonal projections of the lifted characteristics shown in Figures 1.4 onto the x_1x_2-plane coincides with the characteristics shown in Figure 1.3. This is not the case for the semilinear equation. Even though the characteristic equations, given by expression (1.11), are the same for equations (1.8) and (1.38), to obtain the solution of the semilinear equation, we needed to reparametrize the

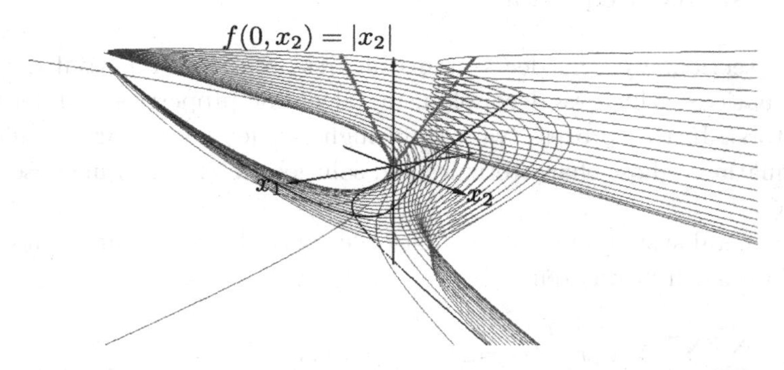

Fig. 1.7 Propagation of the side condition along the characteristics for equation (1.38). The v-shape curve at the origin determines the side condition given at $x_1 = 0$ and the side condition propagated along the characteristics is shown in black. The horizontal axes are the x_1-axis and the x_2-axis, and the vertical axis gives the value of $f(x_1, x_2)$.

characteristics. This reparametrization results in the projection of the lifted characteristics onto the x_1x_2-plane being only a subset of the characteristics, as shown in Figure 1.8. This can be explained by investigating the implicit solution (1.46). Points of the region of the x_1x_2-plane in which the projection of the lifted characteristics does not coincide with the characteristics do not result in real solutions. For example, inserting point $(-1, 0)$ into the implicit solution results in $f^2 = -1$.

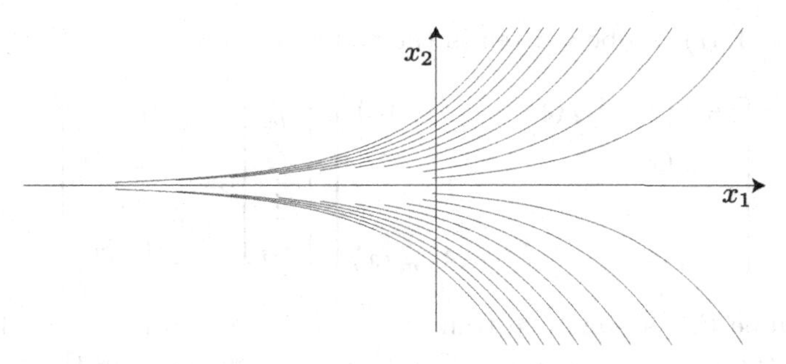

Fig. 1.8 Projection of lifted characteristics of equation (1.38). Compared to Figure 1.3, the only difference is the parametrization of the characteristics, which is given by expressions (1.44) and (1.45). This comparison illustrates the dependence of characteristics on the highest-order derivatives only.

In view of the above example, we conclude that for the first-order equations the characteristics depend only on the terms with the highest-order derivatives, and the solutions depend on all terms.

1.7 Systems of equations

In this section, we consider systems of first-order linear partial differential equations, which we decouple using algebraic properties of differential operators. In this process, we obtain higher-order linear partial differential equations whose characteristics and solutions are studied in subsequent chapters.

A general system of m linear first-order partial differential equations in n variables can be written as

$$\sum_{j=1}^{n}\sum_{k=1}^{m} A_{ijk}(x)\frac{\partial f_k}{\partial x_j} = B_i(x), \quad \text{where} \qquad i = 1, 2, \ldots, m.$$

For facility in subsequent discussions, we rewrite this system using the following notation:

$$\sum_{k=1}^{m} \mathcal{A}_{ik}(x) f_k = B_i(x)\,, \tag{1.47}$$

where

$$\mathcal{A}_{ik}(x) := \sum_{j=1}^{n} A_{ijk}(x)\frac{\partial}{\partial x_j}.$$

Equation (1.47) can be written in the matrix form as

$$\begin{bmatrix} \mathcal{A}_{11}(x) & \mathcal{A}_{12}(x) & \cdots & \mathcal{A}_{1m}(x) \\ \mathcal{A}_{21}(x) & \ddots & & \\ \vdots & & \ddots & \vdots \\ \mathcal{A}_{m1}(x) & & \cdots & \mathcal{A}_{mm}(x) \end{bmatrix} \begin{bmatrix} f_1 \\ f_2 \\ \vdots \\ f_m \end{bmatrix} = \begin{bmatrix} B_1(x) \\ B_2(x) \\ \vdots \\ B_m(x) \end{bmatrix}.$$

To decouple this system, we transform matrix $\mathcal{A}$ into its upper diagonal form by the Gauss elimination. The system of equations in the new form can be expressed as

$$\begin{bmatrix} \mathcal{A}_{11}(x) & \mathcal{A}_{12}(x) & \cdots & \mathcal{A}_{1m}(x) \\ 0 & \mathcal{A}_{22}^2(x) & \cdots & \mathcal{A}_{2m}^2(x) \\ \vdots & & \ddots & \vdots \\ 0 & & \cdots & \mathcal{A}_{mm}^m(x) \end{bmatrix} \begin{bmatrix} f_1 \\ f_2 \\ \vdots \\ f_m \end{bmatrix} = \begin{bmatrix} B_1(x) \\ D^1 B_2(x) \\ \vdots \\ D^{m-1} B_m(x) \end{bmatrix}, \tag{1.48}$$

where $\mathcal{A}_{ij}^{k}(x)$ denotes a linear differential operator of degree k and D^l denotes a differential operator of degree l.

System (1.48) can be decoupled. It is possible to solve the last equation of the system, which is a linear partial differential equation of order m. Once this solution is found, we can substitute it into the second-last equation of the system, which becomes a linear partial differential equation of order $m-1$. We can continue this recursive method until $f_1(x)$ is found. An example of such a system is discussed in Example 1.6 and Exercise 1.12.

We note that by increasing the order of the differential equations, we might generate more solutions than the original equation admits. Also, by increasing the order of differentiation we might lose solutions that are not sufficiently differentiable for the increased order.

Example 1.6. We find the solution of the system of partial differential equations,

$$\frac{\partial f_1}{\partial x_1}+\frac{\partial f_1}{\partial x_2}+\frac{\partial f_2}{\partial x_1}-\frac{\partial f_2}{\partial x_2}=0$$

and

$$\frac{\partial f_1}{\partial x_1}-\frac{\partial f_1}{\partial x_2}+2\frac{\partial f_2}{\partial x_1}+\frac{\partial f_2}{\partial x_2}=x_1,$$

which can be written as

$$\begin{bmatrix} \dfrac{\partial}{\partial x_1}+\dfrac{\partial}{\partial x_2} & \dfrac{\partial}{\partial x_1}-\dfrac{\partial}{\partial x_2} \\ \dfrac{\partial}{\partial x_1}-\dfrac{\partial}{\partial x_2} & 2\dfrac{\partial}{\partial x_1}+\dfrac{\partial}{\partial x_2} \end{bmatrix}\begin{bmatrix} f_1 \\ f_2 \end{bmatrix}=\begin{bmatrix} 0 \\ x_1 \end{bmatrix}. \tag{1.49}$$

Acting on the first row by operator $\partial/\partial x_1-\partial/\partial x_2$, acting on the second row by operator $\partial/\partial x_1+\partial/\partial x_2$, and subtracting the results, we replace the second equation to write

$$\begin{bmatrix} \dfrac{\partial}{\partial x_1}+\dfrac{\partial}{\partial x_2} & \dfrac{\partial}{\partial x_1}-\dfrac{\partial}{\partial x_2} \\ 0 & \left(\dfrac{\partial}{\partial x_1}+\dfrac{\partial}{\partial x_2}\right)\left(2\dfrac{\partial}{\partial x_1}+\dfrac{\partial}{\partial x_2}\right)-\left(\dfrac{\partial}{\partial x_1}-\dfrac{\partial}{\partial x_2}\right)^2 \end{bmatrix}\begin{bmatrix} f_1 \\ f_2 \end{bmatrix}$$
$$=\begin{bmatrix} 0 \\ \left(\dfrac{\partial}{\partial x_1}+\dfrac{\partial}{\partial x_2}\right)x_1 \end{bmatrix}. \tag{1.50}$$

The second equation of this system, which can be written as

$$\left(\frac{\partial^2}{\partial x_1^2}+5\frac{\partial^2}{\partial x_1 \partial x_2}\right) f_2=1, \tag{1.51}$$

is decoupled from unknown function f_1. It is a second-order linear partial differential equation whose characteristics are discussed in Chapter 2. The solution of equation (1.51), obtained by the method of characteristics in Exercise 2.12, is

$$f_2\left(x_1, x_2\right)=\left(x_1-\frac{1}{5} x_2\right) \frac{1}{5} x_2+g\left(x_1-\frac{1}{5} x_2\right)+h\left(\frac{1}{5} x_2\right), \tag{1.52}$$

which—in view of equation (1.51) and without considering distributions, which are discussed in Appendix E—requires f_2 to be at least twice differentiable, whereas the original equations required it to be only once differentiable. After we substitute this solution into the first equation of the original system, this equation becomes

$$\begin{aligned}
\frac{\partial f_1}{\partial x_1}+\frac{\partial f_1}{\partial x_2}= & -\frac{1}{5} x_2-g^{\prime}\left(x_1-\frac{1}{5} x_2\right)+\frac{1}{5} x_1-\frac{2}{25} x_2 \\
& -\frac{1}{5} g^{\prime}\left(x_1-\frac{1}{5} x_2\right)+\frac{1}{5} h^{\prime}\left(\frac{1}{5} x_2\right) \\
= & \frac{1}{5} x_1-\frac{7}{25} x_2-\frac{6}{5} g^{\prime}\left(x_1-\frac{1}{5} x_2\right)+\frac{1}{5} h^{\prime}\left(\frac{1}{5} x_2\right) .
\end{aligned}$$

The solution of this equation is given in Exercise 1.7, and can be written as

$$\begin{aligned}
f_1\left(y_1, y_1-y_2\right)= & \int \frac{1}{5} y_1-\frac{7}{25}\left(y_1-y_2\right)-\frac{6}{5} g^{\prime}\left(y_1-\frac{1}{5}\left(y_1-y_2\right)\right) \\
& +\frac{1}{5} h^{\prime}\left(\frac{1}{5}\left(y_1-y_2\right)\right) \mathrm{d} y_1+c\left(y_2\right),
\end{aligned}$$

where $y_1=x_1$ and $y_2=x_1-x_2$. Integrating, we obtain

$$\begin{aligned}
f_1\left(y_1, y_1-y_2\right)= & \frac{1}{10} y_1^2-\frac{7}{50} y_1^2+\frac{7}{25} y_1 y_2-\frac{3}{2} g\left(y_1-\frac{1}{5}\left(y_1-y_2\right)\right) \\
& +h\left(\frac{1}{5}\left(y_1-y_2\right)\right)+c\left(y_2\right),
\end{aligned}$$

which, in the original coordinates, is

$$f_1\left(x_1, x_2\right)=\frac{6}{25} x_1^2-\frac{7}{25} x_1 x_2-\frac{3}{2} g\left(x_1-\frac{1}{5} x_2\right)+h\left(\frac{1}{5} x_2\right) x_1+c\left(x_1-x_2\right) .$$

Thus, together with f_2 given in expression (1.52), we have the general solution of system (1.49).

The Maxwell equations derived in Appendix C are an important system of first-order linear equations of mathematical physics. In the following example, we use the Gauss elimination method to obtain second-order equations studied in Chapter 2.

Example 1.7. Consider the Maxwell equations, which are formulated in Appendix C.1, namely,

$$\nabla \cdot E = \frac{\rho}{\epsilon_0}, \tag{1.53}$$

$$\nabla \cdot B = 0, \tag{1.54}$$

$$\nabla \times E = -\frac{\partial B}{\partial t} \tag{1.55}$$

and

$$c^2 \nabla \times B = \frac{J}{\epsilon_0} + \frac{\partial E}{\partial t}. \tag{1.56}$$

These equations form a system of eight equations stated explicitly by equations (C.15), which we write in a matrix form as

$$\begin{bmatrix} \partial_1 & \partial_2 & \partial_3 & 0 & 0 & 0 \\ 0 & 0 & 0 & \partial_1 & \partial_2 & \partial_3 \\ 0 & -\partial_3 & \partial_2 & \partial_t & 0 & 0 \\ \partial_3 & 0 & -\partial_1 & 0 & \partial_t & 0 \\ -\partial_2 & \partial_1 & 0 & 0 & 0 & \partial_t \\ -\partial_t & 0 & 0 & 0 & -c^2\partial_3 & c^2\partial_2 \\ 0 & -\partial_t & 0 & c^2\partial_3 & 0 & -c^2\partial_1 \\ 0 & 0 & -\partial_t & -c^2\partial_2 & c^2\partial_1 & 0 \end{bmatrix} \begin{bmatrix} E_1 \\ E_2 \\ E_3 \\ B_1 \\ B_2 \\ B_3 \end{bmatrix} = \begin{bmatrix} \frac{\rho}{\epsilon_0} \\ 0 \\ 0 \\ 0 \\ 0 \\ \frac{J_1}{\epsilon_0} \\ \frac{J_2}{\epsilon_0} \\ \frac{J_3}{\epsilon_0} \end{bmatrix}, \tag{1.57}$$

where $\partial_i := \partial/\partial x_i$ and $\partial_t := \partial/\partial t$. Using the Gauss elimination and invoking the continuity equation (C.1), we get

$$\begin{bmatrix} \partial_1 & \partial_2 & \partial_3 & 0 & 0 & 0 \\ \partial_3 & 0 & -\partial_1 & 0 & \partial_t & 0 \\ 0 & \partial_{13} & -\partial_{12} & 0 & \partial_{2t} & \partial_{3t} \\ 0 & 0 & -\partial_{12t} & c^2\partial_{133} & \partial_{2tt} & -c^2\partial_{113}+\partial_{3tt} \\ 0 & 0 & 0 & \partial_1 & \partial_2 & \partial_3 \\ 0 & 0 & 0 & c^2\partial_{223}+c^2\partial_{333}-\partial_{3tt} & -c^2\partial_{123} & -c^2\partial_{133} \\ 0 & 0 & 0 & 0 & 0 & 0 \\ 0 & 0 & 0 & 0 & 0 & 0 \end{bmatrix} \begin{bmatrix} E_1 \\ E_2 \\ E_3 \\ B_1 \\ B_2 \\ B_3 \end{bmatrix}$$

$$= \left[\frac{\rho}{\epsilon_0}, 0, 0, \partial_{13}\frac{J_2}{\epsilon_0}, 0, \partial_{2t}\frac{\rho}{\epsilon_0} + \partial_{12}\frac{J_1}{\epsilon_0} + \partial_{22}\frac{J_2}{\epsilon_0} + \partial_{33}\frac{J_2}{\epsilon_0}, 0, 0\right]^T,$$

where, say, $\partial_{12t} = \partial_1\partial_2\partial_t := \partial^3/\partial x_1\partial x_2\partial t$. These are six equations for six unknowns. The two rows of zeros together with the last two zeros in the transposed row imply that there is a linear dependence among the Maxwell equations; the last two equations of system (1.57) are the linear combinations of other equations in this system. Proceeding with the elimination, we obtain

$$\begin{bmatrix} \Box & 0 & 0 & 0 & 0 & 0 \\ 0 & \Box & 0 & 0 & 0 & 0 \\ 0 & 0 & \Box & 0 & 0 & 0 \\ 0 & 0 & 0 & \Box & 0 & 0 \\ 0 & 0 & 0 & 0 & \Box & 0 \\ 0 & 0 & 0 & 0 & 0 & \Box \end{bmatrix} \begin{bmatrix} E_1 \\ E_2 \\ E_3 \\ B_1 \\ B_2 \\ B_3 \end{bmatrix} = \begin{bmatrix} c^2\partial_1\frac{\rho}{\epsilon_0} + \partial_t\frac{J_1}{\epsilon_0} \\ c^2\partial_2\frac{\rho}{\epsilon_0} + \partial_t\frac{J_2}{\epsilon_0} \\ c^2\partial_3\frac{\rho}{\epsilon_0} + \partial_t\frac{J_3}{\epsilon_0} \\ \partial_3\frac{J_2}{\epsilon_0} - \partial_2\frac{J_3}{\epsilon_0} \\ \partial_1\frac{J_3}{\epsilon_0} - \partial_3\frac{J_1}{\epsilon_0} \\ \partial_2\frac{J_1}{\epsilon_0} - \partial_1\frac{J_2}{\epsilon_0} \end{bmatrix}, \tag{1.58}$$

where $\Box := c^2(\partial_{11} + \partial_{22} + \partial_{33}) - \partial_{tt}$ is the d'Alembert operator.

The Gauss elimination used in Example 1.7 is equivalent to taking the curl of equations (1.55) and (1.56) and subsequently invoking the identity $\nabla\times\nabla\times X = \nabla(\nabla\cdot X) - \nabla^2 X$, as well as using equations (1.53) and (1.54), as illustrated in Exercises 1.13 and 1.14. Notably, this identity appears also on pages 249 and 261, in formulations leading to wave equations.

This example shows that the Maxwell equations stated in expressions (1.53)–(1.56) constitute eight equations for six unknowns that can be reduced to six linearly independent equations for these unknowns. These six equations, stated in system (1.58), can be written concisely as

$$\Box E = \frac{c^2}{\epsilon_0}\nabla\rho + \frac{1}{\epsilon_0}\frac{\partial J}{\partial t} \tag{1.59}$$

and

$$\Box B = -\frac{1}{\epsilon_0}\nabla \times J\,, \tag{1.60}$$

which are the wave equations with the source terms stated on the right-hand sides. These are second-order linear partial differential equations whose characteristics are discussed in Chapter 2. Notably, these are hyperbolic partial differential equations whose characteristics have a particularly important physical interpretation: they result in wavefronts.

Remark 1.2. To gain an insight into relations between the Maxwell equations and the corresponding wave equations, let us revisit expressions (1.53)–(1.56). Equations (1.53) and (1.54) have no temporal dependence; they are the intrinsic spatial field properties. Equations (1.55) and (1.56) are time-dependent; they are the evolutions of the fields. Thus, equations (1.59) and (1.60) are evolutions of the fields, given the intrinsic field properties. These evolutions are expressed in terms of wave phenomena.

In Appendix C.2, we obtain wave equations from the Maxwell equations using the vector potential, A, and the scalar potential, ϕ. Using the d'Alembert operator defined on page 28, we can write equations (C.23) and (C.24) as

$$\Box A = -\frac{J}{\epsilon_0} \tag{1.61}$$

and

$$\Box \phi = -\frac{c^2\rho}{\epsilon_0}\,, \tag{1.62}$$

respectively.

Equations (1.59) and (1.60) refer to fields, and equations (1.61) and (1.62) to potentials. The wave equations for the fields are related to the wave equations for the potentials, as shown in Exercise 1.15. This relation stems from expressions (C.16) and (C.17), which can be viewed as solutions of equations (1.54) and (1.55). In general, our obtaining wave equations from the Maxwell equations—either by the Gauss elimination or using the vector and scalar potentials—illustrates that electromagnetism is associated with wave phenomena. Each field, B and E, as well as each potential, A and ϕ, satisfy the wave equation independently. However, even though in both cases the equations are decoupled, the electric and magnetic phenomena are intrinsically connected, as described by Maxwell equations. Neither B and E nor A and ϕ are independent of one another.

Closing remarks

In this chapter, we study the concept of characteristics of first-order linear partial differential equations. We find that we cannot set arbitrarily the side conditions along the characteristics since the differential equation itself prescribes restrictions along these hypersurfaces.

The characteristics are given by the differential equation itself, if this equation is at least semilinear. This is not the case for nonlinear equations, discussed in Section 1.3, and for quasilinear equations to be discussed in Section 2.3.6; therein, characteristics depend also on the side conditions.

We can use characteristics to construct nondifferentiable solutions because differentiability is required only along the characteristics, but not along the side condition that must be transverse to characteristics. Furthermore, lifted characteristics allow us to parametrize graphs of multivalued solutions.

1.8 Exercises

Exercise 1.1. Using directional derivatives, solve

$$\frac{\partial f(x_1, x_2)}{\partial x_1} + c\frac{\partial f(x_1, x_2)}{\partial x_2} = 0, \tag{1.63}$$

where c is a constant. Suggest a manner of imposing the side conditions to obtain a particular solution; note that if $c = 0$, this equation reduces to equation (1.1). Equation (1.63) is referred to as the transport equation; justify this name.

Solution. We can rewrite equation (1.63) as

$$[1, c] \cdot \left[\frac{\partial f}{\partial x_1}, \frac{\partial f}{\partial x_2}\right] \equiv [1, c] \cdot \nabla f = 0 \,,$$

where the dot denotes the scalar product. We recognize that this is the directional derivatives of f in direction $[1, c]$. Let us write equation (1.63) as

$$\mathrm{D}_{[1,c]} f\left(x_1, x_2\right) = 0 \,, \tag{1.64}$$

where

$$\mathrm{D}_X := X \cdot \nabla \tag{1.65}$$

stands for the directional-derivative operator along vector X.

Equation (1.64) implies that $f\left(x_1, x_2\right)$ does not change along direction $[1, c]$; in other words, f is constant along this direction—if we choose f at any point on a curve whose tangent is $[1, c]$, equation (1.63) determines the values of f for all points along this curve. As introduced in Section 1.1.2, the curves along which we cannot specify the side conditions are the characteristics. Herein, the characteristics are lines whose tangent is $[1, c]$; in other words,

$$x_2 - c x_1 = C \,, \tag{1.66}$$

with C being a constant that corresponds to the x_2-intercept of a given characteristic.

Since each line is distinguished from the others by the value of $x_2 - c x_1$, we can write the general solution of equation (1.63) as

$$f\left(x_1, x_2\right) = g\left(x_2 - c x_1\right) \,, \tag{1.67}$$

where g is an arbitrary function.

To obtain a particular solution, we can, for instance, specify the value of f along $x_1 = 0$; in other words, we specify it for all points along the x_2-axis. Since $c \neq \infty$, the x_2-axis is not a characteristic, and, hence, we can specify an arbitrary function along this line. However, if $c = 0$ we cannot specify the value of f, since the lines parallel to the x_1-axis are characteristic, as shown for equation (1.1).

If x_1 in equation (1.63) represents time and x_2 represents position, we can view this equation as describing a physical system in which quantity f is being transported with speed c along the x_2-axis. Hence, this equation is referred to as the transport equation.

Exercise 1.2. Find the characteristics of

$$x_2 \frac{\partial f}{\partial x_1} - x_1 \frac{\partial f}{\partial x_2} = x_2, \tag{1.68}$$

using both the directional-derivative method and the incompatibility-of-side-conditions method.

Solution. To express equation (1.68) in terms of directional derivatives, we write

$$[x_2, -x_1] \cdot \left[\frac{\partial f}{\partial x_1}, \frac{\partial f}{\partial x_2}\right] = x_2 ,$$

which, in view of expression (1.64), we can rewrite as

$$\mathrm{D}_{[x_2,-x_1]} f\,(x_1, x_2) = x_2 ;$$

the derivative of f along the curve whose tangent vector is $[x_2, -x_1]$ is equal to x_2. We can write the tangent to this curve as

$$\frac{\mathrm{d}x_2}{\mathrm{d}x_1} = -\frac{x_1}{x_2}, \tag{1.69}$$

which is an ordinary differential equation. Separating the variables and integrating, we get

$$\int x_1 \,\mathrm{d}x_1 = -\int x_2 \,\mathrm{d}x_2 ,$$

which results in

$$x_1^2 + x_2^2 = C , \tag{1.70}$$

where C is the integration constant. Equation (1.70) defines the family of the characteristic curves, which are circles centred at the origin of the x_1x_2-plane whose radii are $\sqrt{C}$. To use the incompatibility of side conditions to find the characteristic curves, we examine equation (1.68) in the context of equation (1.27). We see that $A_1 = x_2$, $A_2 = -x_1$, $B = 0$ and $C = x_2$. Thus, we rewrite equation (1.32) as

$$\frac{\mathrm{d}x_1}{x_2} = -\frac{\mathrm{d}x_2}{x_1} = \frac{\mathrm{d}f}{x_2}. \tag{1.71}$$

Considering the first equality, we write

$$\frac{\mathrm{d}x_2}{\mathrm{d}x_1} = -\frac{x_1}{x_2},$$

which is equation (1.69) whose solution is given by expression (1.70), as expected.

Exercise 1.3. Find the solutions of equation (1.68) along its characteristics.

Solution. We can rewrite equation (1.71) as two equations, namely,

$$\frac{\mathrm{d}x_2}{\mathrm{d}x_1} = -\frac{x_1}{x_2}$$

and

$$\mathrm{d}f = \mathrm{d}x_1 .$$

The corresponding solutions are

$$x_1^2 + x_2^2 = C_1 \tag{1.72}$$

and

$$f = x_1 + C_2 , \tag{1.73}$$

respectively. Equation (1.72) describes the family of the characteristic curves and equation (1.73) describes a plane in the $x_1 x_2 f$-space.

If we view equation (1.72) as an equation for a right circular cylinder, then the intersection of this cylinder with the plane is the graph of the solution along the characteristic, which in this case is an ellipse.

Exercise 1.4. Using equations (1.72) and (1.73), find and verify the general solution of equation (1.68).

Solution. Examining equation (1.72), we see that any constant along a characteristic can be written as a function of $x_1^2 + x_2^2$. If we denote this function by g, we write equation (1.73) as

$$f = x_1 + g\left(x_1^2 + x_2^2\right) .$$

Inserting f into equation (1.68), we verify that it is a solution, namely,

$$\begin{aligned}
&x_2 \frac{\partial}{\partial x_1}\left(x_1 + g\left(x_1^2 + x_2^2\right)\right) - x_1 \frac{\partial}{\partial x_2}\left(x_1 + g\left(x_1^2 + x_2^2\right)\right) \\
&= x_2 \left(1 + \frac{\partial g\left(x_1^2 + x_2^2\right)}{\partial x_1}\right) - x_1 \frac{\partial g\left(x_1^2 + x_2^2\right)}{\partial x_2} \\
&= x_2 \left(1 + g' \frac{\partial\left(x_1^2 + x_2^2\right)}{\partial x_1}\right) - x_1 g' \frac{\partial\left(x_1^2 + x_2^2\right)}{\partial x_2} \\
&= x_2 + 2x_1 x_2 g' - 2x_1 x_2 g' = x_2,
\end{aligned}$$

as required.

Exercise 1.5. Discuss the relation between directional derivatives and change of variables in two dimensions, and exemplify this relation using equation (1.68), namely,

$$x_2 \frac{\partial f(x_1, x_2)}{\partial x_1} - x_1 \frac{\partial f(x_1, x_2)}{\partial x_2} = x_2 \tag{1.74}$$

and its side condition,

$$f(0, x_2) = x_2^2. \tag{1.75}$$

Solution. Let us revisit a general first-order linear partial differential equation (1.5) for a particular case of two independent variables, namely,

$$A_1(x_1, x_2) \frac{\partial f}{\partial x_1} + A_2(x_1, x_2) \frac{\partial f}{\partial x_2} = B(x_1, x_2) f + C(x_1, x_2) . \tag{1.76}$$

Characteristics are solutions of the differential equation obtained using the directional-derivative operator $[A_1, A_2] \cdot \nabla$. At each point, vector $[A_1, A_2]$ must be tangent to the characteristics; in other words, it must be parallel to $[x_1'(s), x_2'(s)]$. This requirement results in the characteristic equation, namely,

$$\mathrm{d}x_1 A_2(x_1, x_2) = \mathrm{d}x_2 A_1(x_1, x_2) . \tag{1.77}$$

Locally, characteristic curves can be represented by level curves of a function, $g(x_1, x_2)$, with different values of d, distinguishing among different characteristics;

$$g(x_1, x_2) = d. \tag{1.78}$$

If g is known, we can write equation (1.76) in a simpler form. Let us take d to be the new independent variable. This variable can replace either x_1 or x_2, depending on the zeros of A_1 and A_2. In particular, if $A_1 \neq 0$ the x_1x_2-plane can be parametrized by d and x_1, which follows from the fact that, in such a case, ∇g is not parallel to the x_1-axis, as illustrated in Figure 1.9. This is the reason for only local representation in terms of values of g, and thus only local change of variables. If $a_1 = 0$ or $a_2 = 0$, equation (1.76) is already reduced to a simpler form.

If we consider x_1 and d as the independent variables, we can express function $f(x_1, x_2)$ as $\tilde{f}(x_1, d)$. The two functions are related by $f(x_1, x_2) = \tilde{f}(x_1, d(x_1, x_2))$. Differentiating f with respect to x_1 and x_2, we write

$$\frac{\partial f(x_1, x_2)}{\partial x_1} = \frac{\partial \tilde{f}(x_1, d)}{\partial x_1} + \frac{\partial \tilde{f}(x_1, d)}{\partial d} \frac{\partial d(x_1, x_2)}{\partial x_1}$$

and

$$\frac{\partial f}{\partial x_2} = \frac{\partial \tilde{f}\left(x_1, d\right)}{\partial d} \frac{\partial d\left(x_1, x_2\right)}{\partial x_2},$$

respectively. Using these two expressions, we rewrite equation (1.76) as

$$\tilde{A}_1\left(x_1, d\right)\left(\frac{\partial \tilde{f}\left(x_1, d\right)}{\partial x_1} + \frac{\partial \tilde{f}\left(x_1, d\right)}{\partial d} \frac{\partial d\left(x_1, x_2\right)}{\partial x_1}\right)$$
$$+\ \tilde{A}_2\left(x_1, d\right) \frac{\partial \tilde{f}\left(x_1, d\right)}{\partial d} \frac{\partial d\left(x_1, x_2\right)}{\partial x_2} = \tilde{B}\left(x_1, d\right) \tilde{f} + \tilde{C}\left(x_1, d\right),$$

where the functions with tilde are related to the original ones in the same way as f is related to $\tilde{f}$; for example, $\tilde{A}_1\left(x_1, d\left(x_1, x_2\right)\right) = A\left(x_1, x_2\right)$. Rearranging the above equation, we get

$$\tilde{A}_1\left(x_1, d\right) \frac{\partial \tilde{f}\left(x_1, d\right)}{\partial x_1} + \left(\tilde{A}_1\left(x_1, d\right) \frac{\partial d}{\partial x_1} + \tilde{A}_2\left(x_1, d\right) \frac{\partial d}{\partial x_2}\right) \frac{\partial \tilde{f}\left(x_1, d\right)}{\partial d}$$
$$= \tilde{B}\left(x_1, d\right) \tilde{f} + \tilde{C}\left(x_1, d\right).$$

Since the term in brackets can be written as $\left[\tilde{A}_1, \tilde{A}_2\right] \cdot \nabla d$ and, as illustrated in Figure 1.9, these two vectors are perpendicular to one another, the term in brackets is zero. Thus, we write the above equation as

$$\tilde{A}_1\left(x_1, d\right) \frac{\partial \tilde{f}\left(x_1, d\right)}{\partial x_1} = \tilde{B}\left(x_1, d\right) \tilde{f} + \tilde{C}\left(x_1, d\right), \tag{1.79}$$

which is a simpler form of the original equation given by expression (1.76).

We change the variables in such a way that if we keep d constant and change x_1, we move along a characteristic. This implies that the derivative in expression (1.79) corresponds to a directional derivatives along a characteristic curve.

To exemplify the coordinate change related to equation (1.79), let us consider equation (1.74) and obtain its general solution. As shown in Exercise 1.2,

$$d = x_1^2 + x_2^2 \tag{1.80}$$

is the equation of characteristics of equation (1.74); the characteristics are circles parametrized by d. For a fixed value of d, we express x_2 as

$$x_2 = \pm\sqrt{d - x_1^2}.$$

In view of equation (1.79), we rewrite equation (1.74) as

$$\sqrt{d - x_1^2} \frac{\partial \tilde{f}\left(x_1, d\right)}{\partial x_1} = \sqrt{d - x_1^2},$$

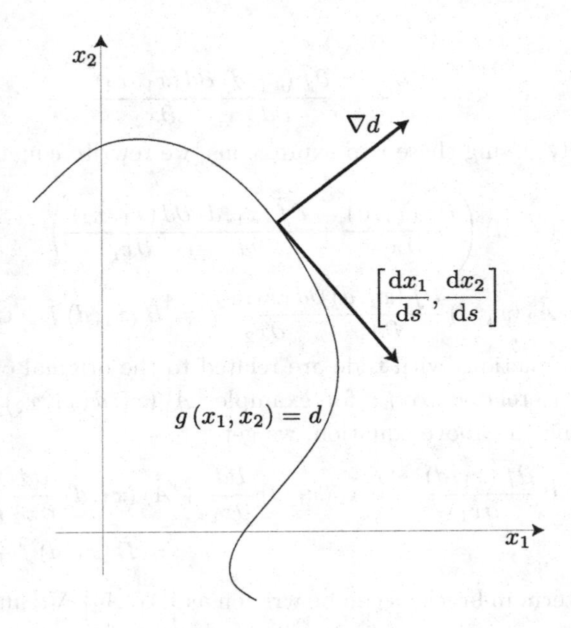

Fig. 1.9 Illustration of perpendicularity of vectors $[A_1, A_2]$ and ∇d used in derivation of equation (1.79): Vectors $[A_1, A_2]$ and $[\mathrm{d}x_1/\mathrm{d}s, \mathrm{d}x_2/\mathrm{d}s]$ are parallel to one another, as given by equation (1.77).

which we write as

$$\frac{\partial \tilde{f}(x_1, d)}{\partial x_1} = 1.$$

This is a simple equation whose solution can be obtained by integration. Integrating, we get

$$\tilde{f}(x_1, d) = x_1 + e(d), \tag{1.81}$$

where $e(d)$ is the integration constant resulting from integration with respect to x_1. Recalling expression (1.80), we write solution (1.81) in the original coordinates as

$$f(x_1, x_2) = x_1 + e\left(x_1^2 + x_2^2\right),$$

which, as verified in Exercise 1.4, is the general solution of equation (1.74).

To obtain a particular solution, we recall condition (1.75). Comparing this condition to the general solution, we conclude that the particular solution is

$$f(x_1, x_2) = x_1 + x_1^2 + x_2^2, \tag{1.82}$$

as expected from Example 1.3.

Exercise 1.6. Find the general solution of

$$a_1 \frac{\partial f}{\partial x_1} + a_2 \frac{\partial f}{\partial x_2} + a_3 \frac{\partial f}{\partial x_3} = b,$$

where a_i and b are constants, such that $a_3 \neq 0$.

Solution. We begin by writing this differential equation as a directional derivative, namely,

$$[a_1, a_2, a_3] \cdot \nabla f = b.$$

Hence, the characteristic curves are the solutions of

$$\begin{aligned} x_1'(s) &= a_1, \\ x_2'(s) &= a_2, \\ x_3'(s) &= a_3. \end{aligned}$$

These solutions are straight lines,

$$\begin{aligned} x_1(s) &= a_1 s + c_1, \\ x_2(s) &= a_2 s + c_2, \\ x_3(s) &= a_3 s + c_3, \end{aligned}$$

where c_i are the integration constants. Along these lines, the original partial differential equation becomes an ordinary differential equation, namely,

$$\frac{\mathrm{d}f}{\mathrm{d}s}(x_1(s), x_2(s), x_3(s)) = b.$$

The solution of this equation is

$$f(x_1(s), x_2(s), x_3(s)) = bs + f_0, \tag{1.83}$$

where f_0 is the integration constant that depends on the choice of a characteristic line along which we integrate. We can distinguish between different characteristic lines by setting $c_3 = 0$ and varying the values of c_1 and c_2. In such a manner, we change the coordinates x_1, x_2 and x_3 to s, c_1 and c_2. The new coordinates are related to the original ones by

$$\begin{aligned} s &= \frac{1}{a_3} x_3, \\ c_1 &= x_1 - \frac{a_1}{a_3} x_3, \\ c_2 &= x_2 - \frac{a_2}{a_3} x_3. \end{aligned}$$

Since the integration constant f_0 depends on the choice of the characteristic line, we can consider it to be an arbitrary function of c_1 and c_2. Thus, in the new coordinates, solution (1.83) becomes

$$f = bs + f_0 (c_1, c_2) .$$

Transforming this expression into the original coordinates, we obtain

$$f (x_1, x_2, x_3) = \frac{b}{a_3} x_3 + f_0 \left(x_1 - \frac{a_1}{a_3} x_3, x_2 - \frac{a_2}{a_3} x_3 \right),$$

which is the general solution of the original partial differential equation.

Exercise 1.7. Find the general solution of

$$\frac{\partial f}{\partial x_1} + \frac{\partial f}{\partial x_2} = g (x_1, x_2) ,$$

where g is an arbitrary function.

Solution. We can rewrite this equation along the characteristic line given by

$$\left[s_1 + x_1^0 (s_2) , s_1 + x_2^0 (s_2) \right]$$

as

$$\frac{\partial f}{\partial s_1} = g (x_1 (s_1, s_2) , x_2 (s_1, s_2)) ,$$

where s_1 is a parameter along the characteristic lines and s_2 specifies the line. We can choose $s_2 = x_1 - x_2$. The solution is

$$f = \int g (x_1 (s_1, s_2) , x_2 (s_1, s_2)) \, \mathrm{d}s_1 + h (s_2) ,$$

where h is an arbitrary differentiable function.

Exercise 1.8. Find the general solution of

$$a_1 \frac{\partial f}{\partial x_1} + a_2 \frac{\partial f}{\partial x_2} = \sin x_1, \tag{1.84}$$

where a_1 and a_2 are nonzero constants, by using the fact that first-order partial differential equations become ordinary differential equations along the characteristic curves.

Solution. To consider an ordinary differential equation along characteristics $[x_1(s), x_2(s)]$, we write the left-hand side of equation (1.84) as

$$\frac{\mathrm{d}f(x_1(s), x_2(s))}{\mathrm{d}s} = \frac{\partial f}{\partial x_1}\frac{\mathrm{d}x_1}{\mathrm{d}s} + \frac{\partial f}{\partial x_2}\frac{\mathrm{d}x_2}{\mathrm{d}s}.$$

Comparing this expression with equation (1.84), we see that

$$\frac{\mathrm{d}x_1}{\mathrm{d}s} = a_1 \tag{1.85}$$

and

$$\frac{\mathrm{d}x_2}{\mathrm{d}s} = a_2 \tag{1.86}$$

are the equations whose solutions are the characteristic curves. Returning to equation (1.84), we write it along these curves as

$$\frac{\mathrm{d}f(x_1(s), x_2(s))}{\mathrm{d}s} = \sin(x_1(s)). \tag{1.87}$$

To solve it, we integrate this equation along the characteristic curves. To integrate along these curves with respect to s, we solve equations (1.85) and (1.86) to get

$$x_1 = a_1 s + x_1^0 \tag{1.88}$$

and

$$x_2 = a_2 s + x_2^0, \tag{1.89}$$

respectively, which is a family of characteristic curves, with $[x_1^0, x_2^0]$ being the point through which a particular curve passes. Herein, these curves are straight lines. We rewrite equation (1.87) as

$$\mathrm{d}f(x_1(s), x_2(s)) = \sin(a_1 s + x_1^0)\,\mathrm{d}s.$$

Integrating both sides, we obtain

$$f(x_1(s), x_2(s)) = -\frac{1}{a_1}\cos(a_1 s + x_1^0) + g, \tag{1.90}$$

where g is the integration constant whose value depends on a particular line. Thus, we have the solution of equation (1.84) along a characteristic line.

We wish to write the solution for the entire x_1x_2-plane. In other words, we wish to write expression (1.90) in terms of x_1 and x_2 only. We return to solutions (1.88) and (1.89). Since $[x_1^0, x_2^0]$ specifies a point in the x_1x_2-plane that lies on a given characteristic line and $a_2 \neq 0$, we can choose this point in such a way that $x_2^0 = 0$. This way, we choose x_1^0 to identify the characteristic lines; in such a case, the integration constant, g, is a function of x_1^0. Thus, letting $x_2^0 = 0$ and solving equations (1.88) and (1.89) for x_1^0 in terms of x_1 and x_2, we get

$$x_1^0 = x_1 - \frac{a_1}{a_2} x_2. \tag{1.91}$$

Using expressions (1.88) and (1.91) in solution (1.90), we write

$$f(x_1, x_2) = -\frac{1}{a_1}\cos(x_1) + g\left(x_1 - \frac{a_1}{a_2} x_2\right), \tag{1.92}$$

which is the general solution of equation (1.84).

To verify solution (1.92), we return to equation (1.84) to get

$$a_1 \frac{\partial f}{\partial x_1} + a_2 \frac{\partial f}{\partial x_2} = a_1 \left(\frac{1}{a_1}\sin x_1 + g'\right) + a_2 \left(-\frac{a_1}{a_2} g'\right) = \sin x_1, \tag{1.93}$$

as required.

Exercise 1.9. Find the general solution of

$$a_1 \frac{\partial f}{\partial x_1} + a_2 \frac{\partial f}{\partial x_2} = \sin x_1, \tag{1.94}$$

where a_1 and a_2 are nonzero constants, by a convenient change of coordinates.

Solution. We would like to express the original differential equation in such a way that the left-hand side is a derivative with respect to a single variable. Hence, we consider

$$\frac{\partial}{\partial y_1} f(x_1(y_1, y_2), x_2(y_1, y_2)) = \frac{\partial f}{\partial x_1}\frac{\partial x_1}{\partial y_1} + \frac{\partial f}{\partial x_2}\frac{\partial x_2}{\partial y_1}. \tag{1.95}$$

Examining this expression together with the left-hand side of equation (1.94), we see that

$$\frac{\partial x_1}{\partial y_1} = a_1$$

and

$$\frac{\partial x_2}{\partial y_1} = a_2,$$

which implies that

$$x_1 = a_1 y_1 + A(y_2)$$

and

$$x_2 = a_2 y_1 + B(y_2).$$

These are equations that relate the original coordinates, x_1 and x_2, to the new coordinates, y_1 and y_2. We require these two equations be linearly independent; otherwise, functions A and B are arbitrary. We set $A(y_2) = a_1 y_2$ and $B(y_2) = -a_2 y_2$. Hence,

$$x_1 = a_1 (y_1 + y_2) \tag{1.96}$$

and

$$x_2 = a_2 (y_1 - y_2). \tag{1.97}$$

Using expressions (1.95) and (1.96), we write equation (1.94) in the new coordinates as

$$\frac{\partial f}{\partial y_1} = \sin(a_1 (y_1 + y_2)).$$

Integrating, we obtain the solution given by

$$f(y_1, y_2) = -\frac{1}{a_1} \cos(a_1 (y_1 + y_2)) + h(y_2),$$

where h is the integration constant whose constancy is with respect to y_1. To express this solution in the original coordinates, we note—in view of expression (1.96)—that the argument of the trigonometric function is x_1. Also, we solve equations (1.96) and (1.97) to get

$$y_2 = \frac{1}{2}\left(\frac{x_1}{a_1} - \frac{x_2}{a_2}\right).$$

Thus, we write the general solution as

$$f(x_1, x_2) = -\frac{1}{a_1} \cos x_1 + h\left(\frac{1}{2}\left(\frac{x_1}{a_1} - \frac{x_2}{a_2}\right)\right). \tag{1.98}$$

To verify solution (1.98), we return to equation (1.94) to get

$$a_1 \frac{\partial f}{\partial x_1} + a_2 \frac{\partial f}{\partial x_2} = a_1 \left(\frac{1}{a_1} \sin x_1 + \frac{1}{2a_1} h'\right) + a_2 \left(-\frac{1}{2a_2} h'\right) = \sin x_1,$$

as required.

Exercise 1.10. Compare and discuss solutions (1.92) and (1.98).

Solution. As shown in Exercises 1.8 and 1.9, both solutions (1.92) and (1.98) satisfy the same differential equation, namely,

$$a_1 \frac{\partial f}{\partial x_1} + a_2 \frac{\partial f}{\partial x_2} = \sin x_1,$$

where a_1 and a_2 are nonzero constants. The apparent difference between these two solutions are the arguments of functions g and h. However, both these arguments represent the same family of characteristic curves. As shown in Exercises 1.8 and 1.9, in the first case we can write the argument as

$$x_1^0 = x_1 - \frac{a_1}{a_2} x_2,$$

while in the second case we can write it as

$$y_2 = \frac{1}{2}\left(\frac{x_1}{a_1} - \frac{x_2}{a_2}\right),$$

which, using the fact that a_1 is a constant, we can rewrite as

$$2a_1 y_2 = x_1 - \frac{a_1}{a_2} x_2 = x_1^0 .$$

Thus, in both cases, the argument of g and h is an expression defining a given characteristic line. In Exercise 1.8, by setting $x_2^0 = 0$, we identify each line of the family of characteristics by its intercept with the x_2-axis, which in such a case is given by x_1^0. In Exercise 1.9, by requiring the linear independence of the two equations that relate the coordinates, we identify each of the characteristics by the value of y_2.

Exercise 1.11. Find the characteristic surfaces of

$$x_2 \frac{\partial g}{\partial x_1} - x_1 \frac{\partial g}{\partial x_2} + x_3^2 = 1.$$

Solution. We can write this equation as

$$[x_2, -x_1, 0] \cdot \nabla g = 1 - x_3^2.$$

The characteristic curves parametrized by s are the solutions of

$$x_1'(s) = x_2,$$
$$x_2'(s) = -x_1,$$
$$x_3'(s) = 0.$$

Since $x_3' = 0$, the solutions are restricted to the planes parallel to the x_1x_2-plane, and hence we can study the solutions only in this plane. We consider x_2 as a function of x_1. The first two equations become

$$\frac{\mathrm{d}x_2}{\mathrm{d}x_1} = -\frac{x_1}{x_2},$$

which is a separable equation whose solution is

$$x_1^2 + x_2^2 = C^2,$$

where C^2 is the integration constant. We conclude that the characteristic surfaces are composed of circles that are parallel to the x_1x_2-plane, and whose radius is C.

Exercise 1.12. Find the general solution of the system composed of

$$\frac{\partial f_1}{\partial x_1} - 2\frac{\partial f_1}{\partial x_2} + 2\frac{\partial f_2}{\partial x_1} + \frac{\partial f_2}{\partial x_2} = x_1$$

and

$$\frac{\partial f_1}{\partial x_1} - \frac{\partial f_1}{\partial x_2} - \frac{\partial f_2}{\partial x_1} + \frac{\partial f_2}{\partial x_2} = 0.$$

Solution. We reduce the system to the upper triangular form by applying

$$\frac{\partial}{\partial x_1} - \frac{\partial}{\partial x_2}$$

to the first equation, applying

$$\frac{\partial}{\partial x_1} - 2\frac{\partial}{\partial x_2}$$

to the second equation, and subtracting the results. Thus, we write

$$\left(\frac{\partial}{\partial x_1} - \frac{\partial}{\partial x_2}\right)\left(\frac{\partial f_1}{\partial x_1} - 2\frac{\partial f_1}{\partial x_2} + 2\frac{\partial f_2}{\partial x_1} + \frac{\partial f_2}{\partial x_2}\right)$$
$$-\left(\frac{\partial}{\partial x_1} - 2\frac{\partial}{\partial x_2}\right)\left(\frac{\partial f_1}{\partial x_1} - \frac{\partial f_1}{\partial x_2} - \frac{\partial f_2}{\partial x_1} + \frac{\partial f_2}{\partial x_2}\right) = 1,$$

and obtain equation for f_2 only:

$$\left(\frac{\partial}{\partial x_1} - \frac{\partial}{\partial x_2}\right)\left(2\frac{\partial f_2}{\partial x_1} + \frac{\partial f_2}{\partial x_2}\right) - \left(\frac{\partial}{\partial x_1} - 2\frac{\partial}{\partial x_2}\right)\left(-\frac{\partial f_2}{\partial x_1} + \frac{\partial f_2}{\partial x_2}\right) = 1.$$

Simplifying, we obtain

$$3\frac{\partial^2 f_2}{\partial x_1^2} - 4\frac{\partial^2 f_2}{\partial x_1 \partial x_2} + \frac{\partial^2 f_2}{\partial x_2^2} = 1,$$

whose characteristic equation is

$$3\left(x_1'\right)^2 - 4x_1'x_2' + \left(x_2'\right)^2 = 0.$$

Instead of eliminating f_1, we could eliminate f_2. We apply

$$-\frac{\partial}{\partial x_1} + \frac{\partial}{\partial x_2}$$

to the first equation, and

$$2\frac{\partial}{\partial x_1} + \frac{\partial}{\partial x_2}$$

to the second one. Subtracting the results, we obtain

$$\left(-\frac{\partial}{\partial x_1} + \frac{\partial}{\partial x_2}\right)\left(\frac{\partial f_1}{\partial x_1} - 2\frac{\partial f_1}{\partial x_2}\right) - \left(2\frac{\partial}{\partial x_1} + \frac{\partial}{\partial x_2}\right)\left(\frac{\partial f_1}{\partial x_1} - \frac{\partial f_1}{\partial x_2}\right) = -1,$$

which we simplify to

$$3\frac{\partial^2 f_1}{\partial x_1^2} - 5\frac{\partial f_1}{\partial x_1 \partial x_2} + \frac{\partial^2 f_1}{\partial x_2^2} = 1,$$

whose characteristic equation is

$$3\left(x_1'\right)^2 - 5x_1'x_2' + \left(x_2'\right)^2 = 0.$$

Dividing by $\left(x_2'\right)^2$, we get

$$3\left(\frac{\mathrm{d}x_1}{\mathrm{d}x_2}\right)^2 - 5\frac{\mathrm{d}x_1}{\mathrm{d}x_2} + 1 = 0,$$

and hence,

$$\frac{\mathrm{d}x_1}{\mathrm{d}x_2} = \frac{5 \pm \sqrt{25 - 12}}{6} = \frac{5 \pm \sqrt{13}}{6}.$$

We denote these values by $a_1$and a_2. Hence, the characteristic curves are straight lines given by

$$x_1 = a_i x_2 + c_i,$$

where c_i are the integration constants with $i = 1, 2$. The change of coordinates along these lines results in

$$\frac{\partial^2 f_1}{\partial y_1 \partial y_2} = -1,$$

where $y_i = x_1 - a_i x_2$ are the new coordinates. In these coordinates the solution is obtained by integrating twice to get

$$\frac{\partial f_1}{\partial y_2} = -y_1 + g\left(y_2\right)$$

and

$$f_1\left(y_1, y_2\right) = -y_1 y_2 + G\left(y_2\right) + H\left(y_1\right),$$

where G and H are arbitrary differentiable functions. In the original coordinates, this solution is

$$f_1\left(x_1, x_2\right) = -\left(x_1 - a_1 x_2\right)\left(x_1 - a_2 x_2\right) + G\left(x_1 - a_2 x_2\right) + H\left(x_1 - a_1 x_2\right).$$

Inserting this solution into the second equation of the original system, we obtain

$$\begin{aligned}\frac{\partial f_2}{\partial x_1} + \frac{\partial f_2}{\partial x_2} &= \frac{\partial f_1}{\partial x_2} - \frac{\partial f_1}{\partial x_1} \\ &= a_1\left(x_1 - a_2 x_2\right) + a_2\left(x_1 - a_1 x_2\right) \\ &\quad - a_2 G'\left(x_1 - a_2 x_2\right) - a_1 H'\left(x_1 - a_1 x_2\right).\end{aligned}$$

If we denote the right-hand side of this equation by $R(x_1, x_2)$, the equation becomes

$$\frac{\partial f_2}{\partial x_1} + \frac{\partial f_2}{\partial x_2} = R(x_1, x_2) . \tag{1.99}$$

To solve this equation, we can consider the characteristic lines of this equation, which are

$$x_1(s) = s + b_1$$

and

$$x_2(s) = s + b_2.$$

Along these lines, equation (1.99) is

$$\frac{\mathrm{d}f_2}{\mathrm{d}s} = R(x_1(s), x_2(s)),$$

and its solution is

$$f_2(x_1(s), x_2(s)) = \int R(x_1(s), x_2(s))\,\mathrm{d}s + C,$$

where the integration constant, C, depends on the choice of the characteristic line along which we integrate. We can identify the lines by the choice of b_1 while setting b_2 to zero. The integral is

$$\begin{aligned}
&\int a_1(x_1(s) - a_2 x_2(s)) + a_2(x_1(s) - a_1 x_2(s)) \\
&\quad - a_2 G'(x_1(s) - a_2 x_2(s)) - a_1 H'(x_1(s) - a_1 x_2(s))\,\mathrm{d}s \\
&= \int a_1(s + b_1 - a_2 s) + a_2(s + b_1 - a_1 s) \\
&\quad - a_2 G'(s + b_1 - a_2 s) - a_1 H'(s + b_1 - a_1 s)\,\mathrm{d}s \\
&= \frac{1}{2}s^2(a_1 - a_2)^2 + s(a_1 b_1 + a_2 b_1) \\
&\quad - \frac{a_2}{1 - a_2} G(s + b_1 - a_2 s) - \frac{a_1}{1 - a_1} H(s + b_1 - a_1 s) .
\end{aligned}$$

Hence,

$$\begin{aligned}
f_2(x_1(s, b_1), x_2(s, b_1)) &= \frac{1}{2}s^2(a_1 - a_2)^2 + s(a_1 b_1 + a_2 b_1) \\
&\quad - \frac{a_2}{1 - a_2} G(s + b_1 - a_2 s) - \frac{a_1}{1 - a_1} H(s + b_1 - a_1 s) \\
&\quad + C(b_1) .
\end{aligned}$$

Since $s = x_2$ and $b_1 = x_1 - x_2$, the solution is

$$\begin{aligned} f_2(x_1,x_2) = {} & \frac{1}{2}x_2^2(a_1-a_2)^2 + x_2(a_1+a_2)(x_1-x_2) \\ & -\frac{a_2}{1-a_2}G(2x_1-x_2-a_2x_2) \\ & -\frac{a_1}{1-a_1}H(2x_1-x_2-a_1x_2) + C(x_1-x_2), \end{aligned}$$

where C, G and H are functions and $a_i = \left(5 \pm \sqrt{13}\right)/6$.

Exercise 1.13. Using the Maxwell equations stated in expressions (1.53)–(1.56) and the vector-calculus identity given by

$$\nabla \times \nabla \times X = \nabla(\nabla \cdot X) - \nabla^2 X, \tag{1.100}$$

where X stands for a vector field, derive equation (1.59).

Solution. Taking the curl of both sides of equation (1.55) and using the equality of mixed partial derivatives, we write

$$\nabla \times \nabla \times E = -\frac{\partial}{\partial t}\nabla \times B.$$

Invoking identity (1.100), we write

$$\nabla(\nabla \cdot E) - \nabla^2 E = -\frac{\partial}{\partial t}\nabla \times B.$$

In view of equation (1.53)—where ρ is a function of position and ϵ_0 is a constant—and using equation (1.56) to express $\nabla \times B$, we write

$$\frac{1}{\epsilon_0}\nabla\rho - \nabla^2 E = -\frac{1}{c^2\epsilon_0}\frac{\partial J}{\partial t} - \frac{1}{c^2}\frac{\partial^2 E}{\partial t^2}.$$

Rearranging, we write

$$\left(c^2\nabla^2 - \frac{\partial^2}{\partial t^2}\right)E = \frac{c^2}{\epsilon_0}\nabla\rho + \frac{1}{\epsilon_0}\frac{\partial J}{\partial t},$$

which in view of the definition of the d'Alembert operator, we state as

$$\Box E = \frac{c^2}{\epsilon_0}\nabla\rho + \frac{1}{\epsilon_0}\frac{\partial J}{\partial t},$$

which is equation (1.59), as required.

Exercise 1.14. Using the Maxwell equations stated in expressions (1.53)–(1.56) and identity (1.100), derive equation (1.60).

Solution. Taking the curl of both sides of equation (1.56) and using the equality of mixed partial derivatives, we write

$$c^2\nabla \times \nabla \times B = \frac{1}{\epsilon_0}\nabla \times J + \frac{\partial}{\partial t}\nabla \times E\,.$$

Invoking identity (1.100), we write

$$c^2\nabla(\nabla \cdot B) - c^2\nabla^2 B = \frac{1}{\epsilon_0}\nabla \times J + \frac{\partial}{\partial t}\nabla \times E\,.$$

In view of equations (1.54) and (1.55), we write

$$-c^2\nabla^2 B = \frac{1}{\epsilon_0}\nabla \times J - \frac{\partial^2 B}{\partial t^2}\,.$$

Rearranging, we write

$$\left(c^2\nabla^2 - \frac{\partial^2}{\partial t^2}\right) B = -\frac{1}{\epsilon_0}\nabla \times J\,,$$

which in view of the definition of the d'Alembert operator, we state as

$$\Box B = -\frac{1}{\epsilon_0}\nabla \times J\,,$$

which is equation (1.60), as required.

Exercise 1.15. Following Example 1.7 and Appendix C, show that equations (1.59) and (1.60) can be derived from equations (1.61) and (1.62).

Solution. Let us write equation (1.61) as

$$\nabla^2 A - \frac{1}{c^2}\frac{\partial^2 A}{\partial t^2} = -\frac{J}{c^2\epsilon_0}. \tag{1.101}$$

Taking the curl of both sides and using the equality of mixed partial derivatives, we get

$$\nabla^2 \nabla \times A - \frac{1}{c^2}\frac{\partial^2 \nabla \times A}{\partial t^2} = -\frac{\nabla \times J}{c^2\epsilon_0}. \tag{1.102}$$

Invoking expression (C.16), namely, $B = \nabla \times A$, we write the above equation as

$$\nabla^2 B - \frac{1}{c^2}\frac{\partial^2 B}{\partial t^2} = -\frac{\nabla \times J}{c^2\epsilon_0},$$

which is equation (1.60), as required.

Let us write equation (1.62) as

$$\nabla^2 \phi - \frac{1}{c^2}\frac{\partial^2 \phi}{\partial t^2} = -\frac{\rho}{\epsilon_0}.$$

Taking the gradient of both sides and using the equality of mixed partial derivatives, we get

$$\nabla^2 \nabla \phi - \frac{1}{c^2} \frac{\partial^2 \nabla \phi}{\partial t^2} = -\frac{\nabla \rho}{\epsilon_0}. \tag{1.103}$$

In view of expression (C.18), namely, $E = -\nabla \phi - \partial A / \partial t$, we add

$$\nabla^2 \frac{\partial A}{\partial t} - \frac{1}{c^2} \frac{\partial^2}{\partial t^2} \frac{\partial A}{\partial t}$$

to both sides of this equation to get

$$\nabla^2 \left(\nabla \phi + \frac{\partial A}{\partial t} \right) - \frac{1}{c^2} \frac{\partial^2}{\partial t^2} \left(\nabla \phi + \frac{\partial A}{\partial t} \right) = -\frac{\nabla \rho}{\epsilon_0} + \nabla^2 \frac{\partial A}{\partial t} - \frac{1}{c^2} \frac{\partial^2}{\partial t^2} \frac{\partial A}{\partial t}.$$

Invoking expression (C.18) and using the equality of mixed partial derivatives, we obtain

$$\nabla^2 E - \frac{1}{c^2} \frac{\partial^2}{\partial t^2} E = \frac{\nabla \rho}{\epsilon_0} - \frac{\partial}{\partial t} \left(\nabla^2 A - \frac{1}{c^2} \frac{\partial^2 A}{\partial t^2} \right).$$

Using equation (1.101), we write

$$\nabla^2 E - \frac{1}{c^2} \frac{\partial^2}{\partial t^2} E = \frac{1}{\epsilon_0} \nabla \rho + \frac{1}{c^2 \epsilon_0} \frac{\partial J}{\partial t},$$

which is equation (1.59), as required.

The fact that equations (1.59) and (1.60) can be derived from equations (1.61) and (1.62) does not imply that the latter are more fundamental than the former. One could argue that the opposite is true, since the potentials appear as solutions of equations stated in terms of fields, as discussed on pages 260–260. Also, examining equations (1.102) and (1.103), we see that they contain third-order derivatives of A and ϕ, respectively. Consequently, a formulation in terms of potentials requires them explicitly to be thrice differentiable, as opposed to equations (1.59) and (1.60), for which E and B need to be differentiable twice.

Chapter 2

Characteristic equations of second-order linear partial differential equations

For a two-dimensional problem the solution $\psi(x, y)$ may be represented by the surface $z = \psi(x, y)$. [...] Cauchy conditions correspond to our specifying not only the line $\psi(s) = z$ but also the normal slope at the edge of the surface $\psi(x, y) = z$. It is as though, instead of a line we had a thin ribbon as a support to the ψ surface, a twisted ribbon which specified slope perpendicular to its axis as well as the height above the z axis.

Morse and Feshbach (1953, Part I)

Preliminary remarks

In the first chapter, we find that characteristics can either be viewed as directional derivatives or as the hypersurfaces along which the side conditions cannot be set arbitrarily. In this chapter, we show that only the second context can be used for second-order equations.

The side conditions for the first-order equations, discussed in Chapter 1, specify the solution along a hypersurface. Several forms of side conditions can be associated with second-order equations. For most of this chapter, we focus our attention on side conditions in the form of the Cauchy data: the solution and its derivative in a transverse direction along a hypersurface.

We begin this chapter with three examples of second-order partial differential equations in two variables for which we find the characteristic curves. In these examples, we use the methods analogous to the ones used in Chapter 1. Subsequently, we formulate a general method to be applied to any second-order semilinear partial differential equation as well as a general method for systems of second-order semilinear partial differential equations. We apply this approach to three equations of mathematical physics: the wave equation, the heat equation, the Laplace equation and to two systems of equations: the elastodynamic equations and the Maxwell equations.

Readers might find it useful to study this chapter together with Appendices B and C, where we formulate the elastodynamic equations and the Maxwell equations, respectively.

2.1 Motivational examples

In contrast to the case of the first-order linear partial differential equations, it is not always possible to find the equations of characteristics for second-order linear partial differential equations using the directional derivatives. There are second-order equations that we cannot express in terms of directional derivatives. An example of such equations is the heat equation discussed in Section 2.1.3. However, we can find their characteristics by the method introduced in Section 1.5, which uses the incompatibility of the side conditions.

2.1.1 *Equation with directional derivative*

2.1.1.1 *Directional derivative*

Let us consider

$$\frac{\partial^2 f}{\partial x_1^2}+2\frac{\partial^2 f}{\partial x_1 \partial x_2}+\frac{\partial^2 f}{\partial x_2^2}=0, \tag{2.1}$$

which we can rewrite as

$$\left(\frac{\partial}{\partial x_1}+\frac{\partial}{\partial x_2}\right)\left(\frac{\partial}{\partial x_1}+\frac{\partial}{\partial x_2}\right)f=0\,,$$

and express as

$$\mathrm{D}_{[1,1]}\mathrm{D}_{[1,1]}f=0\,;$$

$[1,1]$ is a special direction for equation (2.1). The lines parallel to this direction are the characteristics, which herein are straight lines

$$x_1-x_2=C, \tag{2.2}$$

where C is a constant. The solution of equation (2.1) is not constant along the characteristics. Indeed, as shown in Exercise 2.1, this solution is

$$f\left(x_1,x_2\right)=\left(x_1+x_2\right)g\left(x_1-x_2\right)+h\left(x_1-x_2\right)\,, \tag{2.3}$$

where g and h are arbitrary functions of a single variable, x_1-x_2. Along the characteristics, g and h are constant, but the solution f is not. Following an argument analogous to the one on page 7, functions g and h need not be differentiable.

2.1.1.2 *Incompatibility of side conditions*

To illustrate the use of the incompatibility of the side conditions for the second-order differential equations, let us consider equation (2.1) again, together with side conditions

$$f(x_1(s), x_2(s)) = f_0(s) \tag{2.4}$$

and

$$\mathrm{D}_N f(x_1(s), x_2(s)) = f_N(s), \tag{2.5}$$

where $[x_1(s), x_2(s)]$ is a curve parametrized by s, N is a vector normal to this curve, say $[x_2', -x_1']$, and f_0 and f_N are two functions; f_0 specifies the value of f along the curve, and f_N specifies the directional derivative in the direction normal to this curve. Such side conditions, the Cauchy data, are illustrated in Figure 2.1. Since f_N is the derivative along the normal to the curve and f_0 can be used to find the derivative along the curve, these two functions specify the derivative of f in any direction at any point on this curve, as illustrated in Exercise 2.7.

The Cauchy data alone cannot provide us with information about the second derivative in the direction transverse to the hypersurface along which the data are given. To find this derivative, we can invoke the differential equation itself. The differential equation does not provide us with the information about the second derivatives in the transverse direction, if the Cauchy data is given along the characteristics. In such cases, the Cauchy data might contradict the differential equation. The requirement that the side conditions do not contradict the differential equation is given by the compatibility condition, which we discuss below.

By checking if the side conditions are compatible with the differential equation, we find the characteristic curves. We check whether or not the second derivatives along curve $[x_1(s), x_2(s)]$ satisfy the differential equation. In other words, we try to determine the second derivatives along this curve using the given information. We start by expressing the first derivatives in terms of f_0 and f_N.

Taking the derivative of equation (2.4) with respect to s, we get

$$x_1'(s)\frac{\partial f}{\partial x_1}(x_1(s), x_2(s)) + x_2'(s)\frac{\partial f}{\partial x_2}(x_1(s), x_2(s)) = f_0'(s).$$

In view of expression (1.65), we can rewrite expression (2.5) for the normal derivative as

$$[x_2'(s), -x_1'(s)] \cdot \left[\frac{\partial f}{\partial x_1}(x_1(s), x_2(s)), \frac{\partial f}{\partial x_2}(x_1(s), x_2(s))\right] = f_N(s).$$

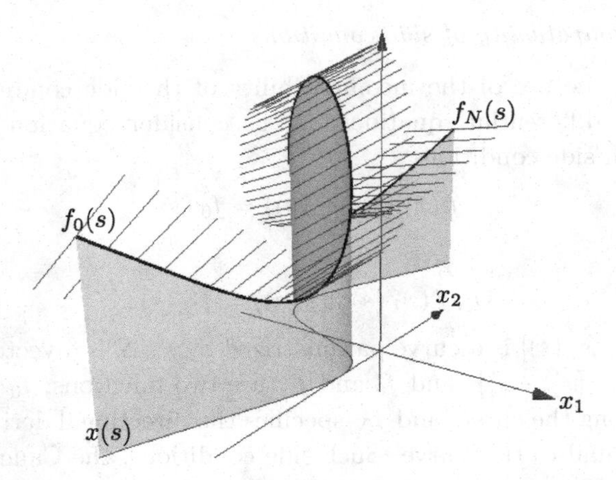

Fig. 2.1 The Cauchy data along curve $[x_1(s), x_2(s)]$ specified by the value of the solution, f_0, and the value of the normal derivative, f_N, along this curve. The value of the normal derivative is indicated by the slopes of the family of straight lines.

The last two equations form a system of linear algebraic equations for the first derivatives, $\partial f/\partial x_1$ and $\partial f/\partial x_2$, along curve $[x_1(s), x_2(s)]$,

$$\begin{bmatrix} x_1'(s) & x_2'(s) \\ x_2'(s) & -x_1'(s) \end{bmatrix} \begin{bmatrix} \dfrac{\partial f}{\partial x_1}(x_1(s), x_2(s)) \\ \dfrac{\partial f}{\partial x_2}(x_1(s), x_2(s)) \end{bmatrix} = \begin{bmatrix} f_0'(s) \\ f_N(s) \end{bmatrix}.$$

Solving this system, we get

$$\frac{\partial f}{\partial x_1}(x_1(s), x_2(s)) = \frac{f_0'(s)\, x_1'(s) + f_N(s)\, x_2'(s)}{(x_1')^2 + (x_2')^2} \tag{2.6}$$

and

$$\frac{\partial f}{\partial x_2}(x_1(s), x_2(s)) = \frac{-f_N(s)\, x_1'(s) + f_0'(s)\, x_2'(s)}{(x_1')^2 + (x_2')^2}, \tag{2.7}$$

where we assume the equality of the mixed partial derivatives. We wish to find the second derivatives of f. For convenience, we denote the right-hand sides of the above expressions by a_1 and a_2, to write

$$\frac{\partial f}{\partial x_1}(x_1(s), x_2(s)) = a_1(s)$$

and

$$\frac{\partial f}{\partial x_2}(x_1(s), x_2(s)) = a_2(s). \tag{2.8}$$

Differentiating these two equations with respect to s, we obtain

$$x_1'(s)\frac{\partial^2 f}{\partial x_1^2}(x_1(s),x_2(s))+x_2'(s)\frac{\partial^2 f}{\partial x_1 \partial x_2}(x_1(s),x_2(s))=a_1'(s) \quad (2.9)$$

and

$$x_1'(s)\frac{\partial^2 f}{\partial x_1 \partial x_2}(x_1(s),x_2(s))+x_2'(s)\frac{\partial^2 f}{\partial x_2^2}(x_1(s),x_2(s))=a_2'(s). \quad (2.10)$$

These are two equations for three unknowns, $\partial^2 f/\partial x_1^2$, $\partial^2 f/\partial x_1 \partial x_2$ and $\partial^2 f/\partial x_2^2$. The third equation, which is necessary to solve for the second derivatives, is the original differential equation. We can write the three equations as

$$\begin{bmatrix} x_1'(s) & x_2'(s) & 0 \\ 0 & x_1'(s) & x_2'(s) \\ 1 & 2 & 1 \end{bmatrix} \begin{bmatrix} \dfrac{\partial^2 f}{\partial x_1^2} \\ \dfrac{\partial^2 f}{\partial x_1 \partial x_2} \\ \dfrac{\partial^2 f}{\partial x_2^2} \end{bmatrix} = \begin{bmatrix} a_1'(s) \\ a_2'(s) \\ 0 \end{bmatrix}, \quad (2.11)$$

where the last equation is equation (2.1). If the determinant of the coefficient matrix of this system is zero, the system has no unique solution. In such a case, the side conditions are given along a curve that does not provide us with the information needed to solve the Cauchy problem. To find this curve, we set the determinant to zero to write

$$(x_1'(s))^2 - 2x_1'(s)\,x_2'(s) + (x_2'(s))^2 = 0,$$

which is

$$(x_1'(s) - x_2'(s))^2 = 0.$$

Hence,

$$x_1'(s) = x_2'(s), \quad (2.12)$$

which means that

$$x_1(s) - x_2(s) = C, \quad (2.13)$$

where C is a constant. This is equation (2.2) parametrized by s.

Equation (2.13) describes the family of curves along which we cannot determine uniquely the second derivatives from the side conditions that are given along these curves. These are the characteristic curves.

The impossibility of obtaining second, and higher, derivatives means that we cannot construct the convergent Taylor series of the solution, which implies that we are not able to solve uniquely the original differential equation. There are two possible cases: either there are no solutions or there are infinitely many solutions. The case of no solutions results from the side conditions that contradict the differential equation. The case of infinitely many solutions results from the side conditions failing to add the needed information about the second derivatives.

2.1.1.3 *Compatibility conditions*

Let us examine the case of infinitely many solutions of system (2.11). This case is tantamount to the original differential equation being a linear combination of the equations resulting from the side conditions. The condition for such a linear combination is called the compatibility condition. There are many ways of establishing such a condition, such as the Gauss-elimination method. To illustrate another way of obtaining the compatibility condition, let us revisit equation (2.1) and the corresponding system (2.11). We see that the condition for linear dependence can be written as

$$b_1\left[x_1', x_2', 0, a_1'\right] + b_2\left[0, x_1', x_2', a_2'\right] = \left[1, 2, 1, 0\right],$$

which is a system of four equations for six unknowns. From the first equation it follows that

$$b_1 = \frac{1}{x_1'}. \tag{2.14}$$

From the third equation it follows that

$$b_2 = \frac{1}{x_2'}. \tag{2.15}$$

Using these equalities, we see that the second equation implies that $x_1' = x_2'$, which is the characteristic equation given by expression (2.12). The fourth equation results in

$$a_1' + a_2' = 0, \tag{2.16}$$

which is the compatibility condition for the side conditions f_0 and f_N along the characteristic curve given by $x_1 - x_2 = C$.

We can use the compatibility condition to propagate the side conditions along the characteristics to obtain the solution of the original Cauchy problem. Even though we motivated our approach by using the Taylor series to find the solution, we see that the characteristics allow us to find solutions that do not require analyticity of the side conditions.

As shown in Exercise 2.8, compatibility condition (2.16) implies that

$$f_0(s) = x_1(s)\,k + l_1\,, \tag{2.17}$$

and, equivalently,

$$f_0(s) = x_2(s)\,k + l_2\,,$$

where k, l_1 and l_2 are constants. We see that the side condition given by expression (2.17) along the characteristic curves does not provide any information that is not contained in the original differential equation. The original differential equation states that the second derivative of a function in the direction of the characteristic curves is zero: the value of the function increases linearly in that direction, which is the statement of expression (2.17). Such a side condition gives redundant information about the solutions. Note that we do not restrict the information given by f_N, which can be arbitrary.

2.1.2 *Wave equation in one spatial dimension*

2.1.2.1 *Directional derivative*

We consider

$$\frac{\partial^2 f(x_1, x_2)}{\partial x_1^2} = c^2 \frac{\partial^2 f(x_1, x_2)}{\partial x_2^2}\,, \tag{2.18}$$

where c is a constant. We can interpret this equation as the wave equation if the independent variables x_1 and x_2 represent time and space, respectively. We can rewrite this equation as

$$\left(c\frac{\partial}{\partial x_2} + \frac{\partial}{\partial x_1}\right)\left(c\frac{\partial}{\partial x_2} - \frac{\partial}{\partial x_1}\right) f(x_1, x_2) = 0\,, \tag{2.19}$$

as shown in Exercise 2.10. Furthermore, we can write equation (2.19) as

$$\left([1, c]\cdot\left[\frac{\partial}{\partial x_1}, \frac{\partial}{\partial x_2}\right]\right)\left([-1, c]\cdot\left[\frac{\partial}{\partial x_1}, \frac{\partial}{\partial x_2}\right]\right) f(x_1, x_2) = 0\,,$$

where each term in parentheses is a directional-derivative operator. Thus, we write

$$\mathrm{D}_{[1,c]}\mathrm{D}_{[-1,c]} f(x_1, x_2) = 0\,,$$

where D_X denotes the directional derivative along vector X. In view of the equality of mixed partial derivatives, we write the above equation as

$$\mathrm{D}_{[-1,c]}\mathrm{D}_{[1,c]} f(x_1, x_2) = 0.$$

We see that any function that is constant along direction $[1, c]$ or $[-1, c]$ is a solution of equation (2.18). Thus,

$$f(x_1, x_2) = g(x_2 - cx_1) + h(x_2 + cx_1) \tag{2.20}$$

is a solution of equation (2.18), and the straight lines given by

$$x_2 \mp cx_1 = C_\pm \tag{2.21}$$

are the characteristics, where C_+ parametrizes the characteristics with the slope c and C_- parametrizes the characteristics with the slope $-c$.

For the wave equation, as shown in Exercise 2.9, the side conditions specify the displacement along a space-time curve at the initial time, as well as the rate of change of displacement with time. If the space-time curve is a characteristic, we know—in view of expression (2.20)—that the displacement remains unchanged along this line. Hence, we are not free to specify its change. Pictorially, we represent characteristics (2.21) in Figure 2.2.

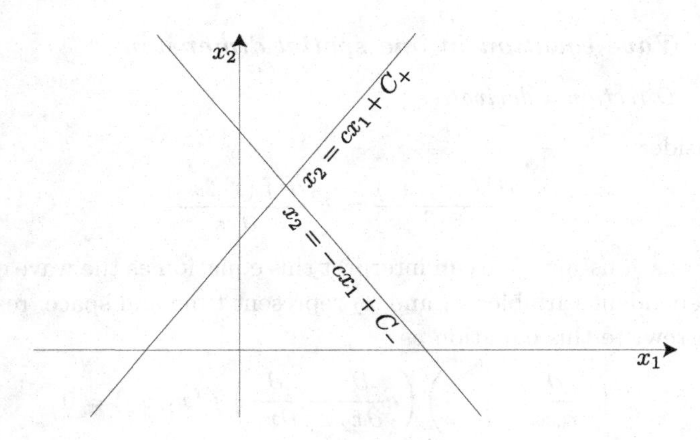

Fig. 2.2 Characteristic curves for the wave equation. In contrast to the first-order linear equations, the characteristic curves cross one another; they are not integral curves of the differential equation.

2.1.2.2 *Incompatibility of side conditions*

Let us find the characteristics using the approach introduced in Section 1.5.

We consider equation (2.18) together with the side conditions along curve $[x_1(s), x_2(s)]$ that are given by

$$f(x_1(s), x_2(s)) = f_0(s) \tag{2.22}$$

and

$$\mathrm{D}_X f\left(x_1(s), x_2(s)\right) = f_X(s), \tag{2.23}$$

where X is a vector that is transverse to curve $[x_1(s), x_2(s)]$, say $[x_2', -x_1']$, and f_0 and f_X are functions: f_0 specifies the value of f along curve $[x_1(s), x_2(s)]$ and f_X specifies the directional derivative in the direction transverse to this curve, which we choose herein to be normal to the curve. The physical insight into this formulation of side conditions is discussed in Exercise 2.9.

Following the approach described in detail in Section 2.1.1.2, we obtain two equations for the three unknown second derivatives, which are equations (2.9) and (2.10). The third equation necessary to solve for the second derivatives is equation (2.18). We can write the three equations as

$$\begin{bmatrix} x_1'(s) & x_2'(s) & 0 \\ 0 & x_1'(s) & x_2'(s) \\ 1 & 0 & -c^2 \end{bmatrix} \begin{bmatrix} \dfrac{\partial^2 f}{\partial x_1^2} \\ \dfrac{\partial^2 f}{\partial x_1 \partial x_2} \\ \dfrac{\partial^2 f}{\partial x_2^2} \end{bmatrix} = \begin{bmatrix} a_1'(s) \\ a_2'(s) \\ 0 \end{bmatrix}. \tag{2.24}$$

If the determinant of this system is zero, there is no unique solution. The determinant equal to zero means that

$$\left(c x_1'(s)\right)^2 = \left(x_2'(s)\right)^2,$$

which implies that

$$x_2'(s) = \pm c x_1'(s).$$

Hence,

$$x_2 \mp c x_1 = C_{\pm} \tag{2.25}$$

are curves along which we cannot determine uniquely the second derivatives from the differential equation and its side conditions that are given along these curves; we are not able to solve uniquely the original differential equation. As expected, equation (2.25) is equation (2.21).

We follow this method to examine the compatibility condition. The side conditions are compatible with the differential equation only if the equations in system (2.24) are linearly dependent on each other. This requirement translates into

$$a_1' x_1' = a_2' x_2',$$

which, considering the expression for the characteristics, can be written as

$$a_1' = \pm c a_2' . \tag{2.26}$$

Integrating this equation with respect to s, we obtain

$$a_1 = \pm c a_2 + k_\pm ,$$

where $k_\pm$ depends on the characteristic along which we integrate expression (2.26); it depends on $C_\pm$. Thus, we write the compatibility conditions at point (x_1, x_2) as

$$a_1 (x_1, x_2) \mp c a_2 (x_1, x_2) = k_\pm (C_\pm (x_1, x_2)) ,$$

and, after substituting for a_1 and a_2 from expressions (2.8), as

$$\frac{\partial f}{\partial x_1} \mp c \frac{\partial f}{\partial x_2} = k_\pm (C_\pm) . \tag{2.27}$$

We can use the compatibility conditions in a manner similar to the one used in Section 1.5 for linear first-order equations, and obtain solutions along the characteristics. We consider side conditions along a noncharacteristic curve. To compare our results with a more standard approach in other books, for example Folland (1995, p. 165) and McOwen (2003, p. 75),[1] we set the side conditions to be

$$f (0, x_2) = f_0 (x_2) \equiv g (x_2)$$

and

$$\frac{\partial f}{\partial x_1} (0, x_2) = f_X (x_2) \equiv h (x_2) . \tag{2.28}$$

Comparing these conditions to condition (2.27), we conclude that

$$k_\pm (C (0, x_2)) = h (x_2) \mp c g' (x_2) .$$

Since $k_\pm$ is constant along each characteristic, we express its dependence on x_1 and x_2 as

$$k_\pm (x_1, x_2) = h (x_2 \mp c x_1) \mp c g' (x_2 \mp c x_1) .$$

Substituting this result into equation (2.27), we see that

$$\begin{aligned} D_{[1,\mp c]} f (x_1, x_2) &= \frac{\partial f}{\partial x_1} (x_1, x_2) \mp c \frac{\partial f}{\partial x_2} (x_1, x_2) \\ &= h (x_2 \mp c x_1) \mp c g' (x_2 \mp c x_1) . \end{aligned}$$

[1]Note that in McOwen (2003) the term "compatibility condition" is used to express selfconsistency of the side conditions, and is not interchangeable with our use of the same term.

This is a linear system for $\partial f/\partial x_1$ and $\partial f/\partial x_2$ whose solution is

$$\frac{\partial f}{\partial x_1}(x_1, x_2) = \frac{1}{2}\left(h(x_2 - cx_1) + h(x_2 + cx_1)\right.$$
$$\left. - cg'(x_2 - cx_1) + cg'(x_2 + cx_1)\right)$$

and

$$\frac{\partial f}{\partial x_2}(x_1, x_2) = \frac{1}{2c}\left(h(x_2 + cx_1) - h(x_2 - cx_1)\right.$$
$$\left. + cg'(x_2 + cx_1) + cg'(x_2 - cx_1)\right).$$

Integrating with respect to x_1 and x_2, we obtain the solution of equation (2.18),

$$f = \frac{1}{2c}\int_{x_2 - cx_1}^{x_2 + cx_1} h(\xi)\,\mathrm{d}\xi + \frac{1}{2}\left(g(x_2 + cx_1) + g(x_2 - cx_1)\right). \tag{2.29}$$

In the context of the wave equation in one spatial dimension, where x_1 stands for time and x_2 for space, this expression is the d'Alembert solution.

This solution is illustrated in Figure 2.3, where we set the initial displacement to be $g(0, x_2) = 3\exp\left\{-(x_2 + 10)^2/4\right\}$ and the initial velocity to be $h(0, x_2) = \exp\left\{-(x_2 - 35)^2/4\right\}$. We observe the behaviour of the solution resulting from the initial displacement and from the initial velocity. The initial displacement propagates along the characteristics only; the initial velocity affects also the region between them. The latter property is the result of the integral in expression (2.29).

To emphasize the fact that along the characteristics of second-order equations we cannot set freely both the values of the function and its transverse derivative along a hypersurface, we examine solution (2.20), namely,

$$f(x_1, x_2) = g(x_2 - cx_1) + h(x_2 + cx_1).$$

We can set f along a characteristic curve, say $cx_1 = x_2$, to any function, say f_0. In such a case, solution (2.20) is

$$f(x_1, x_2) = g(x_2 - cx_1) + f_0(x_2 + cx_1) - g(0).$$

At the first glance it might appear that this result contradicts the defining property of a characteristic curve, namely that we cannot arbitrarily set the side conditions along such a curve. However, we must remember that the side conditions given by expressions (2.22) and (2.23) correspond to both f_0 and f_X, which are functions that do not depend on one another. We see that the derivative of f in the direction of $[c, -1]$ along the characteristic

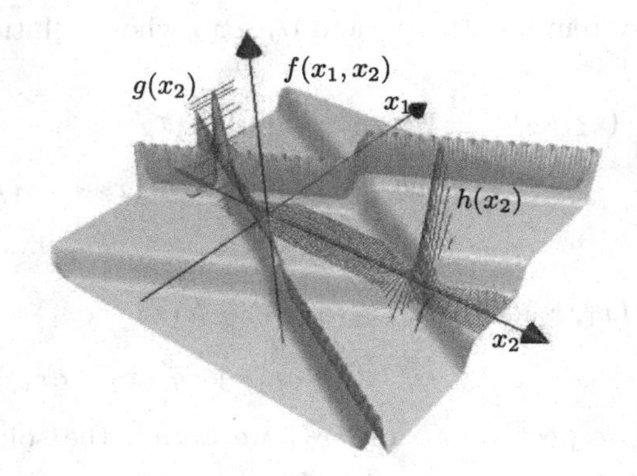

Fig. 2.3 Solution of the wave equation with the Cauchy data. The solution is shown in gray, the Cauchy data along $x_1 = 0$ are shown as a curve for the value of the function and as a family of straight lines for its derivative. The value of the function propagates along the characteristics; the propagation of its derivative is illustrated by steps between characteristics.

$cx_1 = x_2$ is given by

$$\begin{aligned}
&[c, -1] \cdot \left[\frac{\partial f}{\partial x_1}(x_1, cx_1), \frac{\partial f}{\partial x_2}(x_1, cx_1)\right] \\
&= [c, -1] \cdot [-cg'(0) + cf_0'(2cx_1), g'(0) + f_0'(2cx_1)] \\
&= (c^2 - 1) f_0'(2cx_1) - (c^2 + 1) g'(0),
\end{aligned}$$

which must be equal to f_X ; hence, we cannot set the side conditions arbitrarily along these curves. The restrictions on the side conditions along the characteristics are compatibility conditions (2.27).

2.1.3 *Heat equation in one spatial dimension*

We consider

$$\frac{\partial^2 f}{\partial x_1^2} = c\frac{\partial f}{\partial x_2}, \tag{2.30}$$

where c is a constant. We can interpret this equation as the heat equation, if the independent variables x_1 and x_2 represent space and time, respectively. We cannot write this equation using the directional derivative. However, we can investigate the possibility of obtaining the second derivatives from the side conditions.

Let us consider equation (2.30) together with the side conditions along curve $[x_1(s), x_2(s)]$ that are given by

$$f(x_1(s), x_2(s)) = f_0(s) \tag{2.31}$$

and

$$\mathrm{D}_X f\left(x_1(s), x_2(s)\right) = f_X(s), \tag{2.32}$$

where X is a vector that is not tangent to curve $\left[x_1(s), x_2(s)\right]$, say $\left[x_2', -x_1'\right]$, and f_0 and f_X are functions; f_0 specifies the value of f along curve $\left[x_1(s), x_2(s)\right]$ and f_X specifies the directional derivative in the direction transverse to this curve, which herein we choose to be normal to the curve.

As shown in Sections 2.1.1.2 and 2.1.2.2, by differentiating equations (2.31) and (2.32) along the curve $\left[x_1(s), x_2(s)\right]$, we obtain two equations for three unknowns, namely, equations (2.9) and (2.10). The third equation is equation (2.30). We write the three equations as

$$\begin{bmatrix} x_1'(s) & x_2'(s) & 0 \\ 0 & x_1'(s) & x_2'(s) \\ 1 & 0 & 0 \end{bmatrix} \begin{bmatrix} \dfrac{\partial^2 f}{\partial x_1^2} \\ \dfrac{\partial^2 f}{\partial x_1 \partial x_2} \\ \dfrac{\partial^2 f}{\partial x_2^2} \end{bmatrix} = \begin{bmatrix} f_0'' - \dfrac{\partial f}{\partial x_1} x_1'' - \dfrac{\partial f}{\partial x_2} x_2'' \\ f_X' - \dfrac{\partial f}{\partial x_1} x_2'' + \dfrac{\partial f}{\partial x_2} x_1'' \\ c \dfrac{\partial f}{\partial x_2} \end{bmatrix}. \tag{2.33}$$

If the determinant of the coefficient matrix of this system is equal to zero, the system has no unique solution. The determinant is equal to zero if

$$\left(x_2'(s)\right)^2 = 0,$$

which is the characteristic equation of equation (2.30). Hence,

$$x_2(s) = C \tag{2.34}$$

are curves along which we cannot determine uniquely the second derivatives from the side conditions that are given along these curves; we are not able to solve uniquely the original differential equation. Characteristics (2.34) are illustrated in Figure 2.4.

For the heat equation, the side condition represents the temperature along a space-time curve and a change of temperature with time. In this case, the characteristic curve, $x_2 = C$, is a line of constant time. This means that we cannot specify temperature at a particular instant together with the temporal change of temperature.

To understand why curves $x_2 = C$ are characteristics, we consider the Cauchy data along such a curve:

$$f\left(x_1, C\right) = f_0\left(x_1\right)$$

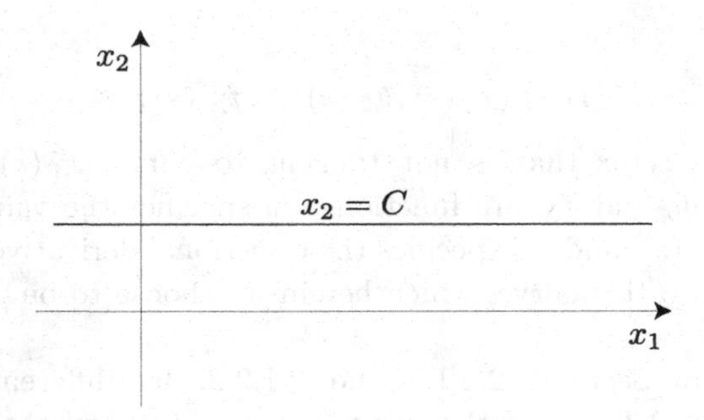

Fig. 2.4 Characteristic curve for the heat equation. In contrast to the wave equation, we cannot propagate the side conditions along the characteristics to obtain the solution.

and

$$D_{[0,1]}f(x_1, C) = f_N(x_1).$$

The second expression states that $\partial f/\partial x_2$ is fixed along $x_2 = C$. The heat equation, however, relates this derivative to the second derivative of f along x_1, and, since these two functions, f_0 and f_N, are related by the heat equation, we cannot choose them arbitrarily.

If we try to find the solution of the heat equation following the method used for the wave equation in Section 2.1.2.2, we fail. The reason for the failure is the fact that the middle equation of system (2.33) is linearly independent of the other two equations. Along the characteristics, which correspond to the determinant being zero, the first and third equations are multiples of one another. As shown in Exercise 2.2, following this method we obtain only a trivial result: the solution has to satisfy the original heat equation. The reason for this is that we cannot integrate the second-order equation along the characteristics to obtain a first-order equation as we did with the wave equation in equation (2.26).

Even though we are not able to find the solution of the heat equation using characteristics, we can find it using the Taylor series, as we discuss in Exercise 2.3. As we show in Exercise 2.4, if we choose the side conditions along the x_1-axis, it is not necessary to have information about the derivative of the solution along the transverse direction to the side condition. Thus, for the heat equation, we need to provide a different type of side condition depending on the curve along which we specify it; there is no unique type of side condition for the heat equation. In Exercise 2.4, we show that along the characteristics we can set the value of the solution arbitrarily and not the value of its derivatives.

In physical applications, the solutions of a differential equation are often restricted to a bounded domain. In the case of the heat equation, the solutions are constrained by a bounded spatial domain, say $x_1 \in \Omega$, and nonegative time, $x_2 \geqslant 0$. Along $x_2 = 0$, the side condition should be given by the value of the solution, say $u(x_1, 0) = g(x_1)$, with $x_1 \in \Omega$—this type of a side condition is the initial condition, since it is given at time $x_2 = 0$. If Ω spans all real space, then this condition, assuming that g is analytical, is sufficient to find the solution. However, if Ω is bounded, then we need to supply extra information about the solution at the boundary of Ω. This side condition is the boundary condition, since it is considered at the boundary of Ω. We note that the distinction between the initial and boundary conditions requires the physical meaning of the differential equation; mathematically, both conditions are given along a geometrical boundary of $\Omega \times [0, \infty)$. Since we are considering the entire boundary of Ω, we should not provide both the value and the derivative of the solution, as we discuss in the next section.

2.1.4 *Laplace equation in two spatial dimensions*

Another important second-order differential equation is the Laplace equation:

$$\frac{\partial^2 f(x_1, x_2)}{\partial x_1^2} + \frac{\partial^2 f(x_1, x_2)}{\partial x_2^2} = 0. \tag{2.35}$$

In spite of its similarity to equation (2.18), the physical and mathematical interpretation of these equations is quite different. In two spatial dimensions, where x_1 and x_2 represent the spatial coordinates, the left-hand side of equation (2.35) is the same as the left-hand sides of both the wave equation and the heat equation. However, with the absence of the time variable, the Laplace equation describes a steady-state case. If we follow the methods described above, we find that the Laplace equation has no real characteristics.

As an aside, we can relate the Laplace equation to the wave equation by considering functions in a complex domain: $z_1 = x_1 + \iota y_1$ and $z_2 = x_2 + \iota y_2$. If the real part satisfies the Laplace equation, then—if we assume that the solution is complex analytic—the solution satisfies the following equation, as shown in Exercise 2.11.

$$\frac{\partial^2 f(x_1, x_2)}{\partial x_1^2} - \frac{\partial^2 f(x_1, x_2)}{\partial y_2^2} = 0,$$

which has real characteristics. Following an approach analogous to the one presented in Section 2.1.2.1, we infer that the characteristics of equation (2.35) are straight lines in the complex plane.

The absence of real characteristics does not mean that we are free to set the side conditions along an arbitrary hypersurface, as illustrated in Exercise 2.5, where we show that the Cauchy data for the Laplace equation in a bounded domain is overdetermined: we can set either the value of the solution along the boundary or the normal derivative, or a combination of the two. Side conditions that specify the value of the solution along the boundary are called the Dirichlet conditions; side conditions that specify the value of the derivative transverse to the boundary are called the Neumann conditions.

The characteristics of the Laplace equation, which are in a complex domain, and the characteristics of the heat equation, which are perpendicular to the time axis, do not have a direct physical meaning. In the remainder of the book, we focus our attention on equations akin to the wave equation; in the next section, we formulate a classification that distinguishes among these equations.

2.2 Hyperbolic, parabolic and elliptic equations

In this section, we classify the second-order linear partial differential equations as hyperbolic, parabolic or elliptic equations, which are exemplified by the wave, heat and Laplace equations, respectively.

To motivate the method of classification, let us write a second-order linear equation in two variables

$$A_{11}\frac{\partial^2 f}{\partial x_1^2}+A_{12}\frac{\partial^2 f}{\partial x_1 \partial x_2}+A_{22}\frac{\partial^2 f}{\partial x_2^2}=B_1\frac{\partial f}{\partial x_1}+B_2\frac{\partial f}{\partial x_2}+Cf+D. \quad (2.36)$$

Equations (2.18), (2.30) and (2.35) can be written in this form. The left-hand side of equation (2.36), which consists of its highest-order derivatives, is associated with the principal symbol,

$$A_{11}\xi_1^2+A_{12}\xi_1\xi_2+A_{22}\xi_2^2, \quad (2.37)$$

which can be obtained by the Fourier transform, as discussed in Appendix E.4. Using the matrix notation, we write this algebraic expression as

$$[\xi_1,\xi_2]\begin{bmatrix} A_{11} & \frac{1}{2}A_{12} \\ \frac{1}{2}A_{21} & A_{22}\end{bmatrix}\begin{bmatrix}\xi_1\\ \xi_2\end{bmatrix}. \quad (2.38)$$

If we denote the 2×2 matrix by A, we say that equation (2.36) is hyperbolic if the eigenvalues of A have the opposite signs. It is parabolic if A is singular—in other words, if one of its eigenvalues is zero. It is elliptic if its eigenvalues have the same sign, which is equivalent to matrix A being positive-definite or negative-definite.

This classification of differential equations can be extended to functions of more than two independent variables. Expression (2.38) can be generalized to

$$[\xi_1, \cdots, \xi_n] \begin{bmatrix} A_{11} & \cdots & \frac{1}{2}A_{1n} \\ \vdots & \ddots & \vdots \\ \frac{1}{2}A_{n1} & \cdots & A_{nn} \end{bmatrix} \begin{bmatrix} \xi_1 \\ \vdots \\ \xi_n \end{bmatrix}, \tag{2.39}$$

with the following eigenvalue criteria. An equation is hyperbolic if one eigenvalue of A has the sign that is opposite to the signs of other eigenvalues. It is called parabolic if one of its eigenvalues is zero and the other eigenvalues are of the same sign. It is called elliptic if its eigenvalues have the same sign, which—in view of A being symmetric—is equivalent to matrix A being definite. Equations whose eigenvalues do not belong to any of these three patterns are rare in physics and we do not discuss them.

Considering the three spatial dimensions and time, we write the corresponding quadratic forms in the matrix form for the wave, heat and Laplace equations, respectively, as

$$\begin{bmatrix} 1 & 0 & 0 & 0 \\ 0 & 1 & 0 & 0 \\ 0 & 0 & 1 & 0 \\ 0 & 0 & 0 & -c^2 \end{bmatrix}, \tag{2.40}$$

$$\begin{bmatrix} 1 & 0 & 0 & 0 \\ 0 & 1 & 0 & 0 \\ 0 & 0 & 1 & 0 \\ 0 & 0 & 0 & 0 \end{bmatrix} \tag{2.41}$$

and

$$\begin{bmatrix} 1 & 0 & 0 \\ 0 & 1 & 0 \\ 0 & 0 & 1 \end{bmatrix}, \tag{2.42}$$

where c is constant. For the hyperbolic equation, one eigenvalue is positive; for the parabolic equation, one eigenvalue is zero; for the elliptic equation, all eigenvalues are negative. Notably, we could scale the variables in such a way that all eigenvalues would be ± 1 or 0. The combination of diagonalization and scaling allow us to study general second-order linear partial differential equations in a convenient form.

If the coefficients are variable—in other words if A is a function of x—the equation can belong to different classes in different regions of domain x, as illustrated in Exercise 2.6.[2]

To justify the nomenclature used to classify differential equations, consider an algebraic equation formed by equating expressions (2.37) and (2.38) to a constant,

$$A_{11}\xi_1^2 + A_{12}\xi_1\xi_2 + A_{22}\xi_2^2 = [\xi_1, \xi_2]\begin{bmatrix} A_{11} & \frac{1}{2}A_{12} \\ \frac{1}{2}A_{21} & A_{22} \end{bmatrix}\begin{bmatrix} \xi_1 \\ \xi_2 \end{bmatrix} = K. \tag{2.43}$$

If $\det A < 0$, which is tantamount to the eigenvalues of the corresponding matrix having the opposite signs, the curve in $\xi_1\xi_2$-space given by equation (2.43) is a hyperbola. If $\det A = 0$, which is tantamount to an eigenvalue being zero, it is a parabola. If $\det A > 0$, which is tantamount to the eigenvalues having the same sign, it is an ellipse. Similarly to each of the three conic sections having distinct properties in terms of boundedness and asymptotics, each of the three types of second-order linear partial differential equations has different properties in terms of characteristics. Since only the hyperbolic equations can be solved by the method of characteristics, they constitute our main focus in this book.

2.3 Characteristics

In this section, we generalize the method based on the incompatibility of side conditions, which allows us to obtain the characteristic curves for all three equations discussed in Section 2.1. This method works not only for linear equations but also for semilinear second-order partial differential equations and for systems of such equations.

[2]Readers interested in an example of a single equation that is hyperbolic, parabolic and elliptic, depending on the region of x, might refer to Strauss (2008, p. 31, Example 2).

2.3.1 *Semilinear equations*

Let us follow the procedure described in Section 2.1.1.2 for the general case of a semilinear second-order equation, namely,

$$\sum_{i,j=1}^{n} A_{ij}(x)\frac{\partial^2 f}{\partial x_i \partial x_j} = B\left(x, f, \frac{\partial f}{\partial x}\right), \tag{2.44}$$

where the right-hand side can depend on the sought-after function and its first derivatives in a nonlinear way. We consider a side condition along hypersurface $x_i = x_i(s_1, \ldots, s_{n-1})$, $i = 1, \ldots, n$, which we write for short as $x = x(s_1, \ldots, s_{n-1})$. The side condition is given by

$$f(x(s_1, \ldots, s_{n-1})) = f_0(s_1, \ldots, s_{n-1}) \tag{2.45}$$

and

$$D_X f(x(s_1, \ldots, s_{n-1})) = f_X(s_1, \ldots, s_{n-1}), \tag{2.46}$$

where X is a vector that is not tangent to hypersurface $x(s_1, \ldots, s_{n-1})$, and f_0 and f_X are functions: f_0 specifies the value of f along hypersurface $x(s_1, \ldots, s_{n-1})$, f_X specifies the value of the directional derivative in the direction transverse to this surface.

To find the second derivatives of f along hypersurface $x(s_1, \ldots, s_{n-1})$, we begin by computing the first derivatives of f along this hypersurface. We differentiate expression (2.45) with respect to all $n-1$ parameters s_j, namely,

$$\sum_{i=1}^{n} \frac{\partial x_i(s_1, \ldots, s_{n-1})}{\partial s_j}\frac{\partial f}{\partial x_i}(x(s_1, \ldots, s_{n-1})) = \frac{\partial f_0(s_1, \ldots, s_{n-1})}{\partial s_j},$$

which gives us $n-1$ equations for n unknowns, $\partial f/\partial x_i$. To get a system of n equations for n unknowns, we use also equation (2.46), which can be written as

$$\sum_{i=1}^{n} X_i(s_1, \ldots, s_{n-1})\frac{\partial f}{\partial x_i}(x(s_1, \ldots, s_{n-1})) = f_X(s_1, \ldots, s_{n-1}).$$

We solve this system to obtain

$$\frac{\partial f}{\partial x_i}(x(s_1, \ldots, s_{n-1})) = a_i(s_1, \ldots, s_{n-1}), \tag{2.47}$$

where $a_i(s_1, \ldots, s_{n-1})$ are the n solutions of the system. Having side conditions (2.45) and (2.46), we can obtain the unique solution for a_i, as illustrated in Exercise 2.7. To find the equations for the second derivatives, we differentiate the n equations (2.47) with respect to all $n-1$ parameters s_k, namely,

$$\sum_{j=1}^{n} \frac{\partial x_j(s_1,\ldots,s_{n-1})}{\partial s_k} \frac{\partial^2 f}{\partial x_i \partial x_j}(x(s_1,\ldots,s_{n-1})) = \frac{\partial a_i(s_1,\ldots,s_{n-1})}{\partial s_k}, \tag{2.48}$$

which provides us with $n(n-1)$ equations for $(n+1)n/2$ second derivatives, $\partial^2 f/\partial x_i \partial x_j$. Among these equations there are $(n+1)n/2-1$ linearly independent equations, since the right-hand side of equation (2.48) contains mixed derivatives, as can be seen from equation (2.47). In the following paragraphs, we give a detailed explanation of this fact by introducing more convenient coordinates.

We choose a new coordinate system, $\tilde{x}_1, \tilde{x}_2, \ldots, \tilde{x}_n$, in such a way, that, at least locally, the hypersurface, $x(s_1,\ldots,s_{n-1})$, is expressed in these coordinates as $\tilde{x}_n = 0$ and the transverse vector along which we know the derivative of f along the hypersurface is expressed in the new coordinates as $[0,\ldots,0,1]$. Hence, we can replace the $n-1$ parameters s_j by $n-1$ coordinates $\tilde{x}_1, \tilde{x}_2, \ldots, \tilde{x}_{n-1}$. In the new coordinates, the original differential equation and the side conditions are

$$\sum_{i,j=1}^{n} \tilde{A}_{ij}(\tilde{x}) \frac{\partial^2 \tilde{f}}{\partial \tilde{x}_i \partial \tilde{x}_j} = \tilde{B}\left(\tilde{x}, \tilde{f}, \frac{\partial \tilde{f}}{\partial \tilde{x}}\right) \tag{2.49}$$

and

$$\begin{aligned} \tilde{f}(\tilde{x}_1,\ldots,\tilde{x}_{n-1},0) &= \tilde{f}_0(\tilde{x}_1,\ldots,\tilde{x}_{n-1}), \\ \mathrm{D}_{[0,\ldots,0,1]}\tilde{f}(\tilde{x}_1,\ldots,\tilde{x}_{n-1},0) &= \tilde{f}_X(\tilde{x}_1,\ldots,\tilde{x}_{n-1}), \end{aligned} \tag{2.50}$$

respectively, where

$$\tilde{f}(\tilde{x}_1,\ldots,\tilde{x}_n) = f(x_1(\tilde{x}_1,\ldots,\tilde{x}_n),\ldots,x_n(\tilde{x}_1,\ldots,\tilde{x}_n)).$$

The form of equation (2.49) is the same as the form of equation (2.44). However, the form of the corresponding side conditions is more convenient herein.

In the new coordinates, the expressions for the first derivatives become

$$\begin{aligned} \frac{\partial \tilde{f}}{\partial \tilde{x}_i}(\tilde{x}_1,\ldots,\tilde{x}_{n-1},0) &= \frac{\partial \tilde{f}_0}{\partial \tilde{x}_i}(\tilde{x}_1,\ldots,\tilde{x}_{n-1}), \quad \text{for } i=1,\ldots,n-1, \\ \frac{\partial \tilde{f}}{\partial \tilde{x}_n}(\tilde{x}_1,\ldots,\tilde{x}_{n-1},0) &= \tilde{f}_X(\tilde{x}_1,\ldots,\tilde{x}_{n-1}). \end{aligned}$$

To find the equations for the second derivatives, we differentiate these equations with respect to the parameters of the hypersurface, which are $\tilde{x}_1, \ldots, \tilde{x}_{n-1}$, to obtain

$$\frac{\partial^2 \tilde{f}}{\partial \tilde{x}_j \partial \tilde{x}_i}(\tilde{x}_1, \ldots, \tilde{x}_{n-1}, 0) = \frac{\partial^2 \tilde{f}_0}{\partial \tilde{x}_j \partial \tilde{x}_i}(\tilde{x}_1, \ldots, \tilde{x}_{n-1}), \tag{2.51}$$

where $i, j = 1, \ldots, n-1$, and

$$\frac{\partial^2 \tilde{f}}{\partial \tilde{x}_j \partial \tilde{x}_n}(\tilde{x}_1, \ldots, \tilde{x}_{n-1}, 0) = \frac{\partial \tilde{f}_X}{\partial \tilde{x}_j}(\tilde{x}_1, \ldots, \tilde{x}_{n-1}), \tag{2.52}$$

where $j = 1, \ldots, n-1$. Expression (2.51) consists of $(n-1)^2$ equations. However, there are only $n(n-1)/2$ independent equations, due to the equality of the mixed partial derivatives. Expression (2.52) consists of $n-1$ independent equations. Combining these expressions, we obtain $(n+1)n/2 - 1$ equations for $(n+1)n/2$ unknowns, $\partial^2 \tilde{f}/\partial \tilde{x}_i \partial \tilde{x}_j$: we are one equation short of completing the system.

We complete this system by taking the original differential equation expressed in the new coordinates, namely equation (2.49). The complete system is

$$\begin{bmatrix} 1 & 0 & \cdots & 0 & 0 \\ 0 & 1 & \cdots & 0 & 0 \\ \vdots & \vdots & \ddots & \vdots & \vdots \\ 0 & 0 & \ldots & 1 & 0 \\ \tilde{A}_{11}(y) & \tilde{A}_{12}(y) & \cdots & \tilde{A}_{n(n-1)}(y) & \tilde{A}_{nn}(y) \end{bmatrix} \begin{bmatrix} \frac{\partial^2 \tilde{f}}{\partial \tilde{x}_1^2} \\ \frac{\partial^2 \tilde{f}}{\partial \tilde{x}_1 \partial \tilde{x}_2} \\ \vdots \\ \frac{\partial^2 \tilde{f}}{\partial \tilde{x}_n \partial \tilde{x}_{n-1}} \\ \frac{\partial^2 \tilde{f}}{\partial \tilde{x}_n^2} \end{bmatrix}$$

$$= \begin{bmatrix} \frac{\partial^2 \tilde{f}_0}{\partial \tilde{x}_1^2}(\tilde{x}_1, \ldots, \tilde{x}_{n-1}) \\ \frac{\partial^2 \tilde{f}_0}{\partial \tilde{x}_1 \partial \tilde{x}_2}(\tilde{x}_1, \ldots, \tilde{x}_{n-1}) \\ \vdots \\ \frac{\partial \tilde{f}_X}{\partial \tilde{x}_{n-1}}(\tilde{x}_1, \ldots, \tilde{x}_{n-1}) \\ \tilde{B}\left(y, \tilde{f}(y), \frac{\partial \tilde{f}}{\partial \tilde{x}}(y)\right) \end{bmatrix}, \tag{2.53}$$

where y stands for $(\tilde{x}_1, \ldots, \tilde{x}_{n-1}, 0)$. The first $n-1$ equations are equations (2.52). The last equation is equation (2.49) whose form on the right-hand side results from the fact that we evaluate it along the hypersurface.

If the determinant of the matrix of system (2.53) is nonzero, we can solve for the second derivatives of the solution along the side conditions. If the coefficients of the differential equation and the side conditions are analytic, we can compute all the higher derivatives, and write the convergent Taylor series as discussed in Section 1.4. To find these derivatives, we can proceed in a manner similar to the one for the second derivatives. We can use f_0 to compute any derivative of the solution in the direction of the side conditions. The derivatives in the transverse direction to the side conditions can be calculated by differentiating the original differential equation in the transverse direction. In the situation where the determinant of the matrix of system (2.53) is zero, we cannot solve uniquely for the derivatives and construct the solution using the Taylor series. In the following paragraphs we turn our attention to the situation of the zero determinant of the above system.

To find the characteristic surface is to find a hypersurface along which we cannot solve uniquely for the second partial derivatives. System (2.53) has no unique solution only if the determinant of the coefficient matrix is zero. Since the determinant of this matrix is $\tilde{A}_{nn}(\tilde{x}_1, \ldots, \tilde{x}_{n-1}, 0)$, we can express this condition as

$$\tilde{A}_{nn}(\tilde{x}_1, \ldots, \tilde{x}_{n-1}, 0) = 0. \tag{2.54}$$

This condition states that the side condition satisfies the differential equation. For this reason we can use the compatibility condition to find solutions along the characteristics if we express the compatibility condition along the characteristics, not along the side conditions. The compatibility condition along the characteristic surface states that the equations in system (2.53) are linearly dependent, namely,

$$\begin{aligned}\tilde{A}_{11}(y)\frac{\partial^2 \tilde{f}_0}{\partial \tilde{x}_1^2}(\tilde{x}_1, \ldots, \tilde{x}_{n-1}) + \tilde{A}_{12}(y)\frac{\partial^2 \tilde{f}_0}{\partial \tilde{x}_1 \partial \tilde{x}_2}(\tilde{x}_1, \ldots, \tilde{x}_{n-1}) \\ + \cdots + \tilde{A}_{n(n-1)}(y)\frac{\partial \tilde{f}_X}{\partial \tilde{x}_{n-1}}(\tilde{x}_1, \ldots, \tilde{x}_{n-1}) \\ = \tilde{B}\left(y, \tilde{f}(y), \frac{\partial \tilde{f}}{\partial \tilde{x}}(y)\right).\end{aligned}$$

To use this equation in the context of the original problem, we express condition (2.54) in terms of the original coordinates. If we assume that the hypersurface can be expressed—at least locally—by $x_n = \psi(x_1, \ldots, x_{n-1})$, then the new coordinates are

$$\tilde{x}_i = x_i, \quad i = 1, \ldots, n-1,$$
$$\tilde{x}_n = x_n - \psi(x_1, \ldots, x_{n-1}),$$

where $\tilde{x}_n = 0$, as required. Expressing the partial derivatives with respect to the original coordinates in terms of the new coordinates, namely,

$$\frac{\partial}{\partial x_i} = \frac{\partial}{\partial \tilde{x}_i} - \frac{\partial \psi}{\partial x_i}\frac{\partial}{\partial \tilde{x}_n}, \quad i = 1, \ldots, n,$$

we write the second derivatives as

$$\begin{aligned}
\frac{\partial^2}{\partial x_j \partial x_i} &= \left(\frac{\partial}{\partial \tilde{x}_j} - \frac{\partial \psi}{\partial x_j}\frac{\partial}{\partial \tilde{x}_n}\right)\left(\frac{\partial}{\partial \tilde{x}_i} - \frac{\partial \psi}{\partial x_i}\frac{\partial}{\partial \tilde{x}_n}\right) \\
&= \frac{\partial^2}{\partial \tilde{x}_j \partial \tilde{x}_i} - \frac{\partial}{\partial \tilde{x}_j}\left(\frac{\partial \psi}{\partial x_i}\frac{\partial}{\partial \tilde{x}_n}\right) - \frac{\partial \psi}{\partial x_j}\frac{\partial^2}{\partial \tilde{x}_n \partial \tilde{x}_i} + \frac{\partial \psi}{\partial x_j}\frac{\partial}{\partial \tilde{x}_n}\left(\frac{\partial \psi}{\partial x_i}\frac{\partial}{\partial \tilde{x}_n}\right) \\
&= \frac{\partial^2}{\partial \tilde{x}_j \partial \tilde{x}_i} - \frac{\partial \psi}{\partial x_j}\frac{\partial^2}{\partial \tilde{x}_n \partial \tilde{x}_i} - \frac{\partial \psi}{\partial x_i}\frac{\partial^2}{\partial \tilde{x}_j \partial \tilde{x}_n} + \frac{\partial \psi}{\partial x_j}\frac{\partial \psi}{\partial x_i}\frac{\partial^2}{\partial \tilde{x}_n^2}.
\end{aligned}$$

Hence, we write

$$\begin{aligned}
&\sum_{i,j=1}^{n} A_{ij}\frac{\partial^2}{\partial x_i \partial x_j} \\
&= \sum_{i,j=1}^{n} A_{ij}\left(\frac{\partial^2}{\partial \tilde{x}_j \partial \tilde{x}_i} - \frac{\partial \psi}{\partial x_j}\frac{\partial^2}{\partial \tilde{x}_n \partial \tilde{x}_i} - \frac{\partial \psi}{\partial x_i}\frac{\partial^2}{\partial \tilde{x}_j \partial \tilde{x}_n} + \frac{\partial \psi}{\partial x_j}\frac{\partial \psi}{\partial x_i}\frac{\partial^2}{\partial \tilde{x}_n^2}\right).
\end{aligned} \tag{2.55}$$

To find the expression for $\tilde{A}_{nn}$, we look for all terms in the above expression that contain $\partial^2/\partial \tilde{x}_n^2$. These terms are

$$\tilde{A}_{nn}\frac{\partial^2}{\partial \tilde{x}_n^2} = \left[\sum_{i,j=1}^{n-1} A_{ij}\left(\frac{\partial \psi}{\partial x_i}\right)\left(\frac{\partial \psi}{\partial x_j}\right) - 2\sum_{i=1}^{n-1} A_{in}\left(\frac{\partial \psi}{\partial x_i}\right) + A_{nn}\right]\frac{\partial^2}{\partial \tilde{x}_n^2},$$

where, in the first two summations, we use the fact that $\partial \psi/\partial x_i = 0$ for $i = n$. Following equation (2.54), we require the term in brackets to be zero. Hence, we write explicitly

$$\begin{aligned}
&\sum_{i,j=1}^{n-1} A_{ij}(x_1, \ldots, x_{n-1}, \psi)\left(\frac{\partial \psi}{\partial x_i}\right)\left(\frac{\partial \psi}{\partial x_j}\right) \\
&- 2\sum_{i=1}^{n-1} A_{in}(x_1, \ldots, x_{n-1}, \psi)\left(\frac{\partial \psi}{\partial x_i}\right) + A_{nn}(x_1, \ldots, x_{n-1}, \psi) = 0,
\end{aligned} \tag{2.56}$$

which is a first-order nonlinear partial differential equation for ψ in $n-1$ independent variables, $x_1, \ldots, x_{n-1}$. The solutions of this equation are functions whose graphs form the characteristic hypersurfaces in the space spanned by $x_1, \ldots, x_n$. Along these hypersurfaces we cannot solve uniquely for the second partial derivatives from the knowledge of the side conditions and the differential equation. For two-dimensional problems, these characteristic hypersurfaces reduce to characteristic curves.

To illustrate this result, we consider equation (2.36), namely,

$$A_{11}\frac{\partial^2 f}{\partial x_1^2}+A_{12}\frac{\partial^2 f}{\partial x_1 \partial x_2}+A_{21}\frac{\partial^2 f}{\partial x_2 \partial x_1}+A_{22}\frac{\partial^2 f}{\partial x_2^2}=B_1\frac{\partial f}{\partial x_1}+B_2\frac{\partial f}{\partial x_2}+Cf+D.$$

If we consider the characteristic curve to be parametrized by x_1, namely $x_2 = \psi(x_1)$, we obtain the characteristic equation given by

$$A_{11}\left(\frac{\mathrm{d}\psi}{\mathrm{d}x_1}\right)^2 - 2A_{12}\frac{\mathrm{d}\psi}{\mathrm{d}x_1} + A_{22} = 0\,, \tag{2.57}$$

where we use the equality of mixed partial derivatives. This is the equation for the derivative of $\psi(x_1)$, which represents the slope of the characteristic curve. Depending on the coefficients, there can be two, one or no real solutions for this slope. This specifies the number of distinct families of characteristic curves, and corresponds to the hyperbolic, parabolic and elliptic equations, respectively. An example of this method is illustrated in Exercise 2.13.

Physical applications of the method presented in this section are illustrated in Sections 2.3.2 and 2.3.3.

2.3.2 *Wave, heat and Laplace equations*

2.3.2.1 *Wave equation*

If we set $n = 3$, $A_{11}(x) = A_{22}(x) = 1$, $A_{33}(x) = -1/v^2$ and we set all the other coefficients to zero in equation (2.44), we obtain the wave equation in two spatial dimensions, namely,

$$\frac{\partial^2 f}{\partial x_1^2}+\frac{\partial^2 f}{\partial x_2^2}=\frac{1}{v^2}\frac{\partial^2 f}{\partial x_3^2}\,, \tag{2.58}$$

with x_3 corresponding to time. Equation (2.58) is a linear case of a general semilinear equation (2.44); the semilinear equation allows us to consider several extensions of the linear wave equation. If $B(x, f, \partial f/\partial x) = C(x)\, f$, we can study linear frequency dispersion. Also, if $B(x, f, \partial f/\partial x) = B_i(x)\,\partial f/\partial x_i$, we can study linear dissipation. Furthermore, if $B(x, f, \partial f/\partial x) = D(x)$ for all x, we can study wave phenomena associated with the wave source.[3]

Let us assume that the characteristic surface can be expressed as

$$x_3 = \psi(x_1, x_2) .$$

Following equation (2.56), we write

$$\left(\frac{\partial \psi}{\partial x_1}\right)^2 + \left(\frac{\partial \psi}{\partial x_2}\right)^2 = \frac{1}{v^2} . \tag{2.59}$$

This is the condition for a surface, $\psi(x_1, x_2)$, in the $x_1x_2x_3$-space along which we cannot set side conditions that would allow us to solve uniquely for the second derivatives of $f(x_1, x_2, x_3)$. In other words, equation (2.59) is the characteristic equation of equation (2.58). This equation is called the eikonal equation. The characteristic surface is at the angle of $\arctan v$ with the x_1x_2-plane. We cannot set arbitrarily the side conditions along such a surface because the disturbance at a point described by the wave equation travels through the x_1x_2-plane with velocity v and forms a "light cone" in the $x_1x_2x_3$-space, which is the envelope of the surfaces in this space that satisfy the eikonal equation and pass through that point.

2.3.2.2 *Heat equation*

We can write the heat equation in three spatial dimensions using the notation of equation (2.44) as follows; $n = 4$, $A_{ij}(x) = \delta_{ij}$, for $i, j = 1, 2, 3$, $A_{44}(x) = 0$, and $B_i(x, f, \partial f/\partial x) = 0$, for $i = 1, 2, 3$, $B_4(x, f, \partial f/\partial x) = 1/k$, with all other coefficients zero. The resulting equation is the linear equation given by

$$\frac{\partial^2 f}{\partial x_1^2} + \frac{\partial^2 f}{\partial x_2^2} + \frac{\partial^2 f}{\partial x_3^2} = \frac{1}{k}\frac{\partial f}{\partial x_4} , \tag{2.60}$$

with k denoting conductivity and x_4 corresponding to time. The equation for the characteristic surface, $x_4 = \psi(x_1, x_2, x_3)$, is

$$\left(\frac{\partial \psi}{\partial x_1}\right)^2 + \left(\frac{\partial \psi}{\partial x_2}\right)^2 + \left(\frac{\partial \psi}{\partial x_3}\right)^2 = 0,$$

---

[3] Readers interested in these extensions might refer to McOwen (2003, pp. 95–97).

which means that

$$\frac{\partial \psi}{\partial x_1} = \frac{\partial \psi}{\partial x_2} = \frac{\partial \psi}{\partial x_3} = 0,$$

and hence, function ψ does not depend on x_1, x_2 or x_3. We conclude that $\psi(x_1, x_2, x_3)$ is constant, which means that the characteristic hypersurfaces in $x_1x_2x_3x_4$-space for the heat equation are hyperplanes given by x_4 being constant.

This result is consistent with our analysis described in Section 2.1.3. If we specify the temperature along the hyperplane perpendicular to the time axis, we can no longer specify the temporal change of the temperature.

2.3.2.3 *Laplace equation*

Let us consider the Laplace equation in three spatial dimensions, namely,

$$\nabla^2 f(x) := \frac{\partial^2 f}{\partial x_1^2} + \frac{\partial^2 f}{\partial x_2^2} + \frac{\partial^2 f}{\partial x_3^2} = 0. \tag{2.61}$$

In the notation used in expression (2.44): $n = 3$, $A_{ij}(x) = \delta_{ij}$ and $B(x, f, \partial f/\partial x) = 0$.

If we choose to parametrize the characteristic surface as

$$x_3 = \psi(x_1, x_2),$$

then, following expression (2.56), we can write the equation for this surface as

$$\left(\frac{\partial \psi}{\partial x_1}\right)^2 + \left(\frac{\partial \psi}{\partial x_2}\right)^2 + 1 = 0.$$

This equation has no real solution, which means that any surface in the three-dimensional space is a noncharacteristic surface.

This conclusion is independent of the particular choice of the parametrization. In the context of the Fourier transform, we can view equation (2.61) as an equation describing a sphere of zero radius.

Since the characteristic surface is determined by the highest derivatives, we reach the same conclusion if B is nonzero.

2.3.3 ***Solution of wave equation***

In this section, we use the characteristics and the side conditions to construct solutions.

Example 2.1. Let us consider the wave equation in two spatial dimensions,

$$\frac{\partial^2 f}{\partial x_1^2}+\frac{\partial^2 f}{\partial x_2^2}-\frac{1}{v^2}\frac{\partial^2 f}{\partial t^2}=0,$$

and its side conditions,

$$f\left(x_1,x_2,0\right)=f_0\left(x_1\right)$$

and

$$\frac{\partial f}{\partial t}\left(x_1,x_2,0\right)=f_N\left(x_1\right).$$

To find the solution of this Cauchy problem, we proceed in a manner described in Section 2.3.1. Letting $\partial f/\partial x_1 =: a_1\left(s_1,s_2\right)$, $\partial f/\partial x_2 =: a_2\left(s_1,s_2\right)$ and $\partial f/\partial t =: a_t\left(s_1,s_2\right)$, we write the system for the second derivatives as

$$\begin{bmatrix} 1 & 0 & \frac{\partial\psi}{\partial s_1} & 0 & 0 & 0 \\ 0 & 1 & \frac{\partial\psi}{\partial s_2} & 0 & 0 & 0 \\ 0 & 1 & 0 & 0 & \frac{\partial\psi}{\partial s_1} & 0 \\ 0 & 0 & 0 & 1 & \frac{\partial\psi}{\partial s_2} & 0 \\ 0 & 0 & 1 & 0 & 0 & \frac{\partial\psi}{\partial s_1} \\ 0 & 0 & 0 & 0 & 1 & \frac{\partial\psi}{\partial s_2} \\ 1 & 0 & 0 & 1 & 0 & -\frac{1}{c^2} \end{bmatrix}\begin{bmatrix} \frac{\partial^2 f}{\partial x_1\partial x_1} \\ \frac{\partial^2 f}{\partial x_1\partial x_2} \\ \frac{\partial^2 f}{\partial x_1\partial t} \\ \frac{\partial^2 f}{\partial x_2\partial x_2} \\ \frac{\partial^2 f}{\partial x_2\partial t} \\ \frac{\partial^2 f}{\partial t^2} \end{bmatrix}=\begin{bmatrix} \frac{\partial a_1}{\partial s_1} \\ \frac{\partial a_1}{\partial s_2} \\ \frac{\partial a_2}{\partial s_1} \\ \frac{\partial a_2}{\partial s_2} \\ \frac{\partial a_t}{\partial s_1} \\ \frac{\partial a_t}{\partial s_2} \\ 0 \end{bmatrix}.$$

The compatibility condition results from adding the first and fourth equations and subtracting from them $\partial\psi/\partial s_1$ times the fifth and $\partial\psi/\partial s_2$ times the sixth equations, and comparing the result with the wave equation. The compatibility condition is

$$\frac{\partial a_1}{\partial s_1}+\frac{\partial a_2}{\partial s_2}-\frac{\partial\psi}{\partial s_1}\frac{\partial a_t}{\partial s_1}-\frac{\partial\psi}{\partial s_2}\frac{\partial a_t}{\partial s_2}=0.$$

Let us choose the characteristic plane that passes through the x_2-axis, namely, $\psi(s_1, s_2) = s_1/v$. This choice is unique—up to the sign—due to ψ having to satisfy the eikonal equation. This choice results in

$$\frac{\partial a_1}{\partial s_1} - \frac{1}{v}\frac{\partial a_t}{\partial s_1} + \frac{\partial a_2}{\partial s_2} = 0\,.$$

Since the problem is invariant along the x_2 direction, the compatibility condition reduces to

$$\frac{\partial a_1}{\partial s_1} - \frac{1}{v}\frac{\partial a_t}{\partial s_1} = 0.$$

Upon integration, we get

$$a_1 - \frac{1}{v}a_t = C.$$

Recalling the expressions for a_i and the notation for the directional derivative, we write

$$D_{\left(1,0,-\frac{1}{v}\right)}f = C,$$

which results in

$$f(x_1, x_2, t) = f_0(x_1 - vt_1, x_2) - vtC.$$

This expression satisfies the Cauchy problem.

The above example illustrates the use of characteristics of the wave equation to construct its solution.

2.3.4 *Systems of semilinear equations*

In this section, we use an approach analogous to the one formulated in Section 2.3.1 to investigate systems of equations.

Consider a system of m semilinear second-order partial differential equations given by

$$\sum_{k=1}^{m}\sum_{i,j=1}^{n} A_{ijkl}(x)\frac{\partial^2 f_k}{\partial x_i \partial x_j} = B_l\left(x, f, \frac{\partial f}{\partial x}\right),$$

where $l = 1, \ldots, m$ and x stands for $x_1, \ldots, x_n$. We want to find a hypersurface along which it is impossible to determine uniquely the second derivatives from the side conditions that are given along this hypersurface. Let the hypersurface be given by

$$x_i = x_i (s_1, \ldots, s_{n-1}) .$$

We express the side conditions as

$$\begin{aligned} f_k (x_1 (s_1, \ldots, s_{n-1}), \ldots, x_n (s_1, \ldots, s_{n-1})) &= f_k^0 (s_1, \ldots, s_{n-1}), \\ \mathrm{D}_X f_k (x_1 (s_1, \ldots, s_{n-1}), \ldots, x_n (s_1, \ldots, s_{n-1})) &= f_k^X (s_1, \ldots, s_{n-1}), \end{aligned}$$

where $k = 1, \ldots, m$. Using the change of coordinates discussed in Section 2.3.1 on page 68, we rewrite these conditions as

$$\tilde{f}_k (\tilde{x}_1, \ldots, \tilde{x}_{n-1}, 0) = \tilde{f}_k^0 (\tilde{x}_1, \ldots, \tilde{x}_{n-1}), \tag{2.62}$$

$$\frac{\partial \tilde{f}_k}{\partial \tilde{x}_n} (\tilde{x}_1, \ldots, \tilde{x}_{n-1}, 0) = \tilde{f}_k^X (\tilde{x}_1, \ldots, \tilde{x}_{n-1}), \tag{2.63}$$

where $k = 1, \ldots, m$. Differentiating expressions (2.62) with respect to $\tilde{x}_1, \ldots, \tilde{x}_{n-1}$, we obtain

$$\frac{\partial \tilde{f}_k}{\partial \tilde{x}_i} (\tilde{x}_1, \ldots, \tilde{x}_{n-1}, 0) = \frac{\partial \tilde{f}_k^0}{\partial \tilde{x}_i} (\tilde{x}_1, \ldots, \tilde{x}_{n-1}), \quad \text{for } i = 1, \ldots, n-1.$$

Differentiating these equations again and differentiating expressions (2.63), we get

$$\begin{aligned} \frac{\partial^2 \tilde{f}_k}{\partial \tilde{x}_j \partial \tilde{x}_i} (\tilde{x}_1, \ldots, \tilde{x}_{n-1}, 0) &= \frac{\partial^2 \tilde{f}_k^0}{\partial \tilde{x}_j \partial \tilde{x}_i} (\tilde{x}_1, \ldots, \tilde{x}_{n-1}), \\ \frac{\partial^2 \tilde{f}_k}{\partial \tilde{x}_j \partial \tilde{x}_n} (\tilde{x}_1, \ldots, \tilde{x}_{n-1}, 0) &= \frac{\partial \tilde{f}_k^X (\tilde{x}_1, \ldots, \tilde{x}_{n-1})}{\partial \tilde{x}_j}, \end{aligned}$$

where $i, j = 1, \ldots, n-1$. This is a set of $m (n-1)^2 + m (n-1)$ equations. However, due to the equality of mixed partial derivatives there are only $mn (n-1)/2 + m (n-1)$ independent equations for $mn (n+1)/2$ unknowns. There are still m equations missing to complete the system. These equations are provided by the original system of m differential equations. In a manner similar to the case of a single equation discussed in Section 2.3.1, we write the complete system for the second derivatives as

$$\begin{bmatrix} 1 & 0 & \cdots & 0 & \cdots & 0 & \cdots & 0 \\ 0 & 1 & \cdots & 0 & \cdots & 0 & \cdots & 0 \\ \vdots & \vdots & \ddots & \vdots & & \vdots & & \vdots \\ 0 & 0 & \cdots & 1 & \cdots & 0 & \cdots & 0 \\ \vdots & \vdots & & \vdots & \ddots & \vdots & & \vdots \\ \tilde{A}_{1111}(y) & \tilde{A}_{1211}(y) & \cdots & \tilde{A}_{ijk1}(y) & \cdots & \tilde{A}_{nn11}(y) & \cdots & \tilde{A}_{nnm1}(y) \\ \vdots & \vdots & & \vdots & & \vdots & \ddots & \vdots \\ \tilde{A}_{11m1}(y) & \tilde{A}_{12m1}(y) & \cdots & \tilde{A}_{ijkm}(y) & \cdots & \tilde{A}_{nn1m}(y) & \cdots & \tilde{A}_{nnmm}(y) \end{bmatrix}$$

$$\begin{bmatrix} \dfrac{\partial^2 \tilde{f}_1}{\partial \tilde{x}_1^2} \\ \dfrac{\partial^2 \tilde{f}_1}{\partial \tilde{x}_1 \partial \tilde{x}_2} \\ \vdots \\ \dfrac{\partial^2 \tilde{f}_k}{\partial \tilde{x}_j \partial \tilde{x}_i} \\ \vdots \\ \dfrac{\partial^2 \tilde{f}_1}{\partial \tilde{x}_n \partial \tilde{x}_n} \\ \vdots \\ \dfrac{\partial^2 \tilde{f}_m}{\partial \tilde{x}_n \partial \tilde{x}_n} \end{bmatrix} = \begin{bmatrix} \dfrac{\partial^2 \tilde{f}_1^0}{\partial \tilde{x}_1^2}(\tilde{x}_1, \ldots, \tilde{x}_{n-1}) \\ \dfrac{\partial^2 \tilde{f}_1^0}{\partial \tilde{x}_1 \partial \tilde{x}_2}(\tilde{x}_1, \ldots, \tilde{x}_{n-1}) \\ \vdots \\ \dfrac{\partial^2 \tilde{f}_k^0}{\partial \tilde{x}_j \partial \tilde{x}_i}(\tilde{x}_1, \ldots, \tilde{x}_{n-1}) \\ \vdots \\ \tilde{B}_1\left(y, \tilde{f}(y), \dfrac{\partial \tilde{f}}{\partial \tilde{x}}(y)\right) \\ \vdots \\ \tilde{B}_m\left(y, \tilde{f}(y), \dfrac{\partial \tilde{f}}{\partial \tilde{x}}(y)\right) \end{bmatrix},$$

where y stands for $(\tilde{x}_1, \ldots, \tilde{x}_{n-1}, 0)$. The determinant of the coefficient matrix of this system is the determinant of the lower-right $m \times m$ matrix, namely,

$$\det\left[\tilde{A}_{nnkl}(\tilde{x}_1, \ldots, \tilde{x}_{n-1}, 0)\right],$$

where n is fixed and $k, l = 1, 2, \ldots, m$. Following the same procedure as in Section 2.3.1 on page 70, we express each entry $\tilde{A}_{nnkl}$ as

$$\tilde{A}_{nnkl}(\tilde{x}_1, \ldots, \tilde{x}_{n-1}, 0) = \sum_{i,j=1}^{n-1} A_{ijkl}(x_1, \ldots, x_{n-1}, \psi)\left(\frac{\partial \psi}{\partial x_i}\right)\left(\frac{\partial \psi}{\partial x_j}\right)$$
$$-2\sum_{i=1}^{n-1} A_{inkl}(x_1, \ldots, x_{n-1}, \psi)\left(\frac{\partial \psi}{\partial x_i}\right) + A_{nnkl}(x_1, \ldots, x_{n-1}, \psi).$$

Using these expressions for $k, l = 1, \ldots, m$, we express the fact that the determinant is zero as

$$\det\left[\sum_{i,j=1}^{n-1} A_{ijkl}\left(\frac{\partial\psi}{\partial x_i}\right)\left(\frac{\partial\psi}{\partial x_j}\right) - 2\sum_{i=1}^{n-1} A_{inkl}\left(\frac{\partial\psi}{\partial x_i}\right) + A_{nnkl}\right] = 0, \tag{2.64}$$

where coefficients A_{ijkl} depend on $x_1, \ldots, x_{n-1}, \psi$. This determinant is a polynomial of degree m. Thus, in general, we obtain m hypersurfaces that satisfy this equation.

2.3.5 *Elastodynamic and Maxwell equations*

In this section, we apply the method discussed in Section 2.3.4 to the elastodynamic equations and Maxwell equations.

2.3.5.1 *Elastodynamic equations*

Consider the elastodynamic equations, which are formulated in Appendix B, namely,

$$\rho(x)\frac{\partial^2 u_i(x,t)}{\partial t^2} = \sum_{j=1}^{3}\sum_{k=1}^{3}\sum_{l=1}^{3}\frac{\partial c_{ijkl}(x)}{\partial x_j}\frac{\partial u_k(x,t)}{\partial x_l} + \sum_{j=1}^{3}\sum_{k=1}^{3}\sum_{l=1}^{3} c_{ijkl}(x)\frac{\partial^2 u_k(x,t)}{\partial x_j \partial x_l}, \tag{2.65}$$

where $i = 1, 2, 3$.

Writing the general system of linear second-order equations as stated in Section 2.3.4, namely,

$$\sum_{k=1}^{m}\sum_{i=1}^{n}\sum_{j=1}^{n} A_{ijkl}(x)\frac{\partial^2 f_k}{\partial x_i \partial x_j} = B_l\left(x, f, \frac{\partial f}{\partial x}\right),$$

we can write the elastodynamic equations by setting $m = 3$, $n = 4$, letting $x_4 = t$ and considering

$$f_i(x) = u_i(x,t), \quad \text{for } i = 1, 2, 3$$
$$A_{ijkl}(x) = c_{ljki}(x), \quad \text{for } i, j, k, l = 1, 2, 3$$

$$A_{i4kl}\left(x\right) = A_{4ikl}\left(x\right) = 0, \quad \text{for } i,k,l = 1,2,3,4$$

$$A_{44kl}\left(x\right) = -\rho\left(x\right)\delta_{kl}, \quad \text{for } k,l = 1,2,3,4$$

$$B_l\left(x,u,\frac{\partial u}{\partial x}\right) = \sum_{j=1}^{3}\sum_{k=1}^{3}\sum_{i=1}^{3}\frac{\partial c_{ljki}\left(x\right)}{\partial x_j}\frac{\partial u_k\left(x,t\right)}{\partial x_i}, \quad \text{for } l = 1,2,3$$

$$B_4\left(x,u,\frac{\partial u}{\partial x}\right) = 0\,.$$

The third equality results from the absence of the mixed partial derivatives containing time, and the last one from the absence of the first-order time derivatives.

Using equation (2.64), we obtain

$$\det\left[\sum_{i=1}^{3}\sum_{j=1}^{3}c_{ljki}\left(x\right)\left(\frac{\partial\psi}{\partial x_i}\right)\left(\frac{\partial\psi}{\partial x_j}\right)-\rho\left(x\right)\delta_{kl}\right]=0\,, \tag{2.66}$$

where $k,l = 1,2,3$ and $t \equiv x_4 = \psi\left(x_1,x_2,x_3\right)$ represents the characteristic surface.[4,5]

In general, this equation has three distinct solutions, which are the eikonal equations corresponding to the three types of waves that propagate in an elastic medium. These equations are discussed in Section 3.4.1.

If we consider homogeneous continua, the form of equation (2.65) changes: the first-derivative coefficients are zero. However, the form of equation (2.66) remains the same, since the characteristic equations depend only on the highest-order derivatives of a given differential equation, as concluded on page 82, below. Furthermore, the form of equation (2.66) remains the same if we exclude anisotropy from equation (2.65). In such a case, the number of coefficients c_{ijkl} is reduced. Notably, an isotropic homogeneous case analogous to equation (2.66) is given by expression (2.67), below, which is the characteristic equation of the Maxwell equations.

2.3.5.2 *Maxwell equations*

Consider the Maxwell equations in their potential form, which are derived in Appendix C, namely,

---

[4]Readers interested in different approaches used to derive equation (2.66) might refer to Bos and Slawinski (2010).

[5]*See also*: Slawinski (2015, Sections 7.2 and 7.3).

$$\frac{\partial^2 A_i}{\partial x_1^2}+\frac{\partial^2 A_i}{\partial x_2^2}+\frac{\partial^2 A_i}{\partial x_3^2}-\frac{1}{c^2}\frac{\partial^2 A_i}{\partial t^2}=-\frac{J_i}{c^2\epsilon_0},\ \text{for } i=1,2,3,$$

$$\frac{\partial^2 \phi}{\partial x_1^2}+\frac{\partial^2 \phi}{\partial x_2^2}+\frac{\partial^2 \phi}{\partial x_3^2}-\frac{1}{c^2}\frac{\partial^2 \phi}{\partial t^2}=-\frac{\rho}{\epsilon_0}.$$

Writing the general system of linear second-order equations as stated in Section 2.3.4, namely,

$$\sum_{k=1}^{m}\sum_{i,j=1}^{n} A_{ijkl}(x)\frac{\partial^2 f_k}{\partial x_i \partial x_j}=B_l\left(x,f,\frac{\partial f}{\partial x}\right),$$

we can write the Maxwell equations by setting $m=n=4$, letting $x_4=t$, and considering

$$\begin{aligned} f_k &= A_k, \quad \text{for } k=1,2,3,\\ f_4 &= \phi \end{aligned}$$

and

$$\begin{aligned} A_{ijkl} &= \delta_{ij}\delta_{kl}, \quad \text{for } i,j,k,l=1,2,3,\\ A_{44kl} &= -\frac{1}{c^2}\delta_{kl}, \quad \text{for } k,l=1,\dots,4,\\ B_k &= -\frac{J_k}{c^2\epsilon_0}, \quad \text{for } k=1,2,3,\\ B_4 &= -\frac{\rho}{\epsilon_0}, \end{aligned}$$

with all other coefficients set to zero.

Using equation (2.64), we obtain

$$\det\left[\sum_{i=1}^{3}\delta_{kl}\left(\frac{\partial\psi}{\partial x_i}\right)^2-\frac{1}{c^2}\delta_{kl}\right]=0, \tag{2.67}$$

where $k,l=1,2,3,4$. This equation has a quadruple root that is given by

$$\sum_{i=1}^{3}\left(\frac{\partial\psi}{\partial x_i}\right)^2=\frac{1}{c^2}, \tag{2.68}$$

which is the eikonal equation for electromagnetic waves.

2.3.6 *Quasilinear equations*

On page 6, studying the first-order equations, we commented on the dependence of characteristics on terms with derivatives only. Herein, studying the second-order equations, we see that that characteristics depend only on terms with second derivatives. In general, characteristics depend only on terms with the highest-order derivatives. If the equation is at least semilinear, we can use the method of characteristics to study its solutions.

In view of these remarks, let us consider a quasilinear second-order equation, which has the general form given by

$$\sum_{i,j=1}^{n} A_{ij}\left(x, f, \frac{\partial f}{\partial x}\right)\frac{\partial^2 f}{\partial x_i \partial x_j} = B\left(x, f, \frac{\partial f}{\partial x}\right),$$

where x stands for $x_1, \ldots, x_n$ and $\partial f/\partial x$ stands for $\partial f/\partial x_i$ with $i \in \{1, \ldots, n\}$. Both the coefficients in front of the second derivatives and the right-hand side of the equation are, generally, nonlinear functions of x, function f and first derivatives. This equation is a generalization of equation (2.44).

Using the approach discussed in Section 2.3.1, we consider a special coordinate system, $\tilde{x}_1, \ldots, \tilde{x}_n$, in which the side conditions can be expressed by equation (2.50), namely

$$\begin{aligned} \tilde{f}\left(\tilde{x}_1, \ldots, \tilde{x}_{n-1}, 0\right) &= \tilde{f}_0\left(\tilde{x}_1, \ldots, \tilde{x}_{n-1}\right), \\ \mathrm{D}_{[0,\ldots,0,1]}\tilde{f}\left(\tilde{x}_1, \ldots, \tilde{x}_{n-1}, 0\right) &= \tilde{f}_X\left(\tilde{x}_1, \ldots, \tilde{x}_{n-1}\right). \end{aligned}$$

In these coordinates, the original equation is

$$\sum_{i\leqslant j} \tilde{A}_{ij}\left(\tilde{x}, \tilde{f}, \frac{\partial \tilde{f}}{\partial \tilde{x}}\right)\frac{\partial^2 \tilde{f}}{\partial \tilde{x}_i \partial \tilde{x}_j} = \tilde{B}\left(\tilde{x}, \tilde{f}, \frac{\partial \tilde{f}}{\partial \tilde{x}}\right).$$

Following the steps discussed in Section 2.3.1, we obtain the system of algebraic equations for the second derivatives along the hypersurface $x_n = 0$. This system is

$$\begin{bmatrix} 1 & 0 & \cdots & 0 \\ 0 & 1 & \cdots & 0 \\ & & \ddots & \\ \tilde{A}_{11}(y) & \tilde{A}_{12}(y) & \cdots & \tilde{A}_{nn}(y) \end{bmatrix} \begin{bmatrix} \dfrac{\partial^2 \tilde{f}}{\partial \tilde{x}_1^2} \\ \dfrac{\partial^2 \tilde{f}}{\partial \tilde{x}_1 \partial \tilde{x}_2} \\ \vdots \\ \dfrac{\partial^2 \tilde{f}}{\partial \tilde{x}_n^2} \end{bmatrix} = \begin{bmatrix} \dfrac{\partial^2 \tilde{f}_0}{\partial \tilde{x}_1^2}\left(\tilde{x}_1, \ldots, \tilde{x}_{n-1}\right) \\ \dfrac{\partial^2 \tilde{f}_0}{\partial \tilde{x}_1 \partial \tilde{x}_2}\left(\tilde{x}_1, \ldots, \tilde{x}_{n-1}\right) \\ \vdots \\ \tilde{B}(y) \end{bmatrix},$$

where y stands for $\left(\tilde{x}_1, \ldots, \tilde{x}_{n-1}, 0, \tilde{f}_0, \partial \tilde{f}/\partial \tilde{x}\left(\tilde{x}_1, \ldots, \tilde{x}_{n-1}, 0\right)\right)$, and where the form of the last entry on the right-hand side is due to the fact that we evaluate the original equation along the hypersurface. This system has no unique solution if

$$\tilde{A}_{nn}\left(\tilde{x}_1,\ldots,\tilde{x}_{n-1},0,\tilde{f}_0,\frac{\partial\tilde{f}}{\partial\tilde{x}}(\tilde{x}_1,\ldots,\tilde{x}_{n-1},0)\right)=0. \tag{2.69}$$

To express condition (2.69) in the original coordinates, we assume that the hypersurface can be expressed by $x_n=\psi(x_1,\ldots,x_{n-1})$. In this case, the new coordinates are

$$\begin{aligned}\tilde{x}_i &= x_i, \quad i=1,\ldots,n-1,\\ \tilde{x}_n &= x_n-\psi(x_1,\ldots,x_{n-1}).\end{aligned}$$

The second derivatives in the original equations are related to the second derivatives in the new coordinates as

$$\begin{aligned}\sum_{i,j=1}^{n} A_{ij}\left(x,f,\frac{\partial f}{\partial x}\right)\frac{\partial^2}{\partial x_i\partial x_j} &= \sum_{i,j=1}^{n} A_{ij}\left(x,f,\frac{\partial f}{\partial x}\right)\\ &\left(\frac{\partial^2}{\partial\tilde{x}_j\partial\tilde{x}_i}-\frac{\partial\psi}{\partial x_j}\frac{\partial^2}{\partial\tilde{x}_n\partial\tilde{x}_i}-\frac{\partial\psi}{\partial x_i}\frac{\partial^2}{\partial\tilde{x}_j\partial\tilde{x}_n}+\frac{\partial\psi}{\partial x_j}\frac{\partial\psi}{\partial x_i}\frac{\partial^2}{\partial\tilde{x}_n^2}\right).\end{aligned}$$

To find the expression for $\tilde{A}_{nn}$, we look for the terms in the above expression that contain $\partial^2/\partial\tilde{x}_n^2$. These terms are

$$\tilde{A}_{nn}\frac{\partial^2}{\partial\tilde{x}_n^2}=\left[\sum_{i,j=1}^{n-1}A_{ij}\left(\frac{\partial\psi}{\partial x_i}\right)\left(\frac{\partial\psi}{\partial x_j}\right)-2\sum_{i=1}^{n-1}A_{in}\left(\frac{\partial\psi}{\partial x_i}\right)+A_{nn}\right]\frac{\partial^2}{\partial\tilde{x}_n^2},$$

where, in the first two summations, we use the fact that $\partial\psi/\partial x_i=0$ for $i=n$. If we evaluate this expression on the hypersurface and set it to zero, we obtain

$$\begin{aligned}&\sum_{i,j=1}^{n-1}A_{ij}\left(x_1,\ldots,x_{n-1},\psi,f_0,\frac{\partial f}{\partial x}(x_1,\ldots,x_{n-1},\psi)\right)\left(\frac{\partial\psi}{\partial x_i}\right)\left(\frac{\partial\psi}{\partial x_j}\right)\\ &\quad-2\sum_{i=1}^{n-1}A_{in}\left(x_1,\ldots,x_{n-1},\psi,f_0,\frac{\partial f}{\partial x}(x_1,\ldots,x_{n-1},\psi)\right)\left(\frac{\partial\psi}{\partial x_i}\right)\\ &\quad+A_{nn}\left(x_1,\ldots,x_{n-1},\psi,f_0,\frac{\partial f}{\partial x}(x_1,\ldots,x_{n-1},\psi)\right)=0.\end{aligned} \tag{2.70}$$

If we take the vector normal to the hypersurface given by $x_n=\psi(x_1,\ldots,x_{n-1})$ to be

$$X=\left[\frac{\partial\psi}{\partial x_1},\ldots,\frac{\partial\psi}{\partial x_{n-1}},-1\right],$$

we can express the normal derivative as

$$\mathrm{D}_X = \left[\frac{\partial \psi}{\partial x_1}, \ldots, \frac{\partial \psi}{\partial x_{n-1}}, -1\right] \cdot \left[\frac{\partial}{\partial x_1}, \ldots, \frac{\partial}{\partial x_n}\right],$$

from which we see that

$$\frac{\partial}{\partial x_n} = \sum_{i=1}^{n-1} \frac{\partial \psi}{\partial x_i} \frac{\partial}{\partial x_i} - \mathrm{D}_X .$$

Thus, we can write the partial derivatives in the arguments of equation (2.70) expressed along the hypersurface as

$$\frac{\partial f}{\partial x_i}(x_1, \ldots, x_{n-1}, \psi) = \frac{\partial f_0}{\partial x_i}(x_1, \ldots, x_{n-1}), \qquad \text{for } i = 1, \ldots, n-1,$$

$$\frac{\partial f}{\partial x_n}(x_1, \ldots, x_{n-1}, \psi) = \sum_{i=1}^{n-1} \frac{\partial \psi}{\partial x_i} \frac{\partial f_0}{\partial x_i} - f_X .$$

Equation (2.70) depends on the side conditions themselves, not only on the hypersurface along which we specify them. We can use this equation to check if the given side conditions allow us to compute the unique second-order derivatives along an *a priori* given hypersurface. However, we cannot obtain the characteristic surface using this equation.

Closing remarks

For first-order linear equations, the characteristics are viewed as curves along which the behaviour of the solutions is determined by the equations; we cannot set side conditions along these curves freely. For second-order linear equations, the characteristics are viewed as surfaces along which we cannot determine uniquely the highest derivatives from the equation and side conditions along these surfaces. The latter definition is valid also for first-order linear equations. It is the definition for characteristics of linear partial differential equations of any order.

The construction of characteristics for nonlinear first-order equations—to which we turn our attention in Chapter 3—is not included explicitly in either of the above definitions.

2.4 Exercises

Exercise 2.1. Find the general solution of equation (2.1), namely,

$$\frac{\partial^2 f}{\partial x_1^2} + 2\frac{\partial^2 f}{\partial x_1 \partial x_2} + \frac{\partial^2 f}{\partial x_2^2} = 0.$$

Solution. We make the change of variables,

$$y_1 = x_1 - x_2,$$
$$y_2 = x_1 + x_2,$$

where y_1 parametrizes the different characteristic curves, and y_2 is a parameter along these curves.

Using this transformation, we express the differential operators in the new variables as

$$\frac{\partial}{\partial x_1} = \frac{\partial y_1}{\partial x_1}\frac{\partial}{\partial y_1} + \frac{\partial y_2}{\partial x_1}\frac{\partial}{\partial y_2} = \frac{\partial}{\partial y_1} + \frac{\partial}{\partial y_2}$$

and

$$\frac{\partial}{\partial x_2} = \frac{\partial y_1}{\partial x_2}\frac{\partial}{\partial y_1} + \frac{\partial y_2}{\partial x_2}\frac{\partial}{\partial y_2} = -\frac{\partial}{\partial y_1} + \frac{\partial}{\partial y_2},$$

which results in the three expressions for the second derivatives:

$$\frac{\partial^2}{\partial x_1^2} = \left(\frac{\partial}{\partial y_1} + \frac{\partial}{\partial y_2}\right)\left(\frac{\partial}{\partial y_1} + \frac{\partial}{\partial y_2}\right) = \frac{\partial^2}{\partial y_1^2} + 2\frac{\partial^2}{\partial y_1 \partial y_2} + \frac{\partial^2}{\partial y_2^2},$$

$$2\frac{\partial^2}{\partial x_1 \partial x_2} = 2\left(\frac{\partial}{\partial y_1} + \frac{\partial}{\partial y_2}\right)\left(-\frac{\partial}{\partial y_1} + \frac{\partial}{\partial y_2}\right) = 2\left(-\frac{\partial^2}{\partial y_1^2} + \frac{\partial^2}{\partial y_2^2}\right)$$

and

$$\frac{\partial^2}{\partial x_2^2} = \left(-\frac{\partial}{\partial y_1} + \frac{\partial}{\partial y_2}\right)\left(-\frac{\partial}{\partial y_1} + \frac{\partial}{\partial y_2}\right) = \frac{\partial^2}{\partial y_1^2} - 2\frac{\partial^2}{\partial y_1 \partial y_2} + \frac{\partial^2}{\partial y_2^2}.$$

Adding the three expressions, we obtain

$$\frac{\partial^2}{\partial x_1^2} + 2\frac{\partial^2}{\partial x_1 \partial x_2} + \frac{\partial^2}{\partial x_2^2} = 4\frac{\partial^2}{\partial y_2^2}.$$

Thus, the original differential equation is

$$\frac{\partial^2 f(y_1, y_2)}{\partial y_2^2} = 0, \tag{2.71}$$

and its solution can be obtained by integrating twice with respect to y_2,

$$f(y_1, y_2) = \int\int 0 \;\, \mathrm{d}y_2 \, \mathrm{d}y_2 = \int g(y_1) \, \mathrm{d}y_2 = y_2 g(y_1) + h(y_1).$$

Expressing this result in the original variables, we obtain the general solution, namely,

$$f(x_1, x_2) = (x_1 + x_2)\, g(x_1 - x_2) + h(x_1 - x_2).$$

Exercise 2.2. Use the incompatibility of the side conditions to look for the solution of the heat equation (2.30),

$$\frac{\partial^2 f}{\partial x_1^2} = c\frac{\partial f}{\partial x_2}.$$

Solution. Examining system (2.33), namely,

$$\begin{bmatrix} x_1'(s) & x_2'(s) & 0 \\ 0 & x_1'(s) & x_2'(s) \\ 1 & 0 & 0 \end{bmatrix} \begin{bmatrix} \frac{\partial^2 f}{\partial x_1^2} \\ \frac{\partial^2 f}{\partial x_1 \partial x_2} \\ \frac{\partial^2 f}{\partial x_2^2} \end{bmatrix} = \begin{bmatrix} f_0'' - \frac{\partial f}{\partial x_1}x_1'' - \frac{\partial f}{\partial x_2}x_2'' \\ f_X' - \frac{\partial f}{\partial x_1}x_2'' + \frac{\partial f}{\partial x_2}x_1'' \\ c\frac{\partial f}{\partial x_2} \end{bmatrix},$$

we see that along $x_1(s) = s$ and $x_2(s) = C$, the compatibility condition reduces to

$$x_1' c\frac{\partial f(x_1, C)}{\partial x_2} = f_0''(x_1, C) - \frac{\partial f(x_1, C)}{\partial x_1}x_1'' - \frac{\partial f(x_1, C)}{\partial x_2}x_2''.$$

Since $x_1' = 1$, $x_1'' = x_2'' = 0$ and

$$f_0''(x_1, C) = \frac{\partial^2 f(x_1, C)}{\partial x_1^2},$$

we rewrite the compatibility condition as

$$c\frac{\partial f(x_1, C)}{\partial x_2} = \frac{\partial^2 f(x_1, C)}{\partial x_1^2},$$

which is a restatement of the heat equation along the characteristic. Hence, a compatibility condition cannot be used to solve the heat equation.

Exercise 2.3. Use the Taylor series at point $[0, 0]$ to solve the heat equation (2.30),

$$\frac{\partial^2 f}{\partial x_1^2} = c\frac{\partial f}{\partial x_2},$$

with the following side conditions:

$$f(0, x_2) = x_2 \tag{2.72}$$

and

$$D_{[1,0]} f(0, x_2) = 0. \tag{2.73}$$

Solution. To construct the series, we find the derivatives of f evaluated at $[0,0]$. We consider condition (2.73) to obtain

$$\frac{\partial f}{\partial x_1}(0,0)=0$$

and condition (2.72) to obtain

$$\frac{\partial f}{\partial x_2}(0,0)=1. \tag{2.74}$$

The second derivatives are obtained by using the heat equation itself to write

$$\frac{\partial^2 f}{\partial x_1^2}=c\frac{\partial f}{\partial x_2}=c,$$

where the second equality is deduced from expression (2.74), by twice differentiating condition (2.72) to get

$$\frac{\partial^2 f}{\partial x_2^2}(0,0)=0$$

and once differentiating condition (2.73) to get

$$\frac{\partial^2 f}{\partial x_1 \partial x_2}(0,0)=0.$$

All higher-order derivatives are zero. Thus, the Taylor series of the solution is

$$\begin{aligned} f(x_1,x_2) &= f(0,0)+\frac{\partial f}{\partial x_1}(0,0)\,x_1+\frac{\partial f}{\partial x_2}(0,0)\,x_2 \\ &\quad+\frac{1}{2!}\frac{\partial^2 f}{\partial x_1^2}(0,0)\,x_1^2+\frac{\partial^2 f}{\partial x_1\partial x_2}(0,0)\,x_1x_2+\frac{1}{2!}\frac{\partial^2 f}{\partial x_2^2}(0,0)\,x_2^2 \\ &= x_2+\frac{c}{2}x_1^2, \end{aligned} \tag{2.75}$$

which satisfies the heat equation and its side conditions.

Given a solution for all times at a particular location, $x_1=0$, together with its spatial change for all instants, we obtain solution (2.75), which is valid for all times and all locations.

Exercise 2.4. Using the Taylor series, construct the solution of the heat equation (2.30), namely,

$$\frac{\partial^2 f}{\partial x_1^2}=c\frac{\partial f}{\partial x_2},$$

given—as the side condition—the value of the solution along the x_1-axis, $f(x_1,0)=g(x_1)$. Obtain the particular solution for

$$g(x_1)=\frac{cx_1^2}{2}. \tag{2.76}$$

Solution. To construct the Taylor series of the solution, we find the derivatives of f at the given point along the x_1-axis. For convenience, we choose this point to be the origin of the x_1x_2-coordinates, even though other points, say $(1,0)$, lead to the same result. The first term of the series is the side condition, $f(x_1,0) = g(x_1)$, evaluated at the origin:

$$f(0,0) = g(0).$$

The first derivative of f with respect to x_1 is the derivative of the side condition evaluated at the origin:

$$\frac{\partial f}{\partial x_1}(0,0) = \frac{\mathrm{d}g}{\mathrm{d}x_1}(0). \tag{2.77}$$

To obtain the first derivative of f with respect to x_2, we write the heat equation along the x_1-axis using the side condition as

$$\frac{\partial f}{\partial x_2} = \frac{1}{c}\frac{\partial^2 f}{\partial x_1^2} = \frac{1}{c}\frac{\mathrm{d}^2 g}{\mathrm{d}x_1^2},$$

and evaluate it at the origin to get

$$\frac{\partial f}{\partial x_2}(0,0) = \frac{1}{c}\frac{\mathrm{d}^2 g}{\mathrm{d}x_1^2}(0). \tag{2.78}$$

To obtain the second derivative with respect to x_1, we differentiate the side condition twice with respect to x_1 and evaluate it at the origin,

$$\frac{\partial^2 f}{\partial x_1^2}(0,0) = \frac{\mathrm{d}^2 g}{\mathrm{d}x_1^2}(0).$$

To obtain the second mixed derivative, we differentiate the heat equation with respect to x_1, use the side condition and evaluate at the origin,

$$\frac{\partial^2 f}{\partial x_1 \partial x_2}(0,0) = \frac{1}{c}\frac{\mathrm{d}^3 g}{\mathrm{d}x_1^3}(0). \tag{2.79}$$

To obtain the second derivative with respect to x_2, we write the heat equation as

$$\frac{\partial f}{\partial x_2} = \frac{1}{c}\frac{\partial^2 f}{\partial x_1^2},$$

differentiate it with respect to x_2 to get

$$\frac{\partial^2 f}{\partial x_2^2} = \frac{1}{c}\frac{\partial}{\partial x_2}\frac{\partial^2 f}{\partial x_1^2},$$

which, using the equality of mixed partial derivatives, we rewrite as

$$\frac{\partial^2 f}{\partial x_2^2} = \frac{1}{c}\frac{\partial}{\partial x_1}\frac{\partial^2 f}{\partial x_1 \partial x_2}.$$

Invoking expression (2.79), we obtain

$$\frac{\partial^2 f}{\partial x_2^2}(0,0) = \frac{1}{c}\frac{\mathrm{d}}{\mathrm{d}x_1}\left(\frac{1}{c}\frac{\mathrm{d}^3 g}{\mathrm{d}x_1^3}\right) = \frac{1}{c^2}\frac{\mathrm{d}^4 g}{\mathrm{d}x_1^4}(0)\,.$$

We can continue in a similar fashion to obtain all higher-order derivatives.

To obtain the particular solution, we use $g(x_1) = cx_1^2/2$ to evaluate the expressions at the origin:

$$f(0,0) = 0,$$

$$\frac{\partial f}{\partial x_1}(0,0) = 0,$$

$$\frac{\partial f}{\partial x_2}(0,0) = 1,$$

$$\frac{\partial^2 f}{\partial x_1^2}(0,0) = c,$$

$$\frac{\partial^2 f}{\partial x_1 \partial x_2}(0,0) = 0$$

and

$$\frac{\partial^2 f}{\partial x_2^2}(0,0) = 0.$$

All other derivatives are equal to zero. Inserting the derivatives into the Taylor series, we see that

$$f(x_1,x_2) = x_2 + \frac{c}{2}x_1^2, \tag{2.80}$$

which is expression (2.75), since we choose the side condition to coincide with solution (2.75) along $x_2 = 0$.

Given a solution at a particular instant, $x_2 = 0$, for all locations x_1, we obtain solution (2.80) for all times and all locations. No information about the derivatives of the solution is needed, since we find the first derivatives from the differential equation together with the side condition.

Exercise 2.5. Solve equation (2.35), namely,

$$\frac{\partial^2 f(x_1,x_2)}{\partial x_1^2} + \frac{\partial^2 f(x_1,x_2)}{\partial x_2^2} = 0, \qquad x_1, x_2 \in (0,\pi)\,, \tag{2.81}$$

where x_1 and x_2 are the two spatial dimensions, with its side conditions,

$$f(0,x_2) = f(\pi,x_2) = 0 \qquad x_2 \in [0,\pi]\,, \tag{2.82}$$

$$f(x_1,0) = 0, \qquad \text{and} \qquad f(x_1,\pi) = F(x_1) \qquad x_1 \in [0,\pi]\,, \tag{2.83}$$

where $F \in C^\infty$, and $F(0) = F(\pi) = 0$. Assume that

$$f(x_1,x_2) = g(x_1)\,h(x_2)\,. \tag{2.84}$$

Solution. Substituting expression $g(x_1)\,h(x_2)$ for $f(x)$ in equation (2.81), we obtain

$$\frac{g''(x_1)}{g(x_1)} = -\frac{h''(x_2)}{h(x_2)}, \tag{2.85}$$

where primes denote the derivative of a function with respect to its argument. Since the sides of the equation are independent of one another, each one must be equal to a constant, which we denote by $-\lambda$. Hence, we can split equation (2.85) into two ordinary differential equations,

$$g''(x_1) + \lambda g(x_1) = 0 \tag{2.86}$$

and

$$h''(x_2) - \lambda h(x_2) = 0. \tag{2.87}$$

Condition (2.82) requires that $g(0) = g(\pi) = 0$; hence, the solution of equation (2.86) is

$$g_n(x_1) = A_n \sin(n x_1),$$

where $n^2 = \lambda_n$, with $n \in \mathbb{N}$. The corresponding solution of equation (2.87), which we write as

$$h''(x_2) - n^2 h(x_2) = 0,$$

is

$$h(x_2) = B_n \sinh(n x_2) + C_n \cosh(n x_2),$$

where the first statement of condition (2.83) requires $C_n = 0$. Invoking assumption (2.84), we write

$$f(x_1, x_2) = g(x_1)\,h(x_2) = \sum_{n=1}^{\infty} a_n \sin(n x_1)\sinh(n x_2), \tag{2.88}$$

where $a_n := A_n B_n$ and the summation is a consequence of the linear combinations of solutions of a linear differential equation being also a solution.

In view of the second statement of condition (2.83), we write solution (2.88) as

$$f(x_1, \pi) = \sum_{n=1}^{\infty} a_n \sin(n x_1)\sinh(n\pi) = F(x_1).$$

Since $F \in C^\infty$ with $F(0) = F(\pi) = 0$, the Fourier sine series converges uniformly and absolutely to F, we can obtain a_n from

$$a_n \sinh(n\pi) = \frac{2}{\pi}\int_0^{\infty} F(x_1)\sin(n x_1)\,\mathrm{d}x_1.$$

Thus, expression (2.88) with a_n given by

$$\frac{2}{\pi \sinh(n\pi)} \int_0^\infty F(x_1) \sin(nx_1)\, dx_1$$

is the solution of the problem stated by expressions (2.81) – (2.83).

Exercise 2.6. Consider

$$\frac{\partial^2 f}{\partial x_1^2} + 2x_1 \frac{\partial^2 f}{\partial x_1 \partial x_2} + \frac{\partial^2 f}{\partial x_2^2} - \sin x_1 \frac{\partial f}{\partial x_1} + \frac{\partial f}{\partial x_2} + 3f = 0.$$

Describe the domain for which it is a hyperbolic equation, for which it is a parabolic equation, and for which it is an elliptic equation.

Solution. Since the classification of partial differential equations depends on the highest derivatives only, in view of expression (2.38), we write

$$\begin{bmatrix} A_{11} & \frac{1}{2}A_{12} \\ \frac{1}{2}A_{21} & A_{22} \end{bmatrix} = \begin{bmatrix} 1 & x_1 \\ x_1 & 1 \end{bmatrix}.$$

To find the eigenvalues, we write

$$\det \begin{bmatrix} 1-\lambda & x_1 \\ x_1 & 1-\lambda \end{bmatrix} = 0,$$

to get

$$\lambda^2 - 2\lambda + 1 - x_1^2 = 0.$$

The discriminant is $\Delta = 4x_1^2$, which is positive. The eigenvalues are

$$\lambda_1 = 1 + x_1$$

and

$$\lambda_2 = 1 - x_1.$$

Hyperbolic equations require the eigenvalues to be of opposite sign; in other words, either $\lambda_1 > 0$, which implies $x_1 > -1$, and $\lambda_2 < 0$, which implies $x_1 > 1$, or $\lambda_1 < 0$, which implies $x_1 < -1$, and $\lambda_2 > 0$, which implies $x_1 < 1$. The first case requires $x_1 > 1$, the second case requires $x_1 < -1$. We can write the requirement for the opposite-sign eigenvalues concisely as $|x_1| > 1$. Elliptic equations require the eigenvalues to be of the same sign. For two positive eigenvalues, we require $x_1 > -1$ and $x_1 < 1$, for two negative eigenvalues, $x_1 < -1$ and $x_1 > 1$. The first case requires

$-1 < x_1 < 1$; the second case cannot be satisfied. We can write the requirement for the same-sign eigenvalues concisely as $|x_1| < 1$. Parabolic equation requires one eigenvalue to be zero; hence, $x_1 = \pm 1$; in other words, $|x_1| = 1$.

In conclusion: the hyperbolic domain is $x_1 \in (-\infty, -1) \cup (1, \infty)$, the elliptic domain is $x_1 \in (-1, 1)$, the parabolic domain is $x_1 = \pm 1$.

Exercise 2.7. Verify that the side conditions along surface $x_3 = \psi(x_1, x_2)$ that are given by

$$f(x_1, x_2, \psi(x_1, x_2)) = g(x_1, x_2) \tag{2.89}$$

and

$$\mathrm{D}_N f(x_1, x_2, \psi(x_1, x_2)) = h(x_1, x_2) \tag{2.90}$$

allow us to obtain the first derivatives, namely, $\partial f / \partial x_1$, $\partial f / \partial x_2$ and $\partial f / \partial x_3$.

Solution. Differentiating condition (2.89) with respect to x_1, we get

$$\frac{\partial}{\partial x_1} f(x_1, x_2, \psi(x_1, x_2)) = \frac{\partial f}{\partial x_1} + \frac{\partial f}{\partial x_3}\frac{\partial \psi}{\partial x_1} = \frac{\partial g}{\partial x_1},$$

where we use the fact that the third argument, ψ, can be denoted by x_3. Similarly, differentiating condition (2.89) with respect to x_2, we get

$$\frac{\partial}{\partial x_2} f(x_1, x_2, \psi(x_1, x_2)) = \frac{\partial f}{\partial x_2} + \frac{\partial f}{\partial x_3}\frac{\partial \psi}{\partial x_2} = \frac{\partial g}{\partial x_2}.$$

A vector in the $x_1 x_2 x_3$-space that is normal to a level set of ψ, namely to a surface given by $\psi(x_1, x_2) - x_3 = const.$, is

$$\begin{aligned} N &= \left[\frac{\partial(\psi(x_1, x_2) - x_3)}{\partial x_1}, \frac{\partial(\psi(x_1, x_2) - x_3)}{\partial x_2}, \frac{\partial(\psi(x_1, x_2) - x_3)}{\partial x_3}\right] \\ &= \left[\frac{\partial \psi}{\partial x_1}, \frac{\partial \psi}{\partial x_2}, -1\right]. \end{aligned}$$

Thus, we can write condition (2.90) as

$$\begin{aligned} N \cdot \left[\frac{\partial f}{\partial x_1}, \frac{\partial f}{\partial x_2}, \frac{\partial f}{\partial x_3}\right] &= \left[\frac{\partial \psi}{\partial x_1}, \frac{\partial \psi}{\partial x_2}, -1\right] \cdot \left[\frac{\partial f}{\partial x_1}, \frac{\partial f}{\partial x_2}, \frac{\partial f}{\partial x_3}\right] \\ &= \frac{\partial f}{\partial x_1}\frac{\partial \psi}{\partial x_1} + \frac{\partial f}{\partial x_2}\frac{\partial \psi}{\partial x_2} - \frac{\partial f}{\partial x_3} = h(x_1, x_2). \end{aligned}$$

We have a system of three linear algebraic equations, which we can write as

$$\begin{bmatrix} 1 & 0 & \dfrac{\partial \psi}{\partial x_1} \\ 0 & 1 & \dfrac{\partial \psi}{\partial x_2} \\ \dfrac{\partial \psi}{\partial x_1} & \dfrac{\partial \psi}{\partial x_2} & -1 \end{bmatrix} \begin{bmatrix} \dfrac{\partial f}{\partial x_1} \\ \dfrac{\partial f}{\partial x_2} \\ \dfrac{\partial f}{\partial x_3} \end{bmatrix} = \begin{bmatrix} \dfrac{\partial g}{\partial x_1} \\ \dfrac{\partial g}{\partial x_2} \\ h \end{bmatrix}.$$

To solve uniquely this system for $\partial f / \partial x_1$, $\partial f / \partial x_2$ and $\partial f / \partial x_3$, we require that

$$\det \begin{bmatrix} 1 & 0 & \dfrac{\partial \psi}{\partial x_1} \\ 0 & 1 & \dfrac{\partial \psi}{\partial x_2} \\ \dfrac{\partial \psi}{\partial x_1} & \dfrac{\partial \psi}{\partial x_2} & -1 \end{bmatrix} \neq 0 .$$

In other words, we require that

$$\left(\frac{\partial \psi}{\partial x_1}\right)^2 + \left(\frac{\partial \psi}{\partial x_2}\right)^2 + 1 \neq 0 . \tag{2.91}$$

Since expression (2.91) is the squared length of a vector normal to surface $\psi(x_1, x_2) - x_3$, it is never zero. Hence, we can solve uniquely for the first partial derivatives of f on this surface.

Exercise 2.8. What are the side conditions for equation (2.1) that allow for infinitely many solutions?

Solution. Recall equation (2.16), namely,

$$a_1' + a_2' = 0, \tag{2.92}$$

where a_1 and a_2 are given by expressions (2.6) and (2.7), namely,

$$a_1(s) = \frac{f_0'(s)\, x_1'(s) + f_N(s)\, x_2'(s)}{(x_1')^2 + (x_2')^2}$$

and

$$a_2(s) = \frac{-f_N(s)\, x_1'(s) + f_0'(s)\, x_2'(s)}{(x_1')^2 + (x_2')^2}.$$

Since the characteristic curves satisfy $x_1' = x_2'$, and we are interested in the side condition along these curves, we reduce the above expressions to write

$$a_1(s) = \frac{f_0'(s) + f_N(s)}{2x_1'}$$

and

$$a_2(s) = \frac{-f_N(s) + f_0'(s)}{2x_1'}.$$

Differentiating these two equations with respect to s , adding them together and using equation (2.92), we get

$$x_1' f_0'' = x_1'' f_0'.$$

This equation can be rewritten as

$$(\ln |f_0'|)' = (\ln |x_1'|)',$$

which implies that

$$\ln |f_0'| = \ln |x_1'| + C.$$

Exponentiating and using the properties of the absolute value, we get

$$f_0(s) = \pm x_1(s) e^C + l,$$

which can be written as

$$f_0(s) = x_1(s) k + l,$$

where k and l are arbitrary constants. This is equation (2.17), as expected.

Exercise 2.9. Set the side conditions for the one-dimensional wave equation written as

$$\frac{\partial^2 f(x,t)}{\partial x^2} = \frac{1}{v^2}\frac{\partial^2 f(x,t)}{\partial t^2}, \tag{2.93}$$

to be the initial conditions giving the displacement f and velocity $\partial f/\partial t$ at time zero.

Solution. Viewing equation (2.93) as the wave equation, we see that the x-axis and the t-axis correspond to space and time, respectively. We set the hypersurface, which in this case is a line, to coincide with the x-axis. In such a case, the first side condition, namely,

$$f(x,0) = f_0(x),$$

is the displacement along this line, which is the displacement at time zero. The second side condition, namely,

$$\mathrm{D}_{[0,1]} f(x,0) = f_X(x),$$

is a derivative in the direction that is not tangent to the hypersurface. Herein, we set this direction to be parallel to the t-axis. Hence, the second side condition provides the information about the rate of change along the x-axis in the direction of time: the velocity of displacement at time zero. In view of expression (1.65), the second side condition is

$$[0,1] \cdot \left[\frac{\partial f}{\partial x}(x,0), \frac{\partial f}{\partial t}(x,0)\right] = \frac{\partial f}{\partial t}(x,0).$$

Exercise 2.10. Show that equation (2.19), namely,

$$\left(c\frac{\partial}{\partial x_2}+\frac{\partial}{\partial x_1}\right)\left(c\frac{\partial}{\partial x_2}-\frac{\partial}{\partial x_1}\right)f\left(x_1,x_2\right)=0\,, \tag{2.94}$$

is equivalent to equation (2.18), namely,

$$\frac{\partial^2 f\left(x_1,x_2\right)}{\partial x_1^2}=c^2\frac{\partial^2 f\left(x_1,x_2\right)}{\partial x_2^2}\,. \tag{2.95}$$

Solution. Using the linearity of differential operators, we can write equation (2.94) as

$$\left(c\frac{\partial}{\partial x_2}+\frac{\partial}{\partial x_1}\right)\left(c\frac{\partial}{\partial x_2}-\frac{\partial}{\partial x_1}\right)f=\left(c\frac{\partial}{\partial x_2}+\frac{\partial}{\partial x_1}\right)\left(c\frac{\partial f}{\partial x_2}-\frac{\partial f}{\partial x_1}\right)=0,$$

which leads to

$$c^2\frac{\partial^2 f}{\partial x_2^2}-c\frac{\partial^2 f}{\partial x_2\partial x_1}+c\frac{\partial^2 f}{\partial x_1\partial x_2}-\frac{\partial^2 f}{\partial x_1^2}=0\,.$$

Since the two middle terms vanish due to the equality of mixed partial derivatives, we obtain

$$c^2\frac{\partial^2 f}{\partial x_2^2}-\frac{\partial^2 f}{\partial x_1^2}=0\,,$$

which is equation (2.95), as required.

Exercise 2.11. Show that a complex analytic function $u(x_1+\iota y_1,x_2+\iota y_2)=f(x_1+\iota y_1,x_2+\iota y_2)+\iota g(x_1+\iota y_1,x_2+\iota y_2)$ whose real part satisfies the Laplace equation,

$$\frac{\partial^2 f}{\partial x_1^2}+\frac{\partial^2 f}{\partial x_2^2}=0\,,$$

satisfies the wave equation,

$$\frac{\partial^2 f}{\partial x_1^2}-\frac{\partial^2 f}{\partial y_2^2}=0.$$

Solution. A function of a complex variable, $z=x+\iota y$, is complex-analytic only if it satisfies the Cauchy-Riemann conditions:[6]

$$\frac{\partial f}{\partial x}=\frac{\partial g}{\partial y}\quad\text{and}\quad\frac{\partial f}{\partial y}=-\frac{\partial g}{\partial x}.$$

[6] Readers interested in a close relation between the Laplace and Cauchy-Riemann equations and operators might refer to Trèves (2006, pp. 6-7 and 34-38).

Substituting $\partial g/\partial y_2$ for the first derivative of f with respect to x_2 in the Laplace equation, we obtain

$$\frac{\partial^2 f}{\partial x_1^2}+\frac{\partial^2 g}{\partial x_2 \partial y_2}=0.$$

Changing the order of differentiation and using the second Cauchy-Riemann condition, we obtain

$$\frac{\partial^2 f}{\partial x_1^2}-\frac{\partial^2 f}{\partial y_2^2}=0,$$

as required.

Exercise 2.12. Using the method of characteristics, find the general solution of

$$\left(\frac{\partial^2}{\partial x_1^2}+5\frac{\partial^2}{\partial x_1 \partial x_2}\right) f\left(x_1, x_2\right)=1.$$

Solution. The algebraic system for the second derivatives is

$$\left[\begin{array}{ccc} x_1'(s) & x_2'(s) & 0 \\ 0 & x_1'(s) & x_2'(s) \\ 1 & 5 & 0 \end{array}\right]\left[\begin{array}{c} \dfrac{\partial^2 f}{\partial x_1^2} \\ \dfrac{\partial^2 f}{\partial x_1 \partial x_2} \\ \dfrac{\partial^2 f}{\partial x_2^2} \end{array}\right]=\left[\begin{array}{c} a_1'(s) \\ a_2'(s) \\ 0 \end{array}\right],$$

where

$$\frac{\partial f}{\partial x_1}\left(x_1(s), x_2(s)\right)=a_1(s)$$

and

$$\frac{\partial f}{\partial x_2}\left(x_1(s), x_2(s)\right)=a_2(s).$$

The determinant of the above matrix is zero if

$$\left(x_2'(s)\right)^2=5 x_1'(s) x_2'(s),$$

which implies that

$$x_2'(s)=0$$

or

$$x_2'(s)=5 x_1'(s).$$

Hence, along the characteristic curves,

$$x_2(s) = C$$

or

$$x_2(s) = 5x_1(s) + D,$$

where either of the values of C or D determines the choice of a characteristic curve.

Along these curves, the original differential equation can be written as

$$\frac{\partial^2}{\partial y_1 \partial y_2} f(x_1(y_1, y_2), x_2(y_1, y_2)) = 1, \tag{2.96}$$

where $y_1 = x_1 - (1/5)x_2$ and $y_2 = (1/5)x_2$. The general solution of equation (2.96) is

$$f(x_1(y_1, y_2), x_2(y_1, y_2)) = y_1 y_2 + g(y_1) + h(y_2),$$

for functions g and h. The solution can be written in the original coordinates as

$$f(x_1, x_2) = \left(x_1 - \frac{1}{5}x_2\right)\frac{1}{5}x_2 + g\left(x_1 - \frac{1}{5}x_2\right) + h\left(\frac{1}{5}x_2\right).$$

Exercise 2.13. Find the characteristic curves of

$$4\frac{\partial^2 f}{\partial x_1^2} + 16x_1\frac{\partial^2 f}{\partial x_1 \partial x_2} + 7x_1^2\frac{\partial^2 f}{\partial x_2^2} = 4\sin(x_2).$$

Solution. Following equation (2.57) and using the fact that $A_{12} + A_{21} = 16x_1$, we write

$$4\left(\frac{\mathrm{d}x_2}{\mathrm{d}x_1}\right)^2 - 16x_1\frac{\mathrm{d}x_2}{\mathrm{d}x_1} + 7x_1^2 = 0,$$

which is the characteristic equation. Solving for $\mathrm{d}x_2/\mathrm{d}x_1$, we obtain

$$\frac{\mathrm{d}x_2}{\mathrm{d}x_1} = \begin{cases} \dfrac{7}{2}x_1 \\ \\ \dfrac{1}{2}x_1 \end{cases}.$$

The solutions of these two equations are

$$x_2 = \frac{7}{4}x_1^2 + C$$

and

$$x_2 = \frac{1}{4}x_1^2 + D,$$

which are the characteristics of the original differential equation.

Exercise 2.14. Consider the wave equation,

$$\frac{\partial^2 u}{\partial x^2}+\frac{\partial^2 u}{\partial y^2}=\frac{1}{v^2}\frac{\partial^2 u}{\partial t^2},$$

and its solution of the form

$$u(x,y,t)=A(x,y)\,F(t-\psi(x,y)), \tag{2.97}$$

where $\psi(x,y)$ is a solution of the related eikonal equation. Find the conditions on $A(x,y)$ given an arbitrary function F.

Solution. Differentiating expression (2.97), we obtain

$$\frac{\partial^2 u}{\partial x^2}=\frac{\partial^2 A}{\partial x^2}F-F'\left(2\frac{\partial A}{\partial x}\frac{\partial \psi}{\partial x}+A\frac{\partial^2 \psi}{\partial x^2}\right)+AF''\left(\frac{\partial \psi}{\partial x}\right)^2,$$

$$\frac{\partial^2 u}{\partial y^2}=\frac{\partial^2 A}{\partial y^2}F-F'\left(2\frac{\partial A}{\partial y}\frac{\partial \psi}{\partial y}+A\frac{\partial^2 \psi}{\partial y^2}\right)+AF''\left(\frac{\partial \psi}{\partial y}\right)^2$$

and

$$\frac{\partial^2 u}{\partial t^2}=AF'',$$

where the primes stand for derivatives with respect to the argument. Substituting these derivatives into the wave equation and rearranging, we obtain

$$F\left(\frac{\partial^2 A}{\partial x^2}+\frac{\partial^2 A}{\partial y^2}\right)-F'\left(2\left(\frac{\partial A}{\partial x}\frac{\partial \psi}{\partial x}+\frac{\partial A}{\partial y}\frac{\partial \psi}{\partial y}\right)+A\left(\frac{\partial^2 \psi}{\partial x^2}+\frac{\partial^2 \psi}{\partial y^2}\right)\right)$$
$$+F''A\left(\left(\frac{\partial \psi}{\partial x}\right)^2+\left(\frac{\partial \psi}{\partial y}\right)^2-\frac{1}{v^2}\right)=0.$$

For this equation to be valid for any function F, we require the three terms in parentheses to be zero; we write them as

$$\nabla^2 A=0 \tag{2.98}$$

and

$$2\nabla A\cdot\nabla\psi+A\nabla^2\psi=0 \tag{2.99}$$

and as the eikonal equation,

$$(\nabla\psi)^2=\frac{1}{v^2}.$$

We assume that the last condition is satisfied; hence, the conditions on A are given by the first two equations. The second equation is called the transport equation and is derived in Chapter 4. According to the first equation, A must be a harmonic function that satisfies the transport equation. Such a solution might not exist; however, we address and resolve this concern in Chapter 4 where we introduce the asymptotic series.

Exercise 2.15. Following Exercise 2.14, find function A that satisfies equations (2.98) and (2.99) for plane waves, $\psi(x, y) = x/v$, where v is a constant velocity.

Solution. Since $\psi = x/v$, we get

$$\nabla\psi(x, y) = \left[\frac{1}{v}, 0\right]$$

and

$$\nabla^2\psi = 0.$$

Inserting these expressions into equation (2.99), we get

$$2\left[\frac{\partial A}{\partial x}, \frac{\partial A}{\partial y}\right] \cdot \left[\frac{1}{v}, 0\right] = \frac{2}{v}\frac{\partial A}{\partial x} = 0,$$

which implies $A(x, y) = f(y)$. Using equation (2.98), we get

$$\frac{\partial^2 f(y)}{\partial y^2} = 0,$$

which implies

$$f(y) = a_0 + a_1 y.$$

Hence, in view of Exercise 2.14,

$$(a_0 + a_1 y)\, F\left(\frac{x}{v} - t\right)$$

is a solution of the wave equation for arbitrary F. If $a_1 = 0$, then we obtain a constant-amplitude solution.

Chapter 3

Characteristic equations of first-order nonlinear partial differential equations

La propriété fondamentale des caractéristiques s'exprime [...] par le fait qu'elles sont les seules surfaces le long desquelles deux solutions de l'équation peuvent se toucher. [...] Les caractéristique ont une signification physique importante; elle sont, en fait, ce que les physiciens appellent des ondes.[1]

Jacques Hadamard (1932)

Preliminary remarks

In this chapter we formulate the characteristic equations of first-order nonlinear partial differential equations. The physical importance of this formulation is exemplified by the eikonal equation, which is derived in the previous chapter.

We return to the concept discussed in Chapter 1, where, for first-order linear equations, we construct the solution by propagating the side conditions along characteristics. However, for nonlinear equations, characteristics depend not only on the differential equation, but also on its side conditions.

We begin this chapter with a motivational example followed by the derivation of the characteristic equations for first-order nonlinear equations. Subsequently, we exemplify the characteristics in the context of elasticity and electromagnetism.

Readers might find it useful to study this chapter together with Appen-

[1] *The fundamental property of characteristics is exhibited by the fact that they are the only surfaces along which two solutions of the equation can touch one another [...] Characteristics have an important physical meaning; they are, in fact, what the physicists refer to as waves.*

dices B and C, where we formulate the elastodynamic equations and the Maxwell equations, respectively.

3.1 Motivational example

To motivate issues discussed in this chapter, we consider a two-dimensional form of the eikonal equation (2.68), namely,

$$\left(\frac{\partial f(x,y)}{\partial x}\right)^2+\left(\frac{\partial f(x,y)}{\partial y}\right)^2=1,$$

which is a first-order nonlinear equation. This equation restricts the magnitude of the gradient of solutions. To solve this equation, we supply a side condition, which we set to be $f=0$ along the straight line given by $[x,0]$. Since we know that the solution is constant along this line, we know also that the gradient of the solution is perpendicular to this line. This information, together with the differential equation, states that the gradient along line $[x,0]$ is $[0,\pm 1]$. Thus we can propagate the side condition along the direction $[0,\pm 1]$. We reduce the original equation to study its solution along the lines tangent to this direction, which are given by

$$\frac{\mathrm{d}x}{\mathrm{d}s}=0,$$
$$\frac{\mathrm{d}y}{\mathrm{d}s}=\pm 1.$$

The solution changes along these curves as

$$\frac{\mathrm{d}f}{\mathrm{d}s}=\frac{\partial f}{\partial x}\frac{\mathrm{d}x}{\mathrm{d}s}+\frac{\partial f}{\partial y}\frac{\mathrm{d}y}{\mathrm{d}s}=[0,\pm 1]\cdot[0,\pm 1]=1,$$

which implies that

$$f(x(s),y(s))=s+s_0.$$

If we choose $[x(0),y(0)]=[0,0]$, then $s_0=0$ and the solution becomes

$$f(x,y)=\pm y.$$

We can view this solution as a propagation of the side condition, $f=0$, along line $[x,0]$ to the neighbouring points by constructing an envelope of circles centred on that line. This envelope corresponds to a level curve of the solution with the value equal to the radius of the circles. Motivated by physics, in particular by Huygens's principle, we might refer to these circles as elementary wavefronts and to curves $[x(s),y(s)]$ as rays.

If the magnitude of the gradient depends on its direction, then the curves given by the length of the gradient are not circles. In such a case, their envelopes do not form the level curves perpendicular to the gradient. Physically, such a case might correspond to anisotropic media—media in which velocities depend on direction—for which rays are not orthogonal to wavefronts, as illustrated in Figure 3.1 and discussed in Section 3.4.1.

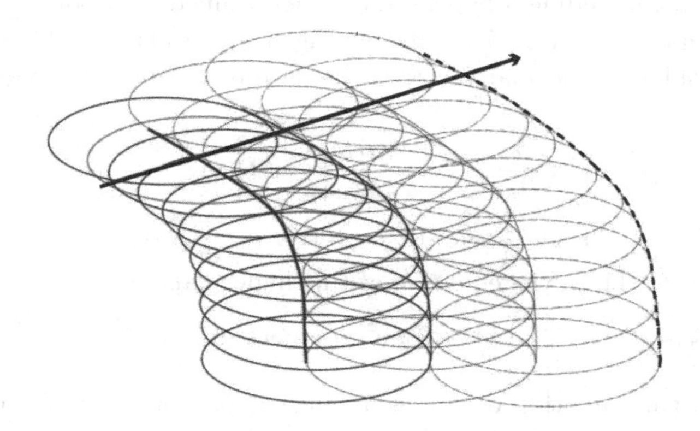

Fig. 3.1 Example of propagation of the side condition along rays that are not perpendicular to the wavefronts.

3.2 Characteristics

Let us consider the general form of an n-dimensional first-order nonlinear partial differential equation for function f,

$$F\left(x_1, x_2, \ldots, x_n, f, \frac{\partial f}{\partial x_1}, \frac{\partial f}{\partial x_2}, \ldots, \frac{\partial f}{\partial x_n}\right) = 0\,, \tag{3.1}$$

where x_i are the n independent variables. The graph of a solution of such an equation is a hypersurface, $w = f(x_1, x_2, \ldots, x_n)$, in the $(n+1)$-dimensional xw-space of vectors $[x_1, \ldots, x_n, w]$. Letting $p_i := \partial f/\partial x_i$, we can rewrite the differential equation as a level surface of a function of $2n+1$ variables given by

$$F(x_1, x_2, \ldots, x_n, w, p_1, p_2, \ldots, p_n) = 0\,. \tag{3.2}$$

We can view this differential equation as a relation among the p_i. Specifically, at a fixed point, $P = \left[x_1^P, \ldots, x_n^P, w^P\right]$, equation (3.2) defines a hypersurface in the $p_1 \ldots p_n$-space, which can be parametrized locally by $n-1$ parameters, $\xi_1, \xi_2, \ldots, \xi_{n-1}$; we write

$$F\left(x_1^P, \ldots, x_n^P, w^P, p_1(\xi_1, \xi_2, \ldots, \xi_{n-1}), \ldots, p_n(\xi_1, \xi_2, \ldots, \xi_{n-1})\right) = 0. \tag{3.3}$$

To find the solution, f, we look for its graph, $w = f(x_1, x_2, \ldots, x_n)$, which can be obtained as the envelope of its tangent hyperplanes in the xw-space.

Any plane, including a hyperplane, is determined by a point it contains and by its normal vector. To find the tangent hyperplane to $f(x)$ at point P, we consider the normal vector to the graph of f in the xw-space; this normal is

$$\left[\left.\frac{\partial f}{\partial x_1}\right|_P, \left.\frac{\partial f}{\partial x_2}\right|_P, \ldots, \left.\frac{\partial f}{\partial x_n}\right|_P, -1\right]$$
$$\equiv \left[p_1(\xi_1, \ldots, \xi_{n-1}), \ldots, p_n(\xi_1, \ldots, \xi_{n-1}), -1\right],$$

since $p := \nabla f$. Thus, the equation of this hyperplane is

$$p_1(\xi_1, \ldots, \xi_{n-1})(x_1 - x_1^P) + \cdots + p_n(\xi_1, \ldots, \xi_{n-1})(x_n - x_n^P) - (w - w^P) = 0. \tag{3.4}$$

As ξ_j vary, this equation describes an $(n-1)$-parameter family of hyperplanes through point P. One of these hyperplanes is tangent to the graph of the solution. The envelope of these hyperplanes is called the Monge cone in honour of Gaspard Monge, and is illustrated in Figure 3.2 for a two-dimensional case, for which the family of hyperplanes is a one-parameter family of planes.

In general, the surfaces given by vectors $[p(\xi), -1]$ and a section of the Monge cone that is perpendicular to the w-axis are not of the same shape, in contrast to Figure 3.2, wherein they are both represented by circles. In the language of differential geometry, the former curve is a figuratrix and the latter is an indicatrix; they are polar reciprocals of one another. A figuratrix corresponds to the wavefront slowness and an indicatrix to the wavefront.[2] In isotropic media only, both figuratrices and indicatrices are spheres; otherwise, they are represented by complicated surfaces.

To study the propagation of side conditions along curves, we reduce the partial differential equation to a set of ordinary differential equations, which is analogous to studying a vector field. Let us consider a particular line from the family of hyperplanes given by equation (3.4). Our choice of the line is the intersection of the given hyperplane with the envelope of all such

[2]Readers interested in figuratrices and indicatrices might refer to Kreyszig (1964). Readers interested in physical meanings of figuratrices and indicatrices might refer to Arnold (1989).

hyperplanes, in other words, the intersection of this hyperplane with the Monge cone. Our motivation for studying the envelope is that it represents

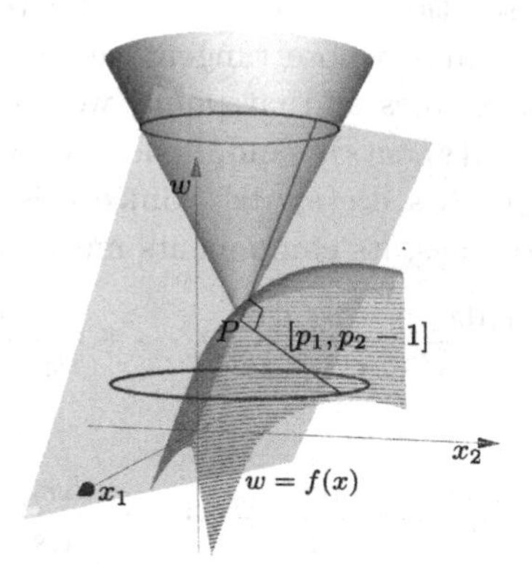

Fig. 3.2 The Monge cone at point $P = \left[x_1^P, x_2^P, w^P\right]$ constructed as an envelope of hyperplanes with normal vectors, $[p_1, p_2, -1]$, where p satisfies $F\left(x^P, p, w^P\right) = 0$. We show one of these hyperplanes that is tangent to both the cone and the graph of $f(x_1, x_2)$. The figuratrix is the larger circle; the indicatrix is the smaller one and is part of the cone.

the set of neighbouring tangent planes, rather than a single plane. We consider the set of points in each hyperplane that remain stationary with respect to an infinitesimal change in ξ_j. For two dimensions, this process is examined in Exercise 3.4.

To write the stationarity condition, we take partial derivatives of equation (3.4) with respect to ξ_j,

$$\frac{\partial p_1}{\partial \xi_j}\left(x_1 - x_1^P\right) + \frac{\partial p_2}{\partial \xi_j}\left(x_2 - x_2^P\right) + \cdots + \frac{\partial p_n}{\partial \xi_j}\left(x_n - x_n^P\right) = 0,$$

which are $n-1$ equations of hyperplanes passing through point P and having normal vectors given by $[\partial p_1/\partial \xi_j, \ldots, \partial p_n/\partial \xi_j, 0]$.

Herein, the intersection of these $n-1$ hyperplanes in the $(n+1)$-dimensional space is a two-dimensional plane. Since vectors $[\partial p_1/\partial \xi_j, \ldots, \partial p_n/\partial \xi_j, 0]$ are tangent to surface $[p(\xi), -1]$, we intersect the two-dimensional plane with the tangent hyperplane to this surface. This hyperplane is given by equation (3.4), which we rewrite as

$$\begin{aligned} p_1(\xi_1, \ldots, \xi_{n-1})\left(x_1 - x_1^P\right) &+ p_2(\xi_1, \ldots, \xi_{n-1})\left(x_2 - x_2^P\right) \\ &+ \cdots + p_n(\xi_1, \ldots, \xi_{n-1})\left(x_n - x_n^P\right) \\ &= w - w^P. \end{aligned}$$

The intersection of these planes is a line.

We have a family of lines at point P, which forms the Monge cone. Each line from this family belongs to one of the possible tangent planes to the graph of a solution of the original differential equation. To find within this graph a curve whose tangent vector at a given point belongs to the family of these lines at that point, we parametrize the curve by s to write $[x_1(s), \ldots, x_n(s), w(s)]$, and, hence, we write its tangent vector as $[\mathrm{d}x_1(s)/\mathrm{d}s, \ldots, \mathrm{d}x_n(s)/\mathrm{d}s, \mathrm{d}w(s)/\mathrm{d}s]$. Since this vector is in the intersection of the n hyperplanes, its components must satisfy

$$\frac{\partial p_1}{\partial \xi_j}\frac{\mathrm{d}x_1}{\mathrm{d}s} + \frac{\partial p_2}{\partial \xi_j}\frac{\mathrm{d}x_2}{\mathrm{d}s} + \cdots + \frac{\partial p_n}{\partial \xi_j}\frac{\mathrm{d}x_n}{\mathrm{d}s} = 0 \tag{3.5}$$

and

$$p_1\frac{\mathrm{d}x_1}{\mathrm{d}s} + p_2\frac{\mathrm{d}x_2}{\mathrm{d}s} + \cdots + p_n\frac{\mathrm{d}x_n}{\mathrm{d}s} = \frac{\mathrm{d}w}{\mathrm{d}s}\,.$$

To gain a geometrical insight, we rewrite these equations in terms of scalar products as

$$\left[\frac{\partial p_1}{\partial \xi_j}, \frac{\partial p_2}{\partial \xi_j}, \ldots, \frac{\partial p_n}{\partial \xi_j}, 0\right] \cdot \left[\frac{\mathrm{d}x_1}{\mathrm{d}s}, \frac{\mathrm{d}x_2}{\mathrm{d}s}, \ldots, \frac{\mathrm{d}x_n}{\mathrm{d}s}, \frac{\mathrm{d}w}{\mathrm{d}s}\right] = 0 \tag{3.6}$$

and

$$[p_1, p_2, \ldots, p_n, -1] \cdot \left[\frac{\mathrm{d}x_1}{\mathrm{d}s}, \frac{\mathrm{d}x_2}{\mathrm{d}s}, \ldots, \frac{\mathrm{d}x_n}{\mathrm{d}s}, \frac{\mathrm{d}w}{\mathrm{d}s}\right] = 0\,, \tag{3.7}$$

respectively. These orthogonality conditions in the $(n+1)$-dimensional space determine vector $[\,\mathrm{d}x_1/\mathrm{d}s, \ldots, \mathrm{d}x_n/\mathrm{d}s, \mathrm{d}w/\mathrm{d}s]$ up to a scalar multiple.

To express the orthogonality conditions in terms of the original differential equation given by function F, and stated in expression (3.3), we differentiate that equation with respect to ξ_j to get

$$\frac{\partial F}{\partial p_1}\frac{\partial p_1}{\partial \xi_j} + \frac{\partial F}{\partial p_2}\frac{\partial p_2}{\partial \xi_j} + \cdots + \frac{\partial F}{\partial p_n}\frac{\partial p_n}{\partial \xi_j} = 0\,,$$

which we rewrite as

$$\left[\frac{\partial F}{\partial p_1}, \frac{\partial F}{\partial p_2}, \ldots, \frac{\partial F}{\partial p_n}, A\right] \cdot \left[\frac{\partial p_1}{\partial \xi_j}, \frac{\partial p_2}{\partial \xi_j}, \ldots, \frac{\partial p_n}{\partial \xi_j}, 0\right] = 0, \tag{3.8}$$

where A is an arbitrary number. We use this arbitrariness to let

$$A = p_1\frac{\partial F}{\partial p_1} + p_2\frac{\partial F}{\partial p_2} + \cdots + p_n\frac{\partial F}{\partial p_n}, \tag{3.9}$$

which we write as

$$\left[\frac{\partial F}{\partial p_1}, \frac{\partial F}{\partial p_2}, \ldots, \frac{\partial F}{\partial p_n}, A\right] \cdot [p_1, p_2, \ldots, p_n, -1] = 0. \tag{3.10}$$

Comparing equations (3.6) and (3.7) with equations (3.8) and (3.10), we see that vector

$$\left[\frac{\mathrm{d}x_1}{\mathrm{d}s}, \frac{\mathrm{d}x_2}{\mathrm{d}s}, \ldots, \frac{\mathrm{d}x_n}{\mathrm{d}s}, \frac{\mathrm{d}w}{\mathrm{d}s}\right]$$

is orthogonal to the same vectors to which vector

$$\left[\frac{\partial F}{\partial p_1}, \frac{\partial F}{\partial p_2}, \ldots, \frac{\partial F}{\partial p_n}, A\right]$$

is orthogonal. Since these are $(n+1)$-dimensional vectors perpendicular to the same linearly independent vectors, it implies that they are parallel to one another,

$$\left[\frac{\mathrm{d}x_1}{\mathrm{d}s}, \frac{\mathrm{d}x_2}{\mathrm{d}s}, \ldots, \frac{\mathrm{d}x_n}{\mathrm{d}s}, \frac{\mathrm{d}w}{\mathrm{d}s}\right]$$

$$= \zeta \left[\frac{\partial F}{\partial p_1}, \frac{\partial F}{\partial p_2}, \ldots, \frac{\partial F}{\partial p_n}, p_1 \frac{\partial F}{\partial p_1} + p_2 \frac{\partial F}{\partial p_2} + \cdots + p_n \frac{\partial F}{\partial p_n}\right], \tag{3.11}$$

where we use expression (3.9) for A. The coefficient of proportionality, ζ, depends on the choice of point P. Hence, ζ is a function of x and w. Expression (3.11) can be written as

$$\frac{\mathrm{d}x_i}{\mathrm{d}s} = \zeta(x, w) \frac{\partial F}{\partial p_i} \tag{3.12}$$

and

$$\frac{\mathrm{d}w}{\mathrm{d}s} = \zeta(x, w) \left(p_1 \frac{\partial F}{\partial p_1} + p_2 \frac{\partial F}{\partial p_2} + \cdots + p_n \frac{\partial F}{\partial p_n}\right). \tag{3.13}$$

The form of ζ can be determined for a given problem by the choice of parameter s, which we choose commonly to be time. Equations (3.12) belong to the characteristic equations of equation (3.1). These equations depend on p, which is restricted by equation (3.2) only.

To restrict p further, we quantify the change of p_i along curve $[x(s), w(s)]$. We differentiate the original differential equation with respect to x_i. Using the chain rule, we write

$$\frac{\partial}{\partial x_i} F(x, f(x), p_1(x), \ldots, p_n(x)) = \frac{\partial F}{\partial x_i} + \frac{\partial F}{\partial w} \frac{\partial f}{\partial x_i} + \sum_{j=1}^{n} \frac{\partial F}{\partial p_j} \frac{\partial p_j}{\partial x_i} = 0. \tag{3.14}$$

Since $p_i := \partial f / \partial x_i$, we write

$$\frac{\partial p_j}{\partial x_i} = \frac{\partial^2 f}{\partial x_i \partial x_j}.$$

Invoking the equality of mixed partial derivatives, we let

$$\frac{\partial p_j}{\partial x_i} = \frac{\partial p_i}{\partial x_j},$$

where we could write explicitly the right-hand side as $\partial p_i / \partial x_j = \partial^2 f / \partial x_j \partial x_i$. We write equation (3.14) as

$$\frac{\partial F}{\partial x_i} + \frac{\partial F}{\partial w}\frac{\partial f}{\partial x_i} + \sum_{j=1}^{n} \frac{\partial F}{\partial p_j}\frac{\partial p_i}{\partial x_j} = 0.$$

Using characteristic equations (3.12), we rewrite this equation as

$$\frac{\partial F}{\partial x_i} + \frac{\partial F}{\partial w}\frac{\partial f}{\partial x_i} + \frac{1}{\zeta}\sum_{j=1}^{n} \frac{\mathrm{d}x_j}{\mathrm{d}s}\frac{\partial p_i}{\partial x_j} = 0. \tag{3.15}$$

In view of the chain rule,

$$\sum_{j=1}^{n} \frac{\mathrm{d}x_j}{\mathrm{d}s}\frac{\partial p_i}{\partial x_j} = \frac{\mathrm{d}p_i(x)}{\mathrm{d}s},$$

we rewrite equation (3.15) as

$$\frac{\partial F}{\partial x_i} + \frac{\partial F}{\partial w}\frac{\partial f}{\partial x_i} + \frac{1}{\zeta}\frac{\mathrm{d}p_i}{\mathrm{d}s} = 0,$$

which can be restated as

$$\frac{\mathrm{d}p_i}{\mathrm{d}s} = -\zeta\left(\frac{\partial F}{\partial x_i} + \frac{\partial F}{\partial w}p_i\right). \tag{3.16}$$

These equations complete the system of $2n + 1$ equations for the same number of unknowns.

To construct the graph of a solution we need only the curves given by $(x_1(s), \ldots, x_n(s), w(s))$. However, to find these curves, we need also functions $p_j(s)$. A geometrical interpretation of the need for functions p to find the solution of the system of linear equations is that the side condition is given by an infinitesimal strip, not a curve; the curve itself is given by $(x_1(s), \ldots, x_n(s), w(s))$ and is referred to as a base characteristic or as a projected characteristic—vector $[-p_1(s), \ldots, -p_n(s), 1]$ is normal to the infinitesimal pieces of the tangent hyperplanes along this curve.

To summarize, we restate characteristic equations (3.12) and (3.13) as

$$\frac{\mathrm{d}x_i}{\mathrm{d}s} = \zeta(x, w)\frac{\partial F}{\partial p_i} \tag{3.17}$$

and

$$\frac{\mathrm{d}w}{\mathrm{d}s} = \zeta\left(x, w\right) \sum_{j=1}^{n} p_j \frac{\partial F}{\partial p_j}, \tag{3.18}$$

respectively, and equations (3.16) as

$$\frac{\mathrm{d}p_i}{\mathrm{d}s} = -\zeta\left(x, w\right) \left(\frac{\partial F}{\partial x_i} + \frac{\partial F}{\partial w} p_i \right), \tag{3.19}$$

where $w = f\left(x_1, x_2, \ldots, x_n\right)$ and $p_i = \partial f / \partial x_i$. Function F is called the Hamiltonian, and equations (3.17–3.19) are the Hamilton equations. They are characteristic equations of a nonlinear first-order partial differential equation. The solutions of this system are characteristic curves. As shown in Exercise 3.1, the Hamilton equations applied to a first-order linear equation result in characteristics obtained in Section 1.1.2.

3.3 Side conditions

The differential equations dictate behaviour of the solutions by restrictions on their derivatives. In the case of the first-order linear equations, these restrictions prescribe the solutions along the characteristic curves and thus the side conditions cannot be set freely along these curves. In the case of the first-order nonlinear equations, the restriction on the derivatives of the solutions is given by the Monge cone; there is no single direction along which the behaviour of solutions is restricted. To define such a direction, we need to take into account the side conditions, which can be propagated along the characteristics to form solutions—the characteristics given by the Hamilton equations.

As discussed in Section 3.1, the side conditions propagate along the characteristic curves. In contrast to linear equations, the direction of propagation depends not only on the equations but also on their side conditions.

3.4 Physical applications

3.4.1 *Elastodynamic equations*

In Section 2.3.5.1, we derive the characteristic equation for the elastodynamic equations—which are formulated in Appendix B—namely, equation (2.66),

$$\det\left[\sum_{i,j=1}^{3} c_{ljki}(x)\left(\frac{\partial\psi}{\partial x_i}\right)\left(\frac{\partial\psi}{\partial x_j}\right)-\rho(x)\,\delta_{kl}\right]=0\,.$$

Denoting $\nabla\psi$ by p, we divide this equation by $\left(p^2\rho(x)\right)^3$, which results in

$$\det\left[\sum_{i,j=1}^{3}\frac{c_{ljki}(x)}{\rho(x)}\frac{p_i}{\|p\|}\frac{p_j}{\|p\|}-\frac{1}{p^2}\delta_{kl}\right]=0\,,$$

an algebraic equation for $1/p^2$; the cubic power of the divisor accounts for the determinant of a 3×3 matrix. The resulting third-degree polynomial has, in general, three real roots, whose values depend on the position, x, and the direction, $n=p/\|p\|$; the roots are real due to the symmetry of the 3×3 matrix, which is a consequence of index symmetries of the elasticity tensor. We denote these roots by $v_i^{-2}(x,n)$, where $i\in\{1,2,3\}$ corresponds to different roots. Subsequently, we write

$$p^2=\frac{1}{v_i^2(x,n)}\,,\tag{3.20}$$

which are the eikonal equations—the characteristic equations of the elastodynamic equations.[3] An eikonal equation describes the slowness of the wavefront propagation whose magnitude depends on direction—an anisotropic dependence.

Level surfaces of ψ correspond to wavefronts at given instants of time. Hence, $p_i:=\partial\psi/\partial x_i$, at a given point, are the components of slowness with which a wavefront propagates at this point. Also, vector p is normal to the wavefront at this point. Hence, its direction is the direction of wavefront propagation at that point. We can express the magnitude of the velocity of the wavefront propagation as

$$v=\frac{1}{\|p\|}\,,$$

where v is the wavefront velocity.

As discussed in Section 3.2, we can solve the eikonal equation using equations (3.17), (3.19) and (3.18). The resulting base characteristics have the physical interpretation of rays. Following expression (3.20), we set $F=p^2v^2$, where v denotes one of the three v_i. Since F does not depend on w, the resulting equations are

$$\frac{\mathrm{d}x_i}{\mathrm{d}s}=\zeta\frac{\partial F}{\partial p_i}\,,$$

[3] *See also*: Slawinski (2015, Section 7.3).

$$\frac{\mathrm{d}p_i}{\mathrm{d}s} = -\zeta \frac{\partial F}{\partial x_i}$$

and

$$\frac{\mathrm{d}w}{\mathrm{d}s} = \zeta \left(p_1 \frac{\partial F}{\partial p_1} + p_2 \frac{\partial F}{\partial p_2} + p_3 \frac{\partial F}{\partial p_3} \right) . \tag{3.21}$$

To facilitate the physical interpretation of these equations and their solutions, we parametrize the characteristics by time. Since the value of ψ corresponds to time, we require that $\mathrm{d}w/\mathrm{d}s = 1$. Since $F = p^2 v^2$ is a homogeneous function of degree two in p, by the Euler homogeneous-function theorem,[4] the term in parentheses in equation (3.21) is equal to $2F$. In view of expression (3.20), $F = 1$, and hence $\zeta = 1/2$. It is common to denote $F/2$ by H, which is called the ray-theory Hamiltonian; in such a case we write

$$\dot{x}_i = \frac{\partial H}{\partial p_i}$$
$$i \in \{1, 2, 3\} , \tag{3.22}$$
$$\dot{p}_i = -\frac{\partial H}{\partial x_i}$$

which are the ray-theory Hamilton equations, and where $\mathrm{d}/\mathrm{d}t$ is denoted by the dot above a symbol.

In the following example we illustrate the use of equations (3.22) in a seismological context.

Example 3.1.[5] In this example, we study the Hamilton ray equations for a particular case of wave propagation that exhibits an elliptical velocity dependence with direction and a linear velocity dependence along one axis. This assumption, together with a consideration of a two-dimensional continuum, allows us to illustrate the meaning of the Hamilton ray equations by obtaining their analytic solutions.

Since the depth is commonly associated with the z-axis, we consider the physical space spanned by the x-axis and the z-axis, which correspond to the horizontal distance and depth, respectively. Considering the xz-plane, we write the wavefront velocity of a wave subjected to the elliptical dependence on direction as

$$v(\vartheta) = \sqrt{v_x^2 \sin^2 \vartheta + v_z^2 \cos^2 \vartheta} , \tag{3.23}$$

[4]Readers interested in this theorem might refer to Courant and Hilbert (1989, Volume 2 p. 11).

[5]This example appears in Rogister and Slawinski (2005).

where v_x and v_z are the magnitudes of wavefront velocity along the x-axis and z-axis, respectively, and ϑ is the wavefront angle measured from the z-axis. We assume that the ratio of the magnitudes of wavefront velocity along the x-axis and z-axis is constant to define

$$c := \frac{v_x}{v_z};$$

this is a dimensionless quantity equal to unity for isotropy. Using this definition, we rewrite expression (3.23) as

$$v(\vartheta) = v_z \sqrt{c^2 \sin^2 \vartheta + \cos^2 \vartheta}\,.$$

We specify the linear dependence of velocity along the z-axis,

$$v(\vartheta, z) = (a + bz)\sqrt{c^2 \sin^2 \vartheta + \cos^2 \vartheta}\,, \tag{3.24}$$

where a and b are constants whose units are velocity and the reciprocal of time, respectively. We refer to the velocity model described by this expression as the *abc* model.

In view of equation (3.20), the eikonal equation for two spatial dimensions is

$$p^2 := p_x^2 + p_z^2 = \frac{1}{v^2}. \tag{3.25}$$

Hence, in view of equation (3.24), the eikonal equation for the *abc* model is

$$p_x^2 + p_z^2 = \frac{1}{(a + bz)^2 (c^2 \sin^2 \vartheta + \cos^2 \vartheta)}\,.$$

To express ϑ in terms of vector $p = [p_x, p_z]$, we write

$$\vartheta = \arctan \frac{p_x}{p_z}\,. \tag{3.26}$$

Inserting this expression into the above equation and using trigonometric identities, we get

$$p_x^2 + p_z^2 = \frac{p_x^2 + p_z^2}{(a + bz)^2 (c^2 p_x^2 + p_z^2)}\,;$$

simplifying, we obtain

$$(a + bz)^2 (c^2 p_x^2 + p_z^2) = 1\,. \tag{3.27}$$

This is the eikonal equation that corresponds to elliptical anisotropy and linear inhomogeneity, and whose solution is the eikonal function, ψ, with its level curves corresponding to wavefronts. Equation (3.27) can be viewed as an expression for a level set of the inverse of Hamiltonian, $H^{-1}(1/2)$, where

$$H := \frac{(a+bz)^2\left(c^2p_x^2+p_z^2\right)}{2}. \tag{3.28}$$

Considering system (3.22) in two dimensions, we let $[x_1, x_2] \equiv [x, z]$ and $[p_1, p_2] \equiv [p_x, p_z]$. Using expression (3.28), we write explicitly this system as

$$\dot{x} = \frac{\partial H(z, p_x, p_z)}{\partial p_x} = (a+bz)^2 c^2 p_x,$$

$$\dot{z} = \frac{\partial H(z, p_x, p_z)}{\partial p_z} = (a+bz)^2 p_z, \tag{3.29}$$

$$\dot{p}_x = -\frac{\partial H(z, p_x, p_z)}{\partial x} = 0$$

and

$$\dot{p}_z = -\frac{\partial H(z, p_x, p_z)}{\partial z} = -b(a+bz)\left(c^2p_x^2+p_z^2\right). \tag{3.30}$$

Since we can write equation (3.27) as

$$c^2p_x^2+p_z^2 = \frac{1}{(a+bz)^2}, \tag{3.31}$$

we rewrite equation (3.30) as

$$\dot{p}_z = -\frac{b}{a+bz}.$$

Since $\dot{p}_x = 0$, it follows that $p_x(t) = \mathfrak{p}$, where $\mathfrak{p}$ denotes a constant. We write the remaining Hamilton equations as a system of ordinary differential equations to be solved for x, z and p_z. These equations are

$$\frac{\mathrm{d}x(t)}{\mathrm{d}t} = (a+bz(t))^2 c^2\mathfrak{p}, \tag{3.32}$$

$$\frac{\mathrm{d}z(t)}{\mathrm{d}t} = (a+bz(t))^2 p_z(t) \tag{3.33}$$

and

$$\frac{\mathrm{d}p_z(t)}{\mathrm{d}t} = -\frac{b}{a+bz(t)}. \tag{3.34}$$

To complete this system, we need to set the side conditions. We choose the values of the unknowns at the initial time: $x(t)$, $z(t)$, $p_x(t)$ and $p_z(t)$ at $t=0$; $p_x(0) = \mathfrak{p}$, and we set $x(0) = 0$, $z(0) = 0$ and $p_z(0) = p_{z0}$.

The condition for $p_z(t)$ is not independent from the condition for $p_x(t)$; they are related by eikonal equation (3.27). Solving this equation for p_z at $t = 0$, which corresponds to $z = 0$, we get

$$p_z(0) = \sqrt{\frac{1}{a^2} - c^2\mathfrak{p}^2}. \tag{3.35}$$

The above system of differential equations accompanied by the side conditions has the following meaning in the context of ray theory. The first two equations of system (3.29) define the vector field in the xz-plane; their solutions, $x(t)$ and $z(t)$, are the integral curves, which describe the path of the propagating wave. These curves are rays. At a given point, $(x(t), z(t))$, we can express the direction of a ray as

$$\theta = \arctan\frac{\mathrm{d}x}{\mathrm{d}z},$$

where θ is the ray angle, which is measured from the z-axis. At any point, we can express the magnitude of the velocity of the disturbance along the ray as

$$V = \sqrt{\left(\frac{\mathrm{d}x}{\mathrm{d}t}\right)^2 + \left(\frac{\mathrm{d}z}{\mathrm{d}t}\right)^2},$$

which is the ray velocity.

Let us examine the above side conditions. Setting $[x(0), z(0)] = [0, 0]$, we fix the origin of the ray at the initial time. In other words, we locate the point source. In view of continuity of wavefronts and considering the inhomogeneity along the z-axis only, we know that

$$\mathfrak{p} = \frac{\sin\vartheta}{v(\vartheta, z)}.$$

Considering $z = 0$ and using expression (3.24), we write

$$p_x(0) = \mathfrak{p} = \frac{\sin\vartheta_0}{a\sqrt{c^2\sin^2\vartheta_0 + \cos^2\vartheta_0}}, \tag{3.36}$$

where ϑ_0 denotes the take-off wavefront angle; in other words, the direction of the wavefront at the source.

We wish to solve system (3.29). Since we already know that $p_x = \mathfrak{p}$, we must solve equations (3.32), (3.33) and (3.34) for $x(t)$, $z(t)$ and $p_z(t)$. Let us write the second equation of system (3.29) as

$$p_z = \frac{\frac{\mathrm{d}z}{\mathrm{d}t}}{(a + bz)^2}.$$

Differentiating with respect to t, we get

$$\frac{\mathrm{d}p_z}{\mathrm{d}t} = \frac{\dfrac{\mathrm{d}^2 z}{\mathrm{d}t^2}(a+bz) - 2b\left(\dfrac{\mathrm{d}z}{\mathrm{d}t}\right)^2}{(a+bz)^3}.$$

Equating this result with the third equation of system (3.29) and rearranging, we get

$$(a+bz)\frac{\mathrm{d}^2 z}{\mathrm{d}t^2} - 2b\left(\frac{\mathrm{d}z}{\mathrm{d}t}\right)^2 + b(a+bz)^2 = 0.$$

Letting $(a+bz)^2 = y$, we get

$$\frac{1}{2b}\frac{\mathrm{d}^2 y}{\mathrm{d}t^2} - \frac{3}{4yb}\left(\frac{\mathrm{d}y}{\mathrm{d}t}\right)^2 + by = 0.$$

Letting $y = a^2 \exp u$, we get

$$\frac{\mathrm{d}^2 u}{\mathrm{d}t^2} - \frac{1}{2}\left(\frac{\mathrm{d}u}{\mathrm{d}t}\right)^2 + 2b^2 = 0.$$

Letting $\mathrm{d}u/\mathrm{d}t = q$, we get

$$\frac{\mathrm{d}q}{\mathrm{d}t} - \frac{1}{2}q^2 + 2b^2 = 0,$$

which we rewrite as

$$\mathrm{d}t = \frac{\mathrm{d}q}{\dfrac{1}{2}q^2 - 2b^2}.$$

Integrating both sides, we get

$$t + A_1 = -\frac{1}{b}\tanh^{-1}\frac{q}{2b},$$

where $\tanh^{-1}$ denotes the inverse function of $\tanh$, and A_1 is an integration constant. Solving for q, we get

$$q = 2b\tanh\left(-b\left(t + A_1\right)\right).$$

To obtain u, we integrate and get

$$u = -2\ln\left(\cosh\left(b\left(t + A_1\right)\right)\right) + A_2,$$

where A_2 is an integration constant. Hence,

$$y = \frac{a^2 \exp(A_2)}{\cosh^2\left(b\left(t + A_1\right)\right)}. \tag{3.37}$$

Since $z = \left(\sqrt{y} - a\right)/b$, we obtain the solution of the second equation of system (3.29), namely,

$$z(t) = \frac{1}{b}\frac{a\exp\left(\frac{A_2}{2}\right)}{\cosh\left(b\left(t + A_1\right)\right)} - \frac{a}{b}.$$

Also, in view of the second equation of system (3.29), we have $p_z = \dot{z}/y$. Thus, we have the solution of the third equation of system (3.29); this solution is

$$p_z(t) = -\frac{\sinh\left(b\left(t + A_1\right)\right)}{a\exp\left(\frac{A_2}{2}\right)}.$$

We write the remaining equation of system (3.29) as

$$\frac{\mathrm{d}x}{\mathrm{d}t} = yc^2\mathfrak{p},$$

where y is given by expression (3.37). Integrating, we get

$$x(t) = \frac{\mathfrak{p}a^2c^2\exp A_2}{b}\tanh\left(b\left(t + A_1\right)\right) + A_3,$$

where A_3 is an integration constant. We write the general solution of system (3.29) concisely as

$$x(t) = \frac{\mathfrak{p}a^2c^2\exp A_2}{b}\tanh\left(b\left(t + A_1\right)\right) + A_3,$$

$$z(t) = \frac{1}{b}\frac{a\exp\left(\frac{A_2}{2}\right)}{\cosh\left(b\left(t + A_1\right)\right)} - \frac{a}{b}, \tag{3.38}$$

$$p_x(t) = \mathfrak{p},$$

$$p_z(t) = -\frac{\sinh\left(b\left(t + A_1\right)\right)}{a\exp\left(\frac{A_2}{2}\right)}.$$

To get the particular solution, we find the three integration constants by invoking the side conditions. At $t = 0$, we rewrite the solutions stated in the first two expressions of system (3.38) as

$$x(0) = \frac{\mathfrak{p}a^2c^2\exp A_2}{b}\tanh\left(bA_1\right) + A_3 = 0$$

and

$$z(0) = \frac{1}{b}\frac{a\exp\left(\frac{A_2}{2}\right)}{\cosh(bA_1)} - \frac{a}{b} = 0\,.$$

Also, considering $z = 0$ and using the second equation of system (3.29) combined with solution $z(t)$ given in system (3.38), and evaluating at $t = 0$, we write

$$p_z(0) = \frac{\dot{z}(0)}{a^2} = -\frac{\sinh(bA_1)}{a\exp\left(\frac{A_2}{2}\right)} = \sqrt{\frac{1}{a^2} - \mathfrak{p}^2 c^2}\,,$$

where the right-hand side is given in expression (3.35).

Considering the last two equations, we have a system of two equations in two unknowns, A_1 and A_2. Solving, we obtain

$$A_1 = -\frac{1}{b}\tanh^{-1}\sqrt{1 - \mathfrak{p}^2 a^2 c^2}$$

and

$$A_2 = -\ln\left(\mathfrak{p}^2 a^2 c^2\right)\,.$$

Inserting $A_1$and A_2 into the equation for $x(0)$, we obtain

$$A_3 = \frac{\sqrt{1 - \mathfrak{p}^2 a^2 c^2}}{\mathfrak{p} b}\,.$$

Examining A_1, A_2 and A_3, we see that the units of A_1 are the units of time, A_2 is dimensionless and the units of A_3 are the units of distance. This is consistent with positions of A_1, A_2 and A_3 in system (3.38).

Having found A_1, A_2 and A_3, we rewrite solutions (3.38) as

$$x(t) = \frac{1}{\mathfrak{p} b}\left(\tanh\left(bt - \tanh^{-1}\sqrt{1 - \mathfrak{p}^2 a^2 c^2}\right) + \sqrt{1 - \mathfrak{p}^2 a^2 c^2}\right)\,, \tag{3.39}$$

$$z(t) = \frac{a}{b}\left(\frac{1}{\mathfrak{p} a c\cosh\left(bt - \tanh^{-1}\sqrt{1 - \mathfrak{p}^2 a^2 c^2}\right)} - 1\right)\,, \tag{3.40}$$

$$p_x(t) = \mathfrak{p}\,,$$

$$p_z(t) = -\frac{1}{\mathfrak{p} a^2 c}\sinh\left(bt - \tanh^{-1}\sqrt{1 - \mathfrak{p}^2 a^2 c^2}\right)\,,$$

where, in view of expression (3.36), we see that

$$\mathfrak{p} = \frac{\sin\vartheta_0}{a\sqrt{c^2\sin^2\vartheta_0 + \cos^2\vartheta_0}}\,,$$

with ϑ_0 being the take-off wavefront angle.

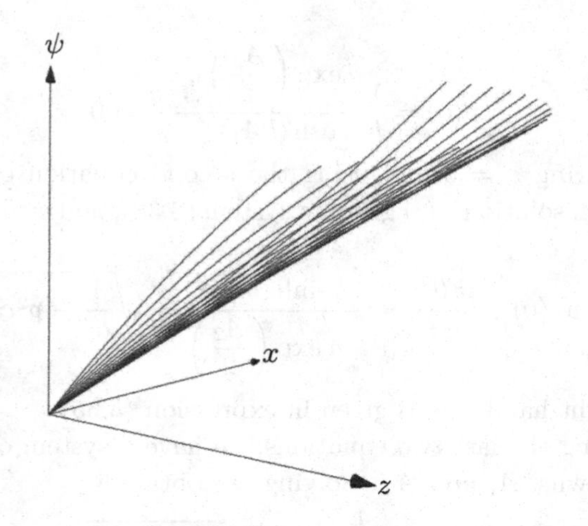

Fig. 3.3 Characteristics of the eikonal equation. Graph of the solution of the Hamilton ray equations (3.29), which is given in expression (3.40) with $a = 2000$, $b = 0.8$ and $c = 1.25$.

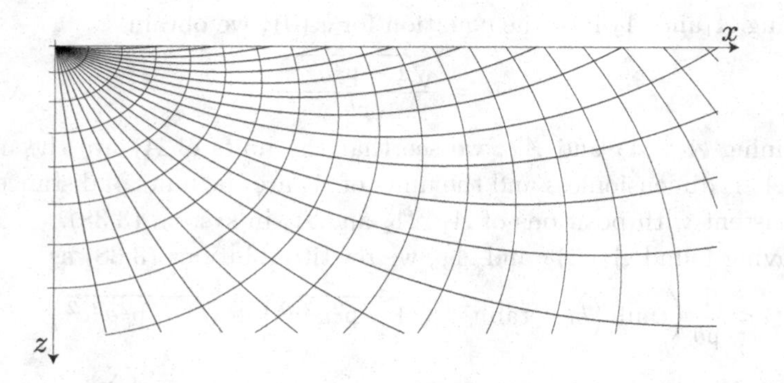

Fig. 3.4 Level curves of the eikonal function, shown in Figure 3.5, and the characteristics of the eikonal equation, shown in Figure 3.3, projected onto the xz-plane. They correspond to wavefronts and rays, respectively, for the *abc* model with $a = 2000$, $b = 0.8$ and $c = 1.25$.

Thus, for the *abc* model, we can choose the wavefront take-off angle, ϑ_0, and using the first two expressions of solutions (3.40) obtain the ray along which the disturbance generated at $(0, 0)$ propagates.

Examination of the first two expressions of solutions (3.40) allows us to learn about the shape of rays for the *abc* model. We can write each of these expressions as

$$\mathfrak{p}bx\left(t\right)-\sqrt{1-\mathfrak{p}^{2}a^{2}c^{2}}=\tanh\left(bt-\tanh^{-1}\sqrt{1-\mathfrak{p}^{2}a^{2}c^{2}}\right)$$

and

$$\left(\frac{b}{a}z\left(t\right)+1\right)\mathfrak{p}ac=\frac{1}{\cosh\left(bt-\tanh^{-1}\sqrt{1-\mathfrak{p}^{2}a^{2}c^{2}}\right)},$$

respectively. Squaring these two equations, adding them together and using standard identities, we obtain

$$\frac{\left(x-\frac{\sqrt{1-\mathfrak{p}^{2}a^{2}c^{2}}}{\mathfrak{p}b}\right)^{2}}{\left(\frac{1}{\mathfrak{p}b}\right)^{2}}+\frac{\left(z+\frac{a}{b}\right)^{2}}{\left(\frac{1}{\mathfrak{p}bc}\right)^{2}}=1\,. \tag{3.41}$$

This is the equation of an ellipse with a centre on the line given by $z=-a/b$. In other words, in the *abc* model, rays are elliptical arcs. In view of $v\left(z\right)=a+bz$, we conclude that the centre of the ellipse corresponds to the level where the velocity vanishes if, as usual, we assume $b>0$ to describe the increase of velocity along the z-axis. As shown in Figure 3.4, we obtain rays and wavefronts. The last two expressions of solutions (3.38) are no longer necessary to obtain rays, traveltimes and wavefronts; however, we need all four equations to solve system (3.29).

For an isotropic case, $c=1$, and equation (3.41) reduces to the equation of a circle whose radius is $1/pb$ and whose centre is at $x=\sqrt{1-p^{2}a^{2}}/pb$ and $z=-a/b$. Derivation of such an equation is shown in Exercise 3.3, where we use a Hamiltonian that is different from expression (3.28). Examining equation (3.41) and Exercise 3.3, we see that two different Hamiltonians rooted in eikonal equation (3.25) result in the same rays since they are the solutions of the characteristic equations of eikonal equation (3.25). For a homogeneous case, $b=0$, and rays are straight lines.

Using solutions (3.40), we can obtain the graph of the solution of the original partial differential equation, namely, eikonal equation (3.27), as a parametric plot, $\left[x\left(t\right),z\left(t\right),t\right]$, for all $\mathfrak{p}$ that are consistent with the original equation, as shown in Figure 3.3. In the present case, following expression (3.36), if we set $\vartheta_{0}\in\left(-\pi/2,\pi/2\right)$, we get

$$\mathfrak{p}\in\left(-\frac{1}{ac},\frac{1}{ac}\right).$$

We can obtain also an explicit form of the solution of equation (3.27). Since $t = \psi(x, z)$, we solve the first equation of system (3.40) for t to get

$$t(x; \mathfrak{p}) = \frac{\tanh^{-1}\left(\mathfrak{p}bx - \sqrt{1 - \mathfrak{p}^2 a^2 c^2}\right) + \tanh^{-1}\sqrt{1 - \mathfrak{p}^2 a^2 c^2}}{b}. \tag{3.42}$$

To express t in terms of x and z, and the parameters of the *abc* model, we solve equation (3.41) for $\mathfrak{p}$ to get

$$\mathfrak{p}(x, z) = \frac{2x}{\sqrt{(x^2 + c^2 z^2)\left((2a + bz)^2 c^2 + b^2 x^2\right)}}. \tag{3.43}$$

Thus, expression (3.42) with $\mathfrak{p}$ given by expression (3.43) is the solution, $\psi = t(x, z)$, of equation (3.27). This solution is shown in Figure 3.5.

Expression (3.42) is not valid for $\mathfrak{p} = 0$, which corresponds to rays along along $x = 0$. In such a case, we cannot invert equation (3.39) for t, since $x(t) = 0$, for all t. Instead—using expression (3.31), with $p_x = 0$—we can write expression (3.33) as

$$\mathrm{d}t = \frac{\mathrm{d}z}{a + bz},$$

to integrate the right-hand side with respect to z—instead of x, as is the case for solution (3.42)—to obtain

$$t = \frac{1}{b}(\ln(a + bz) - \ln a) = \frac{1}{b}\ln\frac{a + bz}{a}, \qquad z \geqslant 0,$$

which is the traveltime, between 0 and z, for rays along $x = 0$.

In Example 3.1, we examine wavefronts and rays for an elastodynamic equation. In Section 3.4.2, we consider rays within the electromagnetic theory.

3.4.2 *Maxwell equations*

The eikonal equation associated with the electromagnetic waves is given by expression (2.68), namely,

$$\sum_{i=1}^{3}\left(\frac{\partial\psi}{\partial x_i}\right)^2 = \frac{1}{c^2},$$

where c is the speed of light in a vacuum. This equation results from the Maxwell equations—given in expressions (C.3), (C.5), (C.7) and (C.10), in Appendix C. These equations describe electromagnetic phenomena in

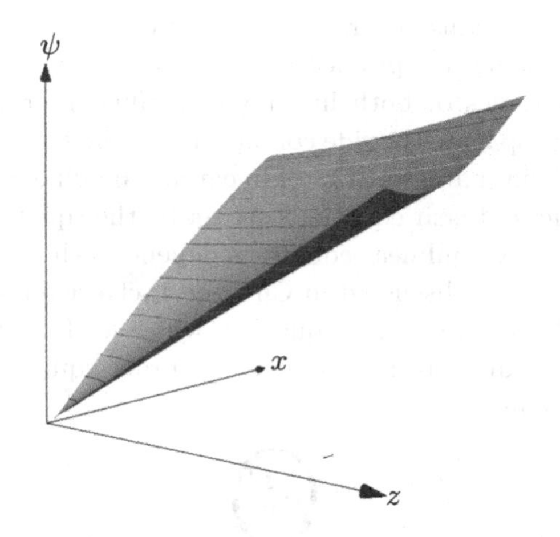

Fig. 3.5 Eikonal function: Surface representing the solution of eikonal equation (3.27) with $a = 2000$, $b = 0.8$ and $c = 1.25$. Note that this surface can be constructed from the characteristic curves illustrated in Figure 3.3.

free space—a space that does not react to these phenomena—consequently, we are dealing with an isotropic and homogeneous medium. Hence, $p = \nabla\psi = 1/c$. Furthermore, F is constant with respect to all variables, and the Hamilton equations (3.17) and (3.19) are reduced to

$$\frac{\mathrm{d}x_i}{\mathrm{d}s} = 0$$

and

$$\frac{\mathrm{d}p_i}{\mathrm{d}s} = 0,$$

respectively, which means that both x and p are constants—the characteristics are straight lines.

Closing remarks

In the previous chapters, we construct characteristics for linear first-order and second-order equations and—in this chapter—for nonlinear first-order equations. Extensions to the higher-order linear equations are straightforward following the method of the Taylor series of solutions for Cauchy problems. Extensions of the higher-order nonlinear equations are difficult due to relations among the $\partial f/\partial x_i$, which depend on higher-order derivatives; in other words, we are not able to relate them using the parameters

ξ_j only. These equations cover the majority of the important differential equations of mathematical physics.

The characteristics of both linear and nonlinear first-order equations also allow us to propagate the side conditions to form the solutions. The difference between the characteristics of linear and nonlinear equations is that the characteristics of linear equations are set by the equation itself, whereas the characteristics of nonlinear equations depend both on the equation and the side condition. As discussed in Chapter 4, characteristics depend only on the highest derivatives of the equation, which leads to similarities of behaviour among linear first-order and second-order equations and nonlinear first-order equations.

3.5 Exercises

Exercise 3.1. Consider equation (1.8), which we rewrite as

$$\frac{\partial \psi\left(x_1, x_2\right)}{\partial x_1}+x_2 \frac{\partial \psi\left(x_1, x_2\right)}{\partial x_2}=0\,. \tag{3.44}$$

Using equations (3.17), show that the characteristic curves of equation (3.44) are given by expression (1.11), namely,

$$x_2=C \exp x_1\,. \tag{3.45}$$

Solution. Denoting the gradient of ψ by p, we write equation (3.44) as

$$p_1+x_2 p_2=0\,.$$

Using Hamiltonian $F=p_1+x_2 p_2$, we write equations (3.17) as

$$\frac{\mathrm{d} x_1}{\mathrm{~d} s}=\zeta \frac{\partial F}{\partial p_1}=\zeta$$

and

$$\frac{\mathrm{d} x_2}{\mathrm{~d} s}=\zeta \frac{\partial F}{\partial p_2}=\zeta x_2\,.$$

We can write the ratio of $\dot{x}_2$ and $\dot{x}_1$ as

$$\frac{\frac{\mathrm{d} x_2}{\mathrm{~d} s}}{\frac{\mathrm{d} x_1}{\mathrm{~d} s}}=\frac{\mathrm{d} x_2}{\mathrm{~d} x_1}=x_2\,,$$

which is expression (1.10). Solving this ordinary differential equation, we obtain

$$x_2 = C \exp x_1 ,$$

where C denotes a constant. This is the characteristic curve given by expression (3.45), as required. Thus, solving the Hamilton equations for x_1 and x_2, we obtain the same result as using the directional-derivative approach in Section 1.1.2.

Exercise 3.2. Find the solution of

$$\frac{\partial f}{\partial x} - \left(\frac{\partial f}{\partial y}\right)^2 = x$$

that satisfies $f(0, y) = f_0(y)$.

Solution. This is a nonlinear first-order partial differential equation. Let the Hamiltonian be given by

$$F = \frac{\partial f}{\partial x} - \left(\frac{\partial f}{\partial y}\right)^2 - x,$$

which can be written as

$$F = p_x - p_y^2 - x.$$

We write the Hamilton equations as

$$\begin{aligned}
x'(s) &= \frac{\partial F}{\partial p_x} = 1, \\
y'(s) &= \frac{\partial F}{\partial p_y} = -2p_y, \\
p_x'(s) &= -\frac{\partial F}{\partial x} - \frac{\partial F}{\partial f} p_x = 1, \\
p_y'(s) &= -\frac{\partial F}{\partial y} - \frac{\partial F}{\partial f} p_y = 0, \\
f'(s) &= p_x \frac{\partial F}{\partial p_x} + p_y \frac{\partial F}{\partial p_y},
\end{aligned}$$

where we choose the parametrization by s to be such that the scaling parameter is equal to one. Integrating, we obtain the solutions of the first four equations,

$$\begin{aligned}
x(s) &= s + x_0, \\
y(s) &= -2p_y s + y_0, \\
p_x(s) &= s + p_{x0}, \\
p_y(s) &= p_{y0},
\end{aligned}$$

where p_{x0}, p_{y0}, x_0 and y_0 are constants. Hence, the fifth equation implies that the unknown function changes along the characteristic curves as

$$f'(s) = s + p_{x0} - 2p_{y0}^2.$$

Integrating, we obtain

$$f(s) = \frac{1}{2}s^2 + \left(p_{x0} - 2p_{y0}^2\right)s + f_0. \tag{3.46}$$

Considering the side condition, $f(0, y) = f_0(y)$, we see that

$$\begin{aligned} x(s) &= s \\ y(s) &= -2f_0's + y_0 \\ p_x(s) &= s + \left(f_0'(y_0)\right)^2 \\ p_y(s) &= f_0'(y_0) \end{aligned}$$

and

$$f(s) = \frac{1}{2}s^2 - \left(f_0'(y_0)\right)^2 s + f_0(y_0). \tag{3.47}$$

We can parametrize the xy-plane by s and y_0. In general, we are unable to express explicitly the dependence of s and y_0 on x and y. Such an expression is possible for certain cases, as we exemplify by considering the following particular form of f_0, namely, $f_0(y) = y$. In such a case, $s = x$ and $y_0 = y + 2x$ and expression (3.47) becomes

$$f(x, y) = \frac{1}{2}x^2 - x + (y + 2x) = \frac{1}{2}x^2 + x + y,$$

which is the required solution.

Exercise 3.3. Recall eikonal equation (3.25), and choose the corresponding Hamiltonian to be

$$H = \frac{1}{2}\left(p_x^2 + p_z^2 - \frac{1}{v^2}\right), \tag{3.48}$$

which is different from the one used in Section 3.1. Considering a vertically inhomogeneous medium and using the Hamilton equations, derive the general expression for a ray as $x = x(z)$. Use the derived expression to find the traveltime along the ray. Also, discuss rays in the vertically inhomogeneous medium where the velocity increases linearly with depth.

Solution. Inserting expression (3.48), which for a vertically inhomogeneous medium is given explicitly by

$$H(p_x, p_z, z) = \frac{1}{2}\left[p_x^2 + p_z^2 - \frac{1}{v^2(z)}\right],$$

into the Hamilton equations, we obtain

$$x' = \frac{\partial H}{\partial p_x} = p_x, \tag{3.49}$$

$$z' = \frac{\partial H}{\partial p_z} = p_z, \tag{3.50}$$

$$p_x' = -\frac{\partial H}{\partial x} = 0 \tag{3.51}$$

and

$$p_z' = -\frac{\partial H}{\partial z} = -\frac{\frac{\mathrm{d}v}{\mathrm{d}z}}{v^3}, \tag{3.52}$$

where the prime denotes a derivative with respect to parameter s.

From equation (3.51), we see that p_x is constant along a ray. Using this fact in equation (3.49), we infer that

$$x(s) = p_x s + x_0, \tag{3.53}$$

which is the x-component of a ray as a function of s.

We recall eikonal equation (3.25), which we write as

$$p_z = \pm\sqrt{\frac{1}{v(z)^2} + p_x^2}.$$

In view of this expression, equation (3.50) implies

$$z' = \pm\sqrt{\frac{1}{v(z)^2} - p_x^2},$$

which is a separable differential equation whose solution is

$$s = \pm\int_0^z \frac{1}{\sqrt{\frac{1}{v(\zeta)^2} - p_x^2}}\,\mathrm{d}\zeta,$$

where we set s in such a way that $z(0) = 0$. Using equation (3.53), we write

$$x(z) = \pm p_x \int_0^z \frac{1}{\sqrt{\frac{1}{v(\zeta)^2} - p_x^2}} \mathrm{d}\zeta + x_0, \tag{3.54}$$

which is the desired expression.

To find the expression for the traveltime along the ray between $[x_0, z_0]$ and $[x, z]$, we use the definitions of p_x and p_z

$$t = \int_{x_0}^{x} p_x \, \mathrm{d}\xi + \int_{z_0}^{z} p_z \, \mathrm{d}\zeta = p_x (x - x_0) \pm \int_{z_0}^{z} \sqrt{\frac{1}{v(\zeta)^2} + p_x^2} \, \mathrm{d}\zeta. \tag{3.55}$$

The $\pm$ sign depends on the direction of the signal. For a downgoing signal, the sign is positive; for an upgoing one the sign is negative.

To discuss rays in a vertically inhomogeneous medium where the velocity increases linearly with depth, we let $v(z) = a + bz$. In such a case, we write expression (3.54) as

$$x(z) = \pm p_x \int_0^z \frac{1}{\sqrt{\frac{1}{(a + b\zeta)^2} - p_x^2}} \mathrm{d}\zeta + x_0.$$

After integration, we get

$$x(z) = \pm \sqrt{\frac{1}{b^2 p_x^2} - \left(z + \frac{a}{b}\right)^2} + x_0 \mp \frac{1}{b}\sqrt{\frac{1}{p_x^2} - a^2},$$

where the $\pm$ sign depends on the direction of propagation.

To interpret this result geometrically, we rewrite it as

$$\left(x - x_0 \pm \frac{1}{b}\sqrt{\frac{1}{p_x^2} - a^2}\right)^2 + \left(z + \frac{a}{b}\right)^2 = \frac{1}{b^2 p_x^2}, \tag{3.56}$$

which is an equation of a circle whose centre is at $\left[x_0 \mp \frac{1}{b}\sqrt{\frac{1}{p_x^2} - a^2}, -a/b\right]$ and whose radius is $1/bp_x$. Thus, in a vertically inhomogeneous medium where the velocity increases linearly with depth, rays are circular arcs. The radius is infinitely long if $b = 0$ or if $p_x = 0$. Thus, the rays are straight lines if the medium is homogeneous or if the ray coincides with the direction of the velocity gradient. As expected, equation (3.56) corresponds to equation (3.41) with $c = 1$; an ellipse is reduced to a circle.

Exercise 3.4. Consider equation

$$\frac{1}{\xi}x_1 + \xi x_2 - 1 = 0\,, \tag{3.57}$$

which represents a one-parameter family of lines in the x_1x_2-plane parametrized by ξ. Find a point that is representative of each line and is parametrized by ξ.

Solution. A choice of the sought-after point is the intersection of the given line with the envelope of all such lines in the x_1x_2-plane. To find this envelope, which results from varying ξ, we consider the $x_1x_2\xi$-space. In other words, we consider the left-hand side of equation (3.57) as a function of x_1, x_2 and ξ; we denote this function by $f(x_1, x_2, \xi)$. The vectors normal to $f = 0$ at the points that project onto the envelope on the x_1x_2-plane have the ξ-component equal to zero, which means that these vectors are parallel to the x_1x_2-plane. In view of equation (3.57), this means that in the $x_1x_2\xi$-space, the ξ-component of the gradient is zero,

$$\frac{\partial}{\partial \xi}\left(\frac{1}{\xi}x_1 + \xi x_2 - 1\right) = 0; \tag{3.58}$$

a point on the envelope belongs to the given tangent line and to the lines that are infinitesimally close to that line, as measured by ξ.

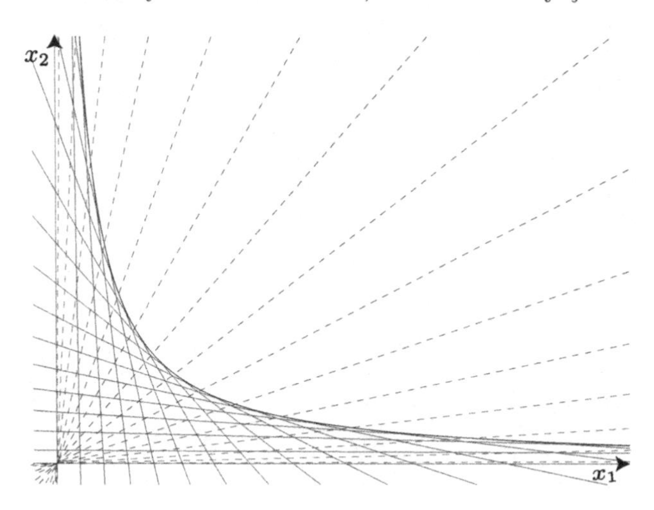

Fig. 3.6 A family of lines $x_1/\xi + \xi x_2 - 1 = 0$ and its envelope constructed by intersecting these lines with the radial lines, $-x_1/\xi^2 + x_2 = 0$, which are obtained by differentiating the family of lines with respect to parameter ξ.

Differentiating expression (3.58), we get

$$-\frac{1}{\xi^2}x_1 + x_2 = 0\,, \tag{3.59}$$

which is another family of lines parametrized by ξ. The point on the envelope parametrized by ξ belongs to this family and to the original family of lines given by equation (3.57). To find this point we combine equations (3.57) and (3.59). Solving these equations for x_1 and x_2, we obtain equations for the envelope parametrized by ξ, namely,

$$x_1 = \frac{1}{2}\xi$$

and

$$x_2 = \frac{1}{2\xi}\,.$$

This envelope is a part of a hyperbola shown in Figure 3.6.

Chapter 4

Propagation of discontinuities for linear partial differential equations

Many physical problems lead to analytic partial differential equations. But the restriction to analytic Cauchy data and solutions is unrealistic. It implies that the solution is determined globally by local conditions near one point, and makes it impossible to describe phenomena that do not interact instantaneously with the whole universe. The Cauchy problem is "ill-posed" except for hyperbolic equations.

Fritz John (1982)

Preliminary remarks

In this chapter we examine the discontinuities of both solutions and their derivatives. To examine discontinuities, we cannot use a method such as the convergent Taylor series for describing a differentiable function at a point. We introduce another series to describe functions at a point: the asymptotic series. In contrast to the convergent Taylor series, the asymptotic series need not converge to a function it represents. Furthermore, in contrast to the convergent Taylor series, this function need not be differentiable at the point of interest. Even if a function is differentiable, the asymptotic series might be preferable to the convergent Taylor series, since the latter one might converge slowly while the initial terms of the former one might be sufficiently close to the values of the function. We refer to the solutions of differential equations that are expressed in terms of asymptotic series as asymptotics—the word used in the title of this book.

The concept of nondifferentiable solutions of differential equations to describe propagation of sharp waveforms was discussed by d'Alembert and Euler in the eighteenth century and elaborated further by Schwartz and

Sobolev in the first half of the twentieth century.[1] A physical motivation to consider discontinuities is the concept of a wavefront. As we show in this chapter, discontinuities of solutions of hyperbolic equations are related to characteristics discussed in Chapters 1, 2 and 3. Indeed, there is a one-to-one relation between wavefronts and characteristics that requires discontinuities in solutions of hyperbolic equations.[2] In the classical theory of partial differential equations, solutions can have discontinuities only along characteristics. We illustrate this using asymptotic series, which rely on the high-frequency content of such discontinuities. For this reason, we present the subject in the frequency domain.

We begin this chapter with an example that motivates the use of the Fourier transform, and consideration of a series expansion of solutions of linear differential equations. Next, we examine the high-frequency behaviour of the Fourier transform for functions exhibiting different types of discontinuities. Finally, we focus our attention on asymptotic series of the Fourier transform around the infinite frequency. From these expansions we conclude that a solution of a linear differential equation can have discontinuities in any of its derivatives only along the characteristics.

Readers might find it useful to study this chapter together with Appendices B, C, D and E, where we formulate the elastodynamic equations and the Maxwell equations, and discuss the Fourier transform and distributions, respectively.

4.1 Motivational example

Discontinuities in a solution are associated with its high-frequency content. Let us consider a solution of the wave equation that is given by a propagation of the Dirac delta,

$$u\left(x,t\right)=A\left(x\right)\delta\left(t-\psi\left(x\right)\right),$$

which we express using the Fourier transform (D.16) as

$$\hat{u}\left(x,\omega\right)=\frac{A\left(x\right)}{\sqrt{2\pi}}e^{-\imath\omega\psi\left(x\right)}. \tag{4.1}$$

[1] Readers interested in the history of the formulation of nondifferentiable solution of the wave equation might refer to Schwartz (2001, Chapter VI).

[2] Readers interested in this requirement for the scalar wave equation might refer to Courant and Hilbert (1989, Volume 2, Chapter 6); for a generalization to the elastodynamic equation to Bos and Slawinski (2010).

Herein, $A(x)$ stands for amplitude of the impulse, and $\psi(x) = t$ represents its location at time t. The Fourier transform of the wave equation is

$$\frac{v^2}{(\iota\omega)^2}\frac{\partial^2 \hat{u}(x,\omega)}{\partial x^2} = \hat{u}(x,\omega) . \tag{4.2}$$

Using expression (4.1) in equation (4.2), we get

$$\frac{v^2}{(\iota\omega)^2}\frac{\partial^2 A(x)\, e^{-\iota\omega\psi(x)}}{\partial x^2} = A(x)\, e^{-\iota\omega\psi(x)} ,$$

which becomes

$$\frac{v^2}{(\iota\omega)^2} e^{-\iota\omega\psi(x)} \left(\frac{\partial^2 A(x)}{\partial x^2} - 2\iota\omega \frac{\partial A(x)}{\partial x}\frac{\partial \psi(x)}{\partial x} - \iota\omega A(x) \frac{\partial^2 \psi}{\partial x^2} \right.$$
$$\left. + (\iota\omega)^2 A(x) \left(\frac{\partial \psi(x)}{\partial x} \right)^2 \right) = A(x)\, e^{-\iota\omega\psi(x)} .$$

Grouping the terms with the same powers of $\iota\omega$, we obtain

$$A(x) \left(v^2 \left(\frac{\partial \psi(x)}{\partial x} \right)^2 - 1 \right) - \frac{1}{\iota\omega} v^2 \left(2 \frac{\partial A(x)}{\partial x}\frac{\partial \psi(x)}{\partial x} + A(x) \frac{\partial^2 \psi}{\partial x^2} \right)$$
$$+ \frac{1}{(\iota\omega)^2} v^2 \frac{\partial^2 A(x)}{\partial x^2} = 0 .$$

For this equation to be satisfied for all ω, the three terms corresponding to the three powers of $\iota\omega$ have to be zero,

$$v^2 \left(\frac{\partial \psi(x)}{\partial x} \right)^2 - 1 = 0 ,$$

$$2 \frac{\partial A(x)}{\partial x}\frac{\partial \psi(x)}{\partial x} + A(x) \frac{\partial^2 \psi}{\partial x^2} = 0$$

and

$$\frac{\partial^2 A(x)}{\partial x^2} = 0 ;$$

these three equations appear in Exercise 2.14. The first equation is the eikonal equation (2.59), but in one spatial dimension. In the following sections, we discuss the meaning of, and the process for obtaining, these equations in the context of propagation of discontinuities. Using expression (4.1) as a trial solution for equation (4.2) requires $A(x)$ to satisfy both the second and the third equations, which might be impossible, as discussed in Exercise 2.15. This difficulty disappears if we replace solution (4.1) by an infinite series. In such a case, each coefficient of the series that corresponds to $A(x)$ satisfies a combination of both equations, as we discuss in Section 4.3.

4.2 Discontinuities and frequency content[3]

The high-frequency behaviour of the Fourier transform gives important information about the smoothness of the function. As discussed in Appendix D, if $f \in L^1(\mathbb{R})$ then we can define the Fourier transform of $f(t)$ to be

$$\widehat{f}(\omega) := \frac{1}{\sqrt{2\pi}} \int_{-\infty}^{\infty} e^{-\iota\omega t} f(t)\,\mathrm{d}t. \tag{4.3}$$

If $g \in L^1(\mathbb{R})$ then we may also define the inverse transform by

$$\check{g}(t) := \frac{1}{\sqrt{2\pi}} \int_{-\infty}^{\infty} e^{\iota\omega t} g(\omega)\,\mathrm{d}\omega \tag{4.4}$$

so that

$$f = \check{\widehat{f}},$$

for $f, \widehat{f} \in L^1(\mathbb{R})$. A connection between the regularity of a function and the growth of its Fourier transform as $|\omega| \to \infty$ is evident in the Riemann-Lebesgue Lemma, below.

Lemma 4.1. *If $f \in L^1(\mathbb{R})$ then*

$$\lim_{|\omega|\to\infty} \widehat{f}(\omega) = 0.$$

In other words, if f has the regularity property of being absolutely integrable over $\mathbb{R}$, then its Fourier transform must diminish to zero as frequency tends to infinity. Conversely, if a Fourier transform does not diminish to zero, it cannot be the transform of an absolutely integrable function.

Integrability is a weak form of regularity; more can be inferred from the Fourier transfom of the function. To study the differentiability, we look at the Fourier transform of derivatives using the property that differentiation in the time domain is replaced by multiplication in the frequency domain. Under suitable conditions on f, the Fourier transform of

$$\frac{\mathrm{d}^n}{\mathrm{d}t^n} f(t) \quad \text{is} \quad (\iota\omega)^n \widehat{f}(\omega) \tag{4.5}$$

and that of

$$t^n f(t) \quad \text{is} \quad \iota^n \frac{\mathrm{d}^n}{\mathrm{d}\omega^n} \widehat{f}(\omega)\,. \tag{4.6}$$

[3] This section is adapted from Bos and Slawinski (2010).

Applying transform (4.4) to relation (4.5), we obtain

$$\frac{\mathrm{d}^n}{\mathrm{d}t^n} f(t) = \int_{-\infty}^{\infty} e^{\iota\omega t} (\iota\omega)^n \widehat{f}(\omega)\, \mathrm{d}\omega.$$

This makes sense only if $\omega^n \widehat{f}(\omega)$ is integrable, which implicitly places a condition on the diminishment of $\omega^n \widehat{f}(\omega)$ as $|\omega| \to \infty$. Conversely, if indeed $\omega^n \widehat{f}(\omega)$ is integrable, then this relation would imply that $f(t)$ has an nth derivative. A formal statement of this behaviour is given by the following theorem.

Theorem 4.1. *Suppose that $f(t) \in L^2(\mathbb{R})$ and that $\omega^k \widehat{f}(\omega) \in L^1(\mathbb{R})$ for $0 \leqslant k \leqslant n$. Then $f(t)$ is n times continuously differentiable for all $t \in \mathbb{R}$.*

A proof may be found, for example, in Zimmer (1990, Cor. 5.2.3). Condition $\omega^k \widehat{f}(\omega) \in L^1(\mathbb{R})$ forces $\omega^k \widehat{f}(\omega)$ to be small for large $|\omega|$, which means that $\widehat{f}(\omega)$ diminishes at least as fast as $1/\omega^k$ as $|\omega| \to \infty$.

If we consider distributions, which are discussed in Appendix E, then the behaviour of the Fourier transform at infinity gives even more specific information. To illustrate this, we consider the two examples below.

Example 4.1. Consider a generalized function that contains a discontinuity given by the Dirac delta,

$$f(t) = A\delta(t) + g(t),$$

where $g \in L^1(\mathbb{R})$, δ is the Dirac delta supported at $t = 0$ and A corresponds to the amplitude of the delta. The Fourier transform of this function is

$$\widehat{f}(\omega) = A\widehat{\delta}(\omega) + \widehat{g}(\omega) = A\frac{1}{\sqrt{2\pi}} + \widehat{g}(\omega).$$

Hence, by the Riemann-Lebesgue Lemma 4.1, the second part vanishes as ω tends to infinity,

$$\lim_{|\omega|\to\infty} \widehat{f}(\omega) = A\frac{1}{\sqrt{2\pi}}.$$

Thus, the amplitude is

$$A = \sqrt{2\pi} \lim_{|\omega|\to\infty} \widehat{f}(\omega).$$

For discontinuities at a location different than zero, we can translate the above expressions from t to $t - a$ by considering

$$f(t) = A\delta(t - a) + g(t - a),$$

which results in

$$\widehat{f}(\omega) = e^{\iota a\omega}\left(A\frac{1}{\sqrt{2\pi}} + \widehat{g}(\omega)\right).$$

We recover A by using Lemma 4.1,

$$A = \sqrt{2\pi}\lim_{|\omega|\to\infty} e^{-\iota\omega a}\,\widehat{f}(\omega).$$

Another type of discontinuity is a step discontinuity, which is discussed in the following example.

Example 4.2. A function with a step discontinuity can be represented by

$$f(t) = A\,\mathrm{sgn}(t) + g(t),$$

where

$$\mathrm{sgn}(t) := \begin{cases} 1 & t > 0 \\ -1 & t < 0 \end{cases},$$

g is such that $g, g' \in L^1(\mathbb{R})$, and A corresponds to the half amplitude of the jump at $t = 0$. The Fourier transform is

$$\widehat{f}(\omega) = A\sqrt{\frac{2}{\pi}}\frac{1}{\iota\omega} + \widehat{g}(\omega).$$

By Lemma 4.1, $\lim_{|\omega|\to\infty}\widehat{g}(\omega) = 0$, and since $g' \in L^1(\mathbb{R})$, by Theorem 4.1, $\lim_{|\omega|\to\infty}\omega\widehat{g}(\omega) = 0$. Hence

$$\lim_{|\omega|\to\infty}\omega\widehat{f}(\omega) = A\sqrt{\frac{2}{\pi}}\frac{1}{\iota} + 0,$$

and we recover the amplitude of the jump,

$$A = \iota\sqrt{\frac{\pi}{2}}\lim_{|\omega|\to\infty}\omega\widehat{f}(\omega).$$

If we translate the jump along the axis to $t = a$, by considering

$$f(t) = A\,\mathrm{sgn}(t-a) + g(t-a),$$

then

$$\widehat{f}(\omega) = e^{\iota a\omega}\left(A\sqrt{\frac{2}{\pi}}\frac{1}{\iota\omega} + \widehat{g}(\omega)\right), \tag{4.7}$$

and we recover A as

$$A = \iota\sqrt{\frac{\pi}{2}}\lim_{|\omega|\to\infty} e^{-\iota a\omega}\omega\widehat{f}(\omega).$$

We can generalize the above example by considering jumps in the derivatives of a function.

Example 4.3. Consider a symmetric jump in all the derivatives at $t = 0$ by examining

$$f(t) = \sum_{n=0}^{\infty} a_n t^n \operatorname{sgn}(t).$$

The kth derivative of the mth term of $f(t)$ is

$$\begin{aligned}(a_m t^m \operatorname{sgn}(t))^{(k)} &= \left(a_m m t^{m-1} \operatorname{sgn}(t) + 2a_m t^m \delta(t)\right)^{(k-1)} \\ &= \left(a_m m t^{m-1} \operatorname{sgn}(t)\right)^{(k-1)},\end{aligned}$$

where the second term in the second expression is zero for $m \neq 0$ due to the Dirac delta being nonzero only for $t = 0$. If $m = 0$, then the above expression becomes

$$(a_m t^m \operatorname{sgn}(t))^{(k)} = (2a_0 \delta(t))^{(k-1)}.$$

Continuing the differentiation, we find

$$\begin{aligned}(a_m t^m \operatorname{sgn}(t))^{(k)} &= \left(a_m m(m-1) t^{m-2} \operatorname{sgn}(t) + 2a_m m t^{m-1} \delta(t)\right)^{(k-2)} \\ &= \left(a_m m(m-1) t^{m-2} \operatorname{sgn}(t)\right)^{(k-2)},\end{aligned}$$

where again the second term in the second expression is zero for $m \neq 1$. If $m = 1$, then

$$(a_m t^m \operatorname{sgn}(t))^{(k)} = (2a_1 \delta(t))^{(k-2)}.$$

We conclude that

$$(a_m t^m \operatorname{sgn}(t))^{(k)} = a_m \frac{m!}{k!} t^{m-k} \operatorname{sgn}(t)$$

for $m > k - 1$. If $m = k - 1$,

$$(a_m t^m \operatorname{sgn}(t))^{(k)} = 2a_m \delta(t).$$

Thus, the kth derivative of $f(t)$ is

$$f^{(k)}(t) = \sum_{n=0}^{k-1} 2a_n \delta^{(k-1-n)}(t) + \sum_{n=k}^{\infty} a_n \frac{n!}{k!} t^{n-k} \operatorname{sgn}(t)$$

and it contains the step discontinuity of magnitude $2a_k$; in particular, the step discontinuity of the function itself at the origin is $2a_0$.

We calculate the Fourier transform of f to see the effect of step discontinuities of different derivatives in the frequency domain:

$$\begin{aligned}\widehat{f}(\omega) &= \sum_{n=0}^{\infty} a_n n! \sqrt{\frac{2}{\pi}} \frac{1}{(\iota\omega)^{n+1}} \\ &=: \sum_{n=1}^{\infty} \frac{A_n}{(\iota\omega)^n}, \end{aligned} \tag{4.8}$$

where the meaning of A_1 is given by

$$A_1 := a_0 \sqrt{\frac{2}{\pi}},$$

and of A_n for $n > 1$ by

$$A_n := a_{n-1}(n-1)! \sqrt{\frac{2}{\pi}};$$

the coefficients A_n are related to coefficients a_{n-1} only. Thus, knowing A_{n+1} is tantamount to knowing the jump $2a_n$. The first nonzero coefficient in the series can be obtained from expression (4.8) by multiplying this expression by the corresponding power of $\iota\omega$; in particular for $n = 1$,

$$A_1 = \iota \lim_{|\omega| \to \infty} \omega \widehat{f}(\omega).$$

If the first nonzero coefficient in the series is a_m, then

$$A_{m+1} = (\iota\omega)^m \lim_{|\omega| \to \infty} \omega^m \widehat{f}(\omega),$$

as expected from Theorem 4.1.

Following the above examples, let us consider a function depending on both space and time,

$$u(x,t), \quad x \in \mathbb{R}^3, t \in \mathbb{R},$$

such that it contains discontinuities at $t = 0$ for all of its temporal derivatives. A model representative of such a function is the generalization of the sum from Example 4.3, where each summand represents a different discontinuity,

$$u(x,t) = a_0(x)\delta(t) + \sum_{n=0}^{\infty} a_{n+1}(x) t^n \operatorname{sgn}(t). \tag{4.9}$$

The Fourier transform of u with respect to t is

$$\widehat{u}(x,\omega) = a_0(x)\frac{1}{\sqrt{2\pi}} + \sum_{n=0}^{\infty} a_{n+1}(x)n!\sqrt{\frac{2}{\pi}}\frac{1}{(\iota\omega)^{n+1}}$$
$$=: \sum_{n=0}^{\infty}\frac{A_n(x)}{(\iota\omega)^n}, \tag{4.10}$$

where the meaning of $A_n(x)$ for $n = 0$ is given by

$$A_0\,(x) := \frac{a_0(x)}{\sqrt{2\pi}},$$

and for $n > 0$ by

$$A_n(x) := a_n(x)(n-1)!\,\sqrt{\frac{2}{\pi}}.$$

We translate the origin, $t = 0$, to a location depending on x by setting $t = \psi(x)$; thus, we obtain $u\,(x, t - \psi(x))$ and its associated Fourier transform with respect to time,

$$\widehat{u}(x,\omega) = e^{-\iota\omega\psi(x)}\sum_{n=0}^{\infty}\frac{A_n(x)}{(\iota\omega)^n}. \tag{4.11}$$

We recover the amplitude, $A_0(x)$, by taking the limit of expression (4.11) as $|\omega| \to \infty$, namely,

$$A_0(x) = \lim_{|\omega|\to\infty} e^{\iota\omega\psi(x)}\,\widehat{u}(x,\omega).$$

If $A_0 = 0$, we use expression (4.11) multiplied by $(\iota\omega)$ to recover $A_1(x)$,

$$A_1(x) = \iota\lim_{|\omega|\to\infty} e^{\iota\omega\psi(x)}\,\omega\,\widehat{u}(x,\omega).$$

Thus, discontinuities can be represented by an infinite series in the Fourier domain given by expression (4.11). In the next section we use such a series to study solutions of differential equations whose solutions include discontinuities.

4.3 Asymptotic series

As discussed in Section 4.2, it is convenient to consider expression (4.11) to represent discontinuities.

Such representations might not have the behaviour of a convergent series. For example, expression (4.11) is not defined for $\omega = 0$. In the following section, we discuss in detail the meaning of these potentially divergent series, and show how the form of expression (4.11) emerges in discussions of differential equations.

4.3.1 *General formulation*

In this section, we discuss the meaning and construction of series (4.11), which is called an asymptotic series, and might not converge in the sense of the Taylor series. The choice of using asymptotic series to represent a function depends on properties of this function and the choice of approximation.

To define the asymptotic approximation, we state that two functions, f and g, are asymptotically equivalent to one another in the Poincaré sense at x_0 if

$$\lim_{x\to x_0}(x-x_0)^{-m}(f(x)-g(x))=0, \qquad \forall m\in\mathbb{N},$$

where f and g are defined on an interval containing x_0. The difference between f and g decreases more rapidly than any power of the difference between x and x_0. We denote this equivalency by $f\sim g$ as $x\to x_0$. To compare functions at infinity, we modify the definition to state that f and g are asymptotically equivalent to one another at infinity, if

$$\lim_{x\to\infty}x^m(f(x)-g(x))=0, \qquad \forall m\in\mathbb{N}. \tag{4.12}$$

The difference between f and g tends to zero more rapidly than any power of x tends to infinity.

To define an asymptotic series of a function, we cannot use directly an asymptotic equivalence of two functions because its expansion into a series might not converge. Thus, we define the asymptotic series as follows.

Definition 4.1. For a function f we say that $\sum_{n=1}^{\infty}a_n(x-x_0)^n$ is its infinite asymptotic series of Poincaré type at x_0 if

$$\lim_{x\to x_0}(x-x_0)^{-n}\left(f(x)-\sum_{j=0}^{n}a_j(x-x_0)^j\right)=0, \qquad \forall n\in\mathbb{N}.$$

We find the coefficients of the asymptotic series by starting with a_0. We compute a_0 from Definition 4.1 by setting $n=0$, namely,

$$\lim_{x\to x_0}(f(x)-a_0)=0\Rightarrow a_0=\lim_{x\to x_0}f(x).$$

Similarly, for $n=1$, we compute a_1 using the computed value of a_0,

$$\lim_{x\to x_0}\frac{(f(x)-a_0)-a_1(x-x_0)}{(x-x_0)}=0\Rightarrow a_1=\lim_{x\to x_0}\frac{f(x)-a_0}{(x-x_0)}.$$

For $n=2$,

$$\lim_{x \to x_0} \frac{(f(x) - a_0) - a_1(x - x_0) - a_2(x - x_0)^2}{(x - x_0)^2} = 0$$
$$\Rightarrow a_2 = \lim_{x \to x_0} \frac{(f(x) - a_0) - a_1(x - x_0)}{(x - x_0)^2}.$$

If $f \in C^2(I)$, where I is open and $x_0 \in I$, then $a_0 = f(x_0)$, $a_1 = f'(x_0)$ and $a_2 = f''(x_0)/2!$.

In general, if $f \in C^\infty(I)$, then the coefficients of its asymptotic series are given uniquely by

$$a_n = \frac{f^{(n)}(x_0)}{n!}$$

and the nth partial sum becomes

$$S_n(x) = \sum_{j=0}^{n} \frac{f^{(j)}(x_0)}{j!(x - x_0)^j}.$$

If f is continuous on I only, then $a_0 = f(x_0)$,

$$a_1 = \lim_{x \to x_0} (f(x) - S_0(x)) / (x - x_0)$$

and, in general,

$$a_n = \lim_{x \to x_0} (f(x) - S_{n-1}(x)) / (x - x_0)^n.$$

Each function has a unique asymptotic sequence, but a given asymptotic sequence corresponds to many functions, as illustrated in the following example.

Example 4.4. Consider

$$g(x) = \begin{cases} \exp\left(-\dfrac{1}{x}\right), & x > 0 \\ 0, & x \leq 0 \end{cases}.$$

We see that $g \in C^\infty(\mathbb{R})$ and $g^{(n)}(0) = 0$. Thus, as $x \to 0$ the asymptotic expansion of this function is a sequence of zeros. This implies that 0 and $g(x)$ have the same expansion around zero; we write $g(x) \sim 0$. This implies that for all functions f that have an asymptotic series around zero, f and $f + g$ have the same asymptotic series around zero; $f(x) + g(x) \sim f(x)$.

Any sequence of real numbers defines an asymptotic series of a function, as shown in the following theorem.

Theorem 4.2. *For any sequence of real numbers* $\{a_n\}_{n=1}^{\infty} \in \mathbb{R}$, *there is a function* f *such that*

$$f(x) \sim \sum_{n=0}^{\infty} a_n (x - x_0)^n ,$$

as $x \to x_0$.

Proof. If $\sum_{n=0}^{N} a_n (x - x_0)^n$ converges uniformly to $f(x)$ as $N \to \infty$, then $f(x) \sim \sum_{n=0}^{\infty} a_n (x - x_0)^n$ as $x \to x_0$. If the series does not converge uniformly to a function, then we have to consider a convergent series that behaves around x_0 in the same way as the original series. This series can be constructed by multiplying each term by a proper cutoff function that is equal to unity in a neighbourhood of x_0 and to zero outside a neighbourhood containing the first neighbourhood. Since the asymptotic series describes only a function around a point, it follows that such a modified series has the same asymptotic behaviour around x_0 as the original series. To make the convergence of the modified series uniform, we choose the neighbourhoods around x_0 to be sufficiently small.[4] □

It is useful to introduce a short-hand notation for the asymptotic behaviour. We express the fact that

$$\lim_{x \to x_0} \frac{f(x) - S_n(x)}{(x - x_0)^n} = 0, \qquad \forall n \in \mathbb{N}$$

by

$$f(x) - S_n(x) = o\left((x - x_0)^n\right), \qquad \forall n \in \mathbb{N}, \tag{4.13}$$

where o is a Landau symbol, commonly referred to as the little-o. It is also convenient to express

$$\lim_{x \to x_0} \left| \frac{f(x) - S_n(x)}{(x - x_0)^{n+1}} \right| < K, \qquad \forall n \in \mathbb{N},$$

where K is a constant, by

$$f(x) - S_n(x) = O\left((x - x_0)^{n+1}\right), \qquad \forall n \in \mathbb{N}, \tag{4.14}$$

[4]Readers interested in a particular construction of the cutoff functions might refer to Guillemin and Sternberg (1990, p. 28).

where O is a Landau symbol, commonly referred to as the big-O. Equations (4.13) and (4.14) are equivalent definitions for the existence of infinite asymptotic series. The proof of this equivalence is given in Exercise 4.1.

Similarly, we can express the asymptotic series of $f(x)$ at infinity, namely

$$f(x) \sim \sum_{n=0}^{\infty} a_n x^{-n},$$

by writing

$$f(x) - S_n(x) = o\left(x^{-n}\right),$$

which is equivalent to

$$f(x) - S_n(x) = O\left(x^{-(n+1)}\right).$$

Series $\sum_{n=0}^{\infty} a_n x^{-n}$ is an asymptotic series of $f(x)$ as $x \to \infty$ if

$$\lim_{x\to\infty} x^n \left(f(x) - S_n(x)\right) = 0, \tag{4.15}$$

for a fixed n, even though, for a fixed x, the following might happen:

$$\lim_{n\to\infty} x^n \left(f(x) - S_n(x)\right) = \infty.$$

To illustrate an asymptotic behaviour at infinity, we consider the following example.

Example 4.5. We find the asymptotic series at infinity of

$$f(x) = xe^x \int_x^{\infty} t^{-1}e^{-t}\, dt. \tag{4.16}$$

If

$$f(x) \sim \sum_{n=0}^{\infty} a_n x^{-n},$$

is the asymptotic series of $f(x)$ as $x \to \infty$, then

$$a_0 = \lim_{x\to\infty} f(x).$$

Explicitly, we write

$$\lim_{x\to\infty} xe^x \int_x^{\infty} t^{-1}e^{-t}\, dt = \lim_{x\to\infty} \frac{x \int_x^{\infty} t^{-1}e^{-t}\, dt}{e^{-x}}.$$

Since $\lim_{x\to\infty} x \int_x^\infty t^{-1}e^{-t}\,\mathrm{d}t = 0$ and $\lim_{x\to\infty} \int_x^\infty t^{-1}e^{-t}\,\mathrm{d}t = 0$, as shown in Exercise 4.2, we can use the de l'Hôpital rule twice to evaluate the limit.

$$\begin{aligned}\lim_{x\to\infty}\frac{\int_x^\infty t^{-1}e^{-t}\,\mathrm{d}t - xx^{-1}e^{-x}}{-e^{-x}} &= 1+\lim_{x\to\infty}\frac{\int_x^\infty t^{-1}e^{-t}\,\mathrm{d}t}{-e^{-x}}\\ &= 1+\lim_{x\to\infty}\frac{-x^{-1}e^{-x}}{e^{-x}} = 1.\end{aligned}$$

Hence, $a_0 = 1$. To compute a_1 we use the formula that follows from the definition of the asymptotic sequence, namely,

$$a_n = \lim_{x\to\infty} x^n\left(f(x) - a_0 - a_1x^{-1} - a_2x^{-2} - \cdots - a_{n-1}x^{-(n-1)}\right). \quad (4.17)$$

Herein, using the de l'Hôpital rule, we compute

$$\begin{aligned}a_1 &= \lim_{x\to\infty}\left(x^2e^x\int_x^\infty t^{-1}e^{-t}\mathrm{d}t - x\right) = \lim_{x\to\infty}\frac{x^2\int_x^\infty t^{-1}e^{-t}\,\mathrm{d}t - xe^{-x}}{e^{-x}}\\ &= \lim_{x\to\infty}\frac{2x\int_x^\infty t^{-1}e^{-t}\,\mathrm{d}t}{-e^{-x}} + \lim_{x\to\infty}\frac{-x^2x^{-1}e^{-x} - e^{-x} + xe^{-x}}{-e^{-x}} = -1.\end{aligned}$$

Using formula (4.17) recursively, we could find all coefficients a_n. A more convenient procedure to find the asymptotic series is to use the integration by parts, namely,

$$xe^x\int_x^\infty t^{-1}e^{-t}\,\mathrm{d}t = xe^x\int_x^\infty t^{-1}\left(-e^{-t}\right)'\,\mathrm{d}t,$$

and proceed to get

$$f(x) = -xe^xt^{-1}e^{-t}\Big|_x^\infty - xe^x\int_x^\infty t^{-2}e^{-t}\,\mathrm{d}t.$$

Continuing the process of integration by parts, we obtain

$$\begin{aligned} f(x) =& 1 + xe^{x}t^{-2}e^{-t}\Big|_{x}^{\infty} + 2!\,xe^{x}\int\limits_{x}^{\infty} t^{-3}e^{-t}\,\mathrm{d}t \\ =& 1 + x^{-1} + 2!\,xe^{x}\int\limits_{x}^{\infty} t^{-3}\left(-e^{-t}\right)'\,\mathrm{d}t \\ =& 1 + x^{-1} + 2!\,x^{-2} + \ldots \\ &+ (-1)^{n}n!\,x^{-n} + (-1)^{n+1}(n+1)!\,xe^{x}\int\limits_{x}^{\infty} t^{-n-1}e^{-t}\,\mathrm{d}t \\ =& \sum_{m=0}^{n}\frac{(-1)^{m}m!}{x^{m}} + (-1)^{n+1}(n+1)!\int\limits_{x}^{\infty} t^{-n-1}e^{-t}\,\mathrm{d}t. \end{aligned}$$

To see that this is the asymptotic series of function f, we have to show that the remainder vanishes faster than x^{n+1} goes to infinity as $x \to \infty$. In other words, we have to show that

$$\lim_{x\to\infty} x^{n+1}(-1)^{n+1}(n+1)!\int\limits_{x}^{\infty} t^{-n-1}e^{-t}\,\mathrm{d}t = 0,$$

which can be accomplished following the steps used in Exercise 4.2. Therefore we write

$$f(x) \sim \sum_{m=0}^{\infty}\frac{(-1)^{m}m!}{x^{m}}, \tag{4.18}$$

as $x \to \infty$. This series is divergent according to the ratio test,

$$\lim_{m\to\infty}\frac{\dfrac{(-1)^{m+1}(m+1)!}{x^{m+1}}}{\dfrac{(-1)^{m}m!}{x^{m}}} = \lim_{m\to\infty}\frac{-m-1}{x} = -\infty, \tag{4.19}$$

for a fixed x.

To understand better the convergence character of the asymptotic series, let us examine Figure 4.1. The graph of function (4.16), which we approximate with the asymptotic series, is displayed by the thick curve. The other five curves represent series (4.18) expanded to the first, sixth, eleventh, sixteenth and twenty-first term, from left to right. All five cases converge to the series with increasing x, as expected from expression (4.18). However, for a fixed x, the series diverges as can be seen by examining the

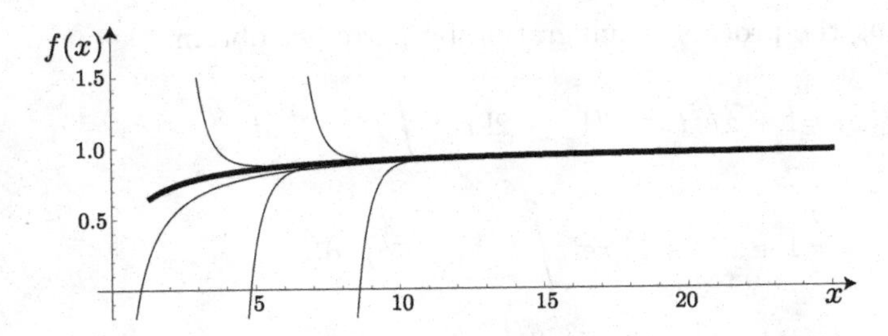

Fig. 4.1 Function (4.16) shown in bold; its asymptotic series (4.18) expanded to the first, sixth, eleventh, sixteenth and twenty-first term. Lower terms approximate the function well in a larger domain than do higher terms.

values of the partial sums at a fixed value, say $x = 5$. As we increase the number of terms, the values at the fixed x are increasing in size and change sign.

For certain functions, the initial terms of their asymptotic series might be sufficiently close to the values of the function, as illustrated by the first term of series (4.18) shown in Figure 4.1.

4.3.2 *Choice of asymptotic sequence*

In the introduction to the asymptotic series on page 137, we use expression (4.11) as the motivation for asymptotic series. However, this expression is not the asymptotic series in ω in the sense of expression (4.15): not only does it contain ι, but it contains also another function of ω, namely $\exp(-\iota\omega\psi(x))$. In the following paragraphs, we generalize the definition of asymptotic series to include functions of the type used in expression (4.11). To do so, we examine the concept of an asymptotic sequence, which we can view as a basis in which we expand a given function.

Definition 4.2. An asymptotic sequence as $x \to x_0$ is a sequence of functions $\{\phi_n(x)\}_{n=0}^{\infty}$ such that $\phi_{n+1}(x) = o(\phi_n(x))$ as $x \to x_0$ for all n.

Using such a sequence, we can develop asymptotic series of functions, as defined below.

Definition 4.3. Let $\{\phi_n\}$ be an asymptotic sequence. We say that f has an infinite asymptotic series with respect to this sequence as $x \to x_0$ if there exists a sequence $\{a_n\} \in \mathbb{R}$ such that

$$\lim_{x \to x_0} \frac{f(x) - \sum_{j=0}^{n} a_j \phi_j(x)}{\phi_n(x)} = 0,$$

which is equivalent to

$$f(x) - S_n(x) = o(\phi_n(x)), \quad \forall n \in \mathbb{N}.$$

For the discussions of vector-valued functions, such as solutions of elastodynamic equations (2.65), this definition can be generalized as follows.

Definition 4.4. For a vector-valued function $f : R \to V$, where V is a normed vector space with norm $\| \cdot \|: V \to R$, function $f(x)$ has an asymptotic series with respect to $\{\phi_n\}$, as $x \to x_0$, if there exists a sequence $a_n \in V$ such that

$$\lim_{x \to x_0} \frac{\| f(x) - \sum_{m=0}^{n} a_m \phi_m \|}{\phi_n} = 0, \qquad \forall n \in \mathbb{N}.$$

As an example of Definition 4.4, we consider the asymptotic series of a vector-valued function with respect to basis $\phi_n = \exp(\iota\omega\psi) / (\iota\omega)^n$ as $\omega \to \infty$,

$$\hat{u}(x, \omega) \sim e^{\iota\omega\psi(x)} \left(A_0(x) + \frac{A_1(x)}{\iota\omega} + \sum_{n=2}^{\infty} \frac{A_n(x)}{(\iota\omega)^n} \right). \tag{4.20}$$

Series (4.20) is an asymptotic series of $\hat{u}$ that we use to study elastodynamic equations in Section 4.7.1. The computation of the first three coefficients is shown in Exercise 4.11.

In subsequent sections, we study equations of the following type:

$$Df = 0, \tag{4.21}$$

where

$$D = \sum_{k \leqslant m} \frac{1}{(\iota\omega)^k} D_k, \tag{4.22}$$

with

$$D_k = \sum_{|\alpha| \leqslant k} B_\alpha^k(x) \frac{\partial^{|\alpha|}}{\partial x^\alpha},$$

where we use the multiindex notation introduced on page 12. To solve such an equation using asymptotic series, we choose the proper asymptotic sequence ϕ_n and the expansion variable. Since the solution is a function of both x and ω, functions ϕ_n should be allowed to depend on both x and ω.

If we choose to expand the solution in ω as $\omega \to \infty$, then the asymptotic sequence must satisfy

$$\lim_{\omega\to\infty} \frac{\phi_{n+1}(x,\omega)}{\phi_n(x,\omega)} = 0. \tag{4.23}$$

Considering the particular form of the differential equation, we require also that $\partial^m \phi_n/\partial x^m$ and $(\iota\omega)^{-k}\phi_n$ be a finite sum of terms $a_k(x)\,\phi_k$. If ϕ_n is of the form

$$\phi_n = \exp\{\iota\omega\psi(x)\}\,\frac{1}{(\iota\omega)^n},$$

these conditions are satisfied, as shown in Exercise 4.12. Thus, we look for solutions in the form given by

$$e^{\iota\omega\psi(x)}\sum_{j=0}^{\infty} A_j(x)\,(\iota\omega)^{-j} \tag{4.24}$$

that satisfy equation (4.21). Substituting expression (4.24) for f in equation (4.21), we write

$$\sum_{k\leqslant m}(\iota\omega)^{-k}\sum_{|\alpha|\leqslant k} B_\alpha^k(x)\,\frac{\partial^{|\alpha|}}{\partial x^\alpha}\left(e^{\iota\omega\psi(x)}\sum_{j=0}^{\infty} A_j(x)\,(\iota\omega)^{-j}\right) = 0, \tag{4.25}$$

where the zero on the right-hand side is represented by its asymptotic series with the coefficients set to zero, namely,

$$0 = e^{\iota\omega\psi}\sum_{l=0}^{\infty} 0\,(\iota\omega)^{-l}.$$

To solve equation (4.21), one has to investigate individual terms of the resulting asymptotic sequence. We compute these terms by applying asymptotic differential operator (4.22) to solution (4.24).

Following the generalized Leibniz rule shown in Exercise 4.5, we write equation (4.25) as

$$\sum_{k\leqslant m}(\iota\omega)^{-k}\sum_{|\alpha|\leqslant k} B_\alpha^k(x)\sum_{\beta+\gamma=\alpha}\frac{\alpha!}{\beta!\gamma!}\frac{\partial^{|\beta|}e^{\iota\omega\psi}}{\partial x^\beta}\sum_{j=0}^{\infty}\frac{\partial^{|\gamma|}A_j}{\partial x^\gamma}(\iota\omega)^{-j} = 0. \tag{4.26}$$

Since the left-hand side of equation (4.26) is an asymptotic series, in Sections 4.4 and 4.4.2, we consider terms with different powers of $\iota\omega$.

4.4 Eikonal equation

4.4.1 *Derivation*

The coefficient corresponding to the zeroth power of $\iota\omega$ on the left-hand side of equation (4.26) results from differentiating k times the exponential terms only, which correspond to $\beta = \alpha$. For multiindex α we can write

$$\begin{aligned}\frac{\partial^{|\alpha|}}{\partial x^{\alpha}} e^{\iota\omega\psi(x)} &= \frac{\partial^{|\alpha|-1}}{\partial x^{\alpha-(0,\ldots,0,1,0,\ldots,0)}} \frac{\partial}{\partial x_i}\left(e^{\iota\omega\psi(x)}\right) \\ &= \frac{\partial^{|\alpha|-1}}{\partial x^{\alpha-(0,\ldots,0,1,0,\ldots,0)}} e^{\iota\omega\psi(x)} (\iota\omega)\frac{\partial\psi}{\partial x_i},\end{aligned}$$

where 1 in multiindex $(0,\ldots,0,1,0,\ldots,0)$ appears in the ith location. Differentiating with respect to x_j, we write

$$\frac{\partial^{|\alpha|}}{\partial x^{\alpha}} e^{\iota\omega\psi(x)} = \frac{\partial^{|\alpha|-2}}{\partial x^{\alpha-(0,\ldots,0,1,0,\ldots,0,1,0,\ldots,0)}} \frac{\partial}{\partial x_j}\left(e^{\iota\omega\psi(x)} (\iota\omega)\frac{\partial\psi}{\partial x_i}\right)$$

as

$$\frac{\partial^{|\alpha|}}{\partial x^{\alpha}} e^{\iota\omega\psi(x)} = \frac{\partial^{|\alpha|-2}\left(e^{\iota\omega\psi(x)}\left((\iota\omega)^2\left(\frac{\partial\psi}{\partial x}\right)^{(0,\ldots,0,1,0,\ldots,0,1,0,\ldots,0)} + (\iota\omega)\frac{\partial^2\psi}{\partial x_i\partial x_j}\right)\right)}{\partial x^{\alpha-(0,\ldots,0,1,0,\ldots,0,1,0,\ldots,0)}},$$

where 1 in multiindex $(0,\ldots,0,1,0,\ldots,0,1,0,\ldots,0)$ appears in the ith and jth locations. If i is equal to j, the multiindex has 2 in the ith location. Continuing the differentiation, we obtain

$$\begin{aligned}\frac{\partial^{|\alpha|} e^{\iota\omega\psi}}{\partial x^{\alpha}} = e^{\iota\omega\psi}\Bigg(&(\iota\omega)^{|\alpha|}\left(\frac{\partial\psi}{\partial x}\right)^{\alpha} \\ &+ (\iota\omega)^{|\alpha|-1}\left(\sum_{i,j\in\alpha}\frac{\partial^2\psi}{\partial x_i\partial x_j}\left(\frac{\partial\psi}{\partial x}\right)^{\alpha-(0,\ldots,0,1,0,\ldots,0,1,0,\ldots,0)}\right) + \cdots\Bigg),\end{aligned} \tag{4.27}$$

where the sum is over all pairs of indices contained in multiindex α as many times as they appear, and the dots indicate the remaining terms with lower powers of $\iota\omega$. Explicitly, the number of times we have to count each term

$$\frac{\partial^2\psi}{\partial x_i\partial x_j}\left(\frac{\partial\psi}{\partial x}\right)^{\alpha-(0,\ldots,0,1,0,\ldots,0,1,0,\ldots,0)}$$

in the summation is the number of times we can choose

$$(0,\ldots,0,1,0,\ldots,0,1,0,\ldots,0)$$

from

$$(0,\ldots,0,\alpha_i,0,\ldots,0,\alpha_j,0,\ldots,0)\,.$$

If $i \neq j$, then this number is $\alpha_i\alpha_j$. If $i = j$, then this number is $\alpha_i\,(\alpha_i - 1)$. Using the Kronecker delta, we rewrite the summation as

$$\sum_{i,j=1}^{n} \frac{1}{2}\left(\alpha_i\alpha_j\,(1-\delta_{ij}) + \alpha_i\,(\alpha_j-1)\,\delta_{ij}\right) \frac{\partial^2\psi}{\partial x_i\partial x_j} \left(\frac{\partial\psi}{\partial x}\right)^{\alpha-(0,\ldots,0,1,0,\ldots,0,1,0,\ldots,0)}, \tag{4.28}$$

where the factor of $1/2$ offsets the result of summing over both i and j, and thus counting each term twice. The first term on the right-hand side of expression (4.28) is derived in Exercise 4.4.

Equating the coefficients corresponding to the zeroth power of $(\iota\omega)$ on both sides of equation (4.26), we see that

$$\sum_{k\leqslant m}\sum_{|\alpha|=k} B_\alpha^k\,(x)\left(\frac{\partial\psi}{\partial x}\right)^{|\alpha|} A_0(x) = 0. \tag{4.29}$$

In this equation there are two unknown functions, $\psi\,(x)$ and $A_0\,(x)$. If we consider only nonzero scalar functions A_0, then this equation reduces to

$$\sum_{k\leqslant m}\sum_{|\alpha|=k} B_\alpha^k\,(x)\left(\frac{\partial\psi}{\partial x}\right)^{|\alpha|} = 0. \tag{4.30}$$

This is a first-order nonlinear partial differential equation for $\psi\,(x)$. We refer to it as the eikonal equation, as illustrated in Example 4.6.

If A_0 is a vector, then we cannot reduce equation (4.29) to equation (4.30), since a linear operator acting on a vector resulting in zero does not imply that the vector is zero. In such a case, equation (4.29) represents a zero-eigenvalue problem.

Example 4.6. Consider the reduced wave equation, namely,

$$v^2\nabla^2\hat{u}\,(x,\omega) = (\iota\omega)^2\,\hat{u}\,(x,\omega)\,. \tag{4.31}$$

In this case, equation (4.30) reduces to

$$\sum_i\left(\frac{\partial\psi}{\partial x_i}\right)^2 - \frac{1}{v^2} = 0, \tag{4.32}$$

as shown in Exercise 4.7. This is eikonal equation (2.59), derived in Section 2.3.2.1.

As we discuss in Appendix E.3, equation (4.29) can be written as

$$\sigma(D)(\mathrm{d}\psi) A_0(x) = 0. \tag{4.33}$$

If we consider only nonzero scalar functions A_0, then this equation reduces to

$$\sigma(D)(\mathrm{d}\psi) = 0. \tag{4.34}$$

We can view equation (4.33) as a set of n equations of the type of equation (4.34). These n equations correspond to equating the eigenvalues of $\sigma(D)(d\psi)$ to zero. Solutions, $\psi(x)$, of first-order nonlinear differential equation (4.34) are discussed within the context of characteristics in Chapter 3, which we summarize in the following section.

4.4.2 *Solution*

In this section, we discuss the solutions of the eikonal equation. The solution using the general method of characteristics for the first-order nonlinear partial differential equations is given in Chapter 3. To use the results from that chapter in the context of eikonal equation (4.34), we express the eikonal equation in the notation of Chapter 3 as

$$0 = F\left(x_1, x_2, x_3, \psi, \frac{\partial\psi}{\partial x_1}, \frac{\partial\psi}{\partial x_2}, \frac{\partial\psi}{\partial x_3}\right) \equiv \sigma(D)(\mathrm{d}\psi).$$

For vector-valued equations, we treat this expression as a set of n equations corresponding to the zero eigenvalues of $\sigma(D)(\mathrm{d}\psi)$. In the case of the eikonal equation, $\sigma(D)(\mathrm{d}\psi)$ does not depend explicitly on ψ. Consequently, the characteristic equations (3.17), (3.19) and (3.18) become

$$\begin{aligned}
\frac{\mathrm{d}x_i}{\mathrm{d}t} &= \zeta\frac{\partial\sigma(D)(p)}{\partial p_i}, \\
\frac{\mathrm{d}p_i}{\mathrm{d}t} &= -\zeta\frac{\partial\sigma(D)(p)}{\partial x_i}, \\
\frac{\mathrm{d}\psi}{\mathrm{d}t} &= \zeta\sum_{i=1}^{3}\frac{\partial\sigma(D)(p)}{\partial p_i}p_i.
\end{aligned} \tag{4.35}$$

The solutions of this system depend on the original partial differential equation, which manifests itself in the form of $\sigma(D)(p)$. As an example, we return to the reduced wave equation (4.31), and consider it in two spatial dimensions. In such a case,

$$\sigma(D)(p) = p_1^2 + p_2^2 - \frac{1}{v^2}.$$

The solution is discussed in Exercise 3.3, in which we include $\zeta = 1/2$ in the Hamiltonian, in accordance with discussion leading to equation (3.22), where the reason for choosing $\zeta = 1/2$ is due to the physical interpretation of ψ as traveltime. In the context of the asymptotic series, we view t as a parameter only, since the asymptotic series is in the frequency domain, not in its Fourier-transformed time domain.

4.5 Transport equation

4.5.1 *Derivation*

Returning to equation (4.26), we focus our attention on the terms with $(\iota\omega)^{-1}$. There are four types of such terms in this series.

The first type results from setting $j = 1$ and $|\beta| = |\alpha| = k$:

$$\sum_{k\leqslant m}\sum_{|\alpha|=k} B_{\alpha}^{k}(x)\left(\frac{\partial\psi}{\partial x}\right)^{|\alpha|} e^{\iota\omega\psi} A_1(x)(\iota\omega)^{-1}.$$

The second type results from setting $j = 0$ and $|\beta| = |\alpha| = k - 1$:

$$\sum_{k\leqslant m}\sum_{|\alpha|=k-1} B_{\alpha}^{k}(x)\left(\frac{\partial\psi}{\partial x}\right)^{|\alpha|} e^{\iota\omega\psi} A_0(x)(\iota\omega)^{-1}.$$

The third type results from setting $j = 0$, $|\beta| = |\alpha| = k$ and considering expression (4.28):

$$\sum_{k\leqslant m}\sum_{|\alpha|=k} B_{\alpha}^{k}\sum_{i,j=1}^{n}\frac{1}{2}\left(\alpha_i\alpha_j\left(1-\delta_{ij}\right)+\alpha_i\left(\alpha_j-1\right)\delta_{ij}\right)$$
$$\cdot\frac{\partial^2\psi}{\partial x_i\partial x_j}\left(\frac{\partial\psi}{\partial x}\right)^{\alpha-(0,\ldots,0,1,0,\ldots,0,1,0,\ldots,0)} e^{i\omega\psi} A_0(\iota\omega)^{-1}.$$

The fourth type results from setting $j = 0$, $|\beta| + 1 = |\alpha| = k$:

$$\sum_{k\leqslant m}\sum_{|\alpha|=k} B_{\alpha}^{k}\sum_{i=1}^{n}\alpha_i\left(\frac{\partial\psi}{\partial x}\right)^{\alpha-(0,\ldots,0,1,0,\ldots,0)} e^{i\omega\psi}\frac{\partial A_0}{\partial x_i}(\iota\omega)^{-1},$$

where 1 in $(0,\ldots,0,1,0,\ldots,0)$ appears in the ith location.

Gathering these four expressions and factoring out the common terms, we write

$$e^{\iota\omega\psi}(\iota\omega)^{-1}\left(\sum_{k\leqslant m}\sum_{|\alpha|=k}B_\alpha^k(x)\left(\frac{\partial\psi}{\partial x}\right)^{|\alpha|}A_1(x)\right.$$
$$+\sum_{k\leqslant m}\sum_{|\alpha|=k-1}B_\alpha^k(x)\left(\frac{\partial\psi}{\partial x}\right)^{|\alpha|}A_0$$
$$+\sum_{k\leqslant m}\sum_{|\alpha|=k}B_\alpha^k\sum_{i,j=1}^{n}\frac{1}{2}\left(\alpha_i\alpha_j\left(1-\delta_{ij}\right)+\alpha_i\left(\alpha_j-1\right)\delta_{ij}\right)$$
$$\frac{\partial^2\psi}{\partial x_i\partial x_j}\left(\frac{\partial\psi}{\partial x}\right)^{\alpha-(0,\ldots,0,1,0,\ldots,0,1,0,\ldots,0)}A_0$$
$$\left.+\sum_{k\leqslant m}\sum_{|\alpha|=k}B_\alpha^k\sum_{i=1}^{n}\alpha_i\left(\frac{\partial\psi}{\partial x}\right)^{\alpha-(0,\ldots,0,1,0,\ldots,0)}\frac{\partial A_0}{\partial x_i}\right).$$

In an asymptotic series that is equal to zero all the coefficients are zero; hence, we conclude that

$$\sum_{k\leqslant m}\sum_{|\alpha|=k}B_\alpha^k(x)\left(\frac{\partial\psi}{\partial x}\right)^{|\alpha|}A_1(x)+\left(\sum_{k\leqslant m}\sum_{|\alpha|=k-1}B_\alpha^k(x)\left(\frac{\partial\psi}{\partial x}\right)^{|\alpha|}\right.$$
$$+\sum_{k\leqslant m}\sum_{|\alpha|=k}B_\alpha^k\sum_{i,j=1}^{n}\frac{1}{2}\left(\alpha_i\alpha_j\left(1-\delta_{ij}\right)+\alpha_i\left(\alpha_j-1\right)\delta_{ij}\right)$$
$$\frac{\partial^2\psi}{\partial x_i\partial x_j}\left(\frac{\partial\psi}{\partial x}\right)^{\alpha-(0,\ldots,0,1,0,\ldots,0,1,0,\ldots,0)}$$
$$\left.+\sum_{k\leqslant m}\sum_{|\alpha|=k}B_\alpha^k\sum_{i=1}^{n}\alpha_i\left(\frac{\partial\psi}{\partial x}\right)^{\alpha-(0,\ldots,0,1,0,\ldots,0)}\frac{\partial}{\partial x_i}\right)A_0(x)=0. \tag{4.36}$$

In the following example we discuss this equation in the context of the wave equation. Following the discussion, we refer to this equation as the transport equation.

Example 4.7. Consider the reduced wave equation stated in expression (4.31), namely,

$$v^2\nabla^2\hat{u}(x,\omega)=(i\omega)^2\,\hat{u}(x,\omega). \tag{4.37}$$

In this case, equation (4.36) reduces to

$$\left(\sum_i\left(\frac{\partial\psi}{\partial x_i}\right)^2-\frac{1}{v^2}\right)A_1(x)+\left(\sum_i\frac{\partial^2\psi}{\partial x_i^2}+\sum_i 2\frac{\partial\psi}{\partial x_i}\frac{\partial}{\partial x_i}\right)A_0(x)=0, \tag{4.38}$$

as shown in Exercise 4.8. Furthermore, if we assume that ψ satisfies the eikonal equation given by expression (4.32), equation (4.38) becomes

$$\left(\sum_i \frac{\partial^2 \psi}{\partial x_i^2} + \sum_i 2 \frac{\partial \psi}{\partial x_i} \frac{\partial}{\partial x_i}\right) A_0(x) = 0,$$

which is a first-order linear partial differential equation for $A_0(x)$. A_0 is the first coefficient in the asymptotic series of the wave equation solution, and represents the first-order approximation of the amplitude of the wave. This coefficient is found by propagating the side condition along rays, which justifies the name of transport equation.

Using the symbols of the differential operators discussed in Appendix E.3, we rewrite equation (4.36) as

$$\sigma(D)(\mathrm{d}\psi) A_1(x) + \left(\sum_{k \leqslant m} \sum_{|\alpha|=k-1} B_\alpha^k(x) \left(\frac{\partial \psi}{\partial x}\right)^{|\alpha|}\right. \tag{4.39}$$

$$\left. + \sum_{k \leqslant m} \frac{1}{2} \sum_{i,j=1}^{n} \frac{\partial^2 \sigma(D_k)(p)}{\partial p_i \partial p_j} \frac{\partial^2 \psi}{\partial x_i \partial x_j} + \sum_{k \leqslant m} \sum_{i=1}^{n} \frac{\partial \sigma(D_k)(p)}{\partial p_i} \frac{\partial}{\partial x_i}\right) A_0(x) = 0,$$

where $p = \mathrm{d}\psi$, as shown in Exercise 4.9.

If $\psi(x)$ satisfies the eikonal equation stated in expression (4.30) and if either $A_j(x)$ are scalars or we assume that $A_1(x)$ is in the kernel of $\sigma(D)(d\psi)$, the first term of the above equation is zero. Subsequently, equation (4.39) reduces to

$$\left(\sum_{k \leqslant m} \sum_{|\alpha|=k-1} B_\alpha^k(x) \left(\frac{\partial \psi}{\partial x}\right)^{|\alpha|} + \sum_{k \leqslant m} \frac{1}{2} \sum_{i,j=1}^{n} \frac{\partial^2 \sigma(D_k)(p)}{\partial p_i \partial p_j} \frac{\partial^2 \psi}{\partial x_i \partial x_j}\right.$$

$$\left. + \sum_{k \leqslant m} \sum_{i=1}^{n} \frac{\partial \sigma(D_k)(p)}{\partial p_i} \frac{\partial}{\partial x_i}\right) A_0(x) = 0, \tag{4.40}$$

which is a first-order partial differential equation for $A_0(x)$ and is referred to as the reduced transport equation.

4.5.2 *Solution*

In this section we discuss the solutions of the reduced transport equation given by expression (4.40). We write this equation as

$$\sum_{k\leqslant m}\sum_{i=1}^{n}\frac{\partial\sigma\left(D_{k}\right)\left(p\right)}{\partial p_{i}}\frac{\partial}{\partial x_{i}}A_{0}\left(x\right)=-\left(\sum_{k\leqslant m}\sum_{|\alpha|=k-1}B_{\alpha}^{k}\left(x\right)\left(\frac{\partial\psi}{\partial x}\right)^{|\alpha|}\right. \tag{4.41}$$
$$\left.+\sum_{k\leqslant m}\frac{1}{2}\sum_{i,j=1}^{n}\frac{\partial^{2}\sigma\left(D_{k}\right)\left(p\right)}{\partial p_{i}\partial p_{j}}\frac{\partial^{2}\psi}{\partial x_{i}\partial x_{j}}\right)A_{0}\left(x\right).$$

We can solve this equation by the method of characteristics, discussed in Chapter 1. In view of the solutions of the eikonal equation discussed in Section 4.4.2, the characteristics of the transport and the eikonal equations are the same. Along these characteristics, given by equations (4.35), the left-hand side of equation (4.41) is

$$\frac{1}{\zeta}\sum_{i=1}^{n}\frac{\mathrm{d}}{\mathrm{d}t}x_{i}\left(t\right)\frac{\partial}{\partial x_{i}}A_{0}\left(x\left(t\right)\right),$$

where $x\left(t\right)$ is a characteristic. As discussed in Section 4.4.2, we view t as a parameter only, since we work in the frequency domain. For vector valued functions, we view this expressions as a set of n equations for each eigenvalue of $\sigma\left(D\right)\left(p\right)$. In such a case, we get the solution of the transport equation not as a vector, A_0, but only as its length. The direction of A_0 has to be determined directly from eikonal equation (4.33), as the direction of the eigenvector corresponding to the given eigenvalue.

Following the chain rule, we write the transport equation along the characteristics as

$$\frac{\mathrm{d}}{\mathrm{d}t}A_{0}\left(x\left(t\right)\right)=-\zeta\left(\sum_{k\leqslant m}\sum_{|\alpha|=k-1}B_{\alpha}^{k}\left(x\right)\left(\frac{\partial\psi}{\partial x}\right)^{|\alpha|}\right.$$
$$\left.+\sum_{k\leqslant m}\frac{1}{2}\sum_{i,j=1}^{n}\frac{\partial^{2}\sigma\left(D_{k}\right)\left(p\right)}{\partial p_{i}\partial p_{j}}\frac{\partial^{2}\psi}{\partial x_{i}\partial x_{j}}\right)A_{0}\left(x\left(t\right)\right).$$

Integrating along the characteristic, as shown in Exercise 4.10, we obtain the solution of the transport equation,

$$A_{0}\left(x\left(t\right)\right)=\exp\left\{-\int_{0}^{t}\left(\zeta\sum_{k\leqslant m}\sum_{|\alpha|=k-1}B_{\alpha}^{k}\left(x\right)\left(\frac{\partial\psi}{\partial x}\right)^{|\alpha|}\right.\right.$$
$$\left.\left.+\sum_{k\leqslant m}\frac{\zeta}{2}\sum_{i,j=1}^{n}\frac{\partial^{2}\sigma\left(D_{k}\right)\left(p\right)}{\partial p_{i}\partial p_{j}}\frac{\partial^{2}\psi}{\partial x_{i}\partial x_{j}}\right)\mathrm{d}s\right\}A_{0}\left(x\left(0\right)\right),$$

where $A_0\left(x\left(0\right)\right)$ is the zero-order amplitude at time $t=0$.

If $\psi(x(t))$ is a solution of the eikonal equation, then, in view of Hamilton equations (4.35), we write the second part of the above integrand as

$$\sum_{k\leqslant m}\frac{\zeta}{2}\sum_{i,j=1}^{n}\frac{\partial^2\sigma(D_k)(p)}{\partial p_i\partial p_j}\frac{\partial^2\psi}{\partial x_i\partial x_j}=\sum_{k\leqslant m}\frac{\zeta}{2}\sum_{i,j=1}^{n}\frac{\partial}{\partial p_j}\frac{\partial\sigma(D_k)(p)}{\partial p_i}\frac{\partial^2\psi}{\partial x_i\partial x_j}$$

$$=\sum_{k\leqslant m}\frac{\zeta}{2}\sum_{i,j=1}^{n}\frac{\partial\frac{1}{\zeta}\dot{x}_i}{\partial p_j}\frac{\partial^2\psi}{\partial x_i\partial x_j}$$

$$=\frac{\zeta}{2}\sum_{i,j=1}^{n}\frac{\partial\frac{1}{\zeta}\dot{x}_i}{\partial p_j}\frac{\partial p_j}{\partial x_i},$$

where $\dot{x}\equiv \mathrm{d}x/\mathrm{d}t$ and we use the fact that $p_j=\partial\psi/\partial x_j$. Using the chain rule, we write the above concisely

$$\frac{\zeta}{2}\nabla\cdot\left(\frac{1}{\zeta}\dot{x}(x(t))\right).$$

Consequently, solution $A_0(x(t))$ becomes

$$A_0(x(t))=\exp\left\{-\int_{t_0}^{t}\zeta\sum_{k\leqslant m}\sum_{|\alpha|=k-1}B_\alpha^k(x)\left(\frac{\partial\psi}{\partial x}\right)^{|\alpha|}\mathrm{d}s\right\}$$

$$\cdot\exp\left\{-\int_{t_0}^{t}\frac{\zeta}{2}\nabla\cdot\left(\frac{1}{\zeta}\dot{x}(x(s))\right)\mathrm{d}s\right\}A_0(x(0)). \tag{4.42}$$

The form of solution $A_0(x(t))$ depends on the form of solution of the eikonal equation, $\psi(x(t))$.

Let us explain the meaning of the second exponential. Since the integrand in the second exponent consists of a divergence, we invoke the Divergence Theorem stated on page 215. We consider the volume integral given by

$$\iiint_{V(t)}\nabla\cdot\dot{x}(x(s))\,\mathrm{d}V, \tag{4.43}$$

where volume $V(t)$ is spanned by area S_0 that is normal to the characteristics, $x(s), s\in[t_0,t]$, along which this area propagates, as illustrated in Figure 4.2. Using Theorem A.1, we write this integral as

$$\iiint_{V(t)}\nabla\cdot\dot{x}\;\mathrm{d}V=\iint_{S(t)}\dot{x}\cdot N\;\mathrm{d}S, \tag{4.44}$$

where N is the outward normal to the surface of the raytube, $S(t)$, and also is the outward normal to the tube ends. Along the tube, N is normal to $\dot{x}$; hence, $\dot{x} \cdot N = 0$. On the two tube ends, N is parallel to $\dot{x}$; hence, since N is a unit vector, $\dot{x} \cdot N = \pm |\dot{x}|$. Consequently, we rewrite equation (4.44) as

$$\iiint_{V(t)} \nabla \cdot \dot{x} \, \mathrm{d}V = -\iint_{S_0} |\dot{x}| \, \mathrm{d}S + \iint_{S_t} |\dot{x}| \, \mathrm{d}S, \tag{4.45}$$

where the negative sign results from the fact that at S_0, $\dot{x}$ and N have opposite directions. To study the propagation along the raytube, we differentiate both sides of equation (4.45) with respect to t to write

$$\frac{\mathrm{d}}{\mathrm{d}t} \iiint_{V(t)} \nabla \cdot \dot{x} \, \mathrm{d}V = -\frac{\mathrm{d}}{\mathrm{d}t} \iint_{S_0} |\dot{x}| \, \mathrm{d}S + \frac{\mathrm{d}}{\mathrm{d}t} \iint_{S_t} |\dot{x}| \, \mathrm{d}S. \tag{4.46}$$

Invoking the definition of derivative, we write the left-hand side of equation (4.46) as

$$\lim_{\mathrm{d}t \to 0} \frac{\iiint_{V(t+\mathrm{d}t)} \nabla \cdot \dot{x} \, \mathrm{d}V - \iiint_{V(t)} \nabla \cdot \dot{x} \, \mathrm{d}V}{\mathrm{d}t} = \lim_{\mathrm{d}t \to 0} \frac{\iiint_{V(t+\mathrm{d}t)-V(t)} \nabla \cdot \dot{x} \, \mathrm{d}V}{\mathrm{d}t}. \tag{4.47}$$

Let us consider the numerator. According to the Mean-value Theorem for integrals, there exists a surface, $S_{t+\mathrm{d}t}$, such that

$$\iiint_{V(t+\mathrm{d}t)-V(t)} \nabla \cdot \dot{x} \, \mathrm{d}V = \mathrm{d}t \iint_{S_{t+dt}} \nabla \cdot \dot{x} \, \mathrm{d}S,$$

Hence, returning to expression (4.47), we get

$$\begin{aligned} \lim_{\mathrm{d}t \to 0} \frac{\iiint_{V(t+\mathrm{d}t)-V(t)} \nabla \cdot \dot{x} \, \mathrm{d}V}{\mathrm{d}t} &= \lim_{\mathrm{d}t \to 0} \frac{\mathrm{d}t \iint_{S_{t+dt}} \nabla \cdot \dot{x} \, \mathrm{d}S}{\mathrm{d}t} \\ &= \lim_{\mathrm{d}t \to 0} \iint_{S_{t+dt}} \nabla \cdot \dot{x} \, \mathrm{d}S = \iint_{S_t} \nabla \cdot \dot{x} \, \mathrm{d}S. \end{aligned}$$

Thus, we write the left-hand side of equation (4.46) as

$$\frac{\mathrm{d}}{\mathrm{d}t} \iiint_{V(t)} \nabla \cdot \dot{x} \, \mathrm{d}V = \iint_{S_t} \nabla \cdot \dot{x} \, \mathrm{d}S. \tag{4.48}$$

Examining the right-hand side of equation (4.46), we see that the integral over S_0 is a constant and, hence, its derivative is zero. Let us consider the integral over S_t. According to the Mean-value Theorem for integrals, there exists a point, $y(t)$, on surface S_t such that

$$\iint_{S_t} |\dot{x}| \, \mathrm{d}S = S_t \, |\dot{x}(y(t))| \, .$$

Using this expression and invoking the product rule and the chain rule, we write

$$\frac{\mathrm{d}}{\mathrm{d}t} \iint_{S_t} |\dot{x}| \, \mathrm{d}S = \frac{\mathrm{d}}{\mathrm{d}t} \left(S_t \, |\dot{x}(y(t))| \right) = \frac{\mathrm{d}S_t}{\mathrm{d}t} \, |\dot{x}(y(t))| + S_t \nabla \, |\dot{x}(y(t))| \cdot \dot{y}(t) \, , \tag{4.49}$$

which is the right-hand side of equation (4.46). Equating the left-hand and right-hand sides of equation (4.46), which are given in expressions (4.48) and (4.49) respectively, we write equation (4.46) as

$$\iint_{S_t} \nabla \cdot \dot{x} \, \mathrm{d}S = \frac{\mathrm{d}S_t}{\mathrm{d}t} \, |\dot{x}(y(t))| + S_t \nabla \, |\dot{x}(y(t))| \cdot \dot{y}(t) \, , \tag{4.50}$$

which is the time derivative of integral (4.43). To find the solution of equation (4.50), we restrict the parametrization by t to the arclength parametrization. In such a case, $|\dot{x}| = 1$ and the second term on the right-hand side of equation (4.50) is zero. Hence, equation (4.50) becomes

$$\iint_{S_t} \nabla \cdot \dot{x} \, \mathrm{d}S = \frac{\mathrm{d}S_t}{\mathrm{d}t} . \tag{4.51}$$

To use this result in the examination of solution (4.42), we remove the surface integral. Let us consider the left-hand side of equation (4.51). According to the Mean-value theorem for integrals, there exists a point, $z(t)$, on surface S_t such that

$$\iint_{S_t} \nabla \cdot \dot{x} \, \mathrm{d}S = S_t \nabla \cdot \dot{x}(z(t)) \, .$$

Thus we write equation (4.51) as

$$S_t \nabla \cdot \dot{x}(z(t)) = \frac{\mathrm{d}S_t}{\mathrm{d}t} .$$

Using the chain rule, we rewrite it as

$$\frac{\frac{\mathrm{d}S_t}{\mathrm{d}t}}{S_t} = \frac{\mathrm{d}}{\mathrm{d}t} \ln S_t = \nabla \cdot \dot{x}(y(t)) \, .$$

For an infinitesimally small surface, S_t, we can replace point $y(t)$ with $x(t)$. Thus, we write

$$\nabla \cdot \dot{x}(x(t)) = \frac{\mathrm{d}}{\mathrm{d}t} \ln S_t, \tag{4.52}$$

which is another form of equation (4.51) and, consequently, equation (4.50) for the arclength parametrization.

If parameter t corresponds to the arclength, then the change in the surface area corresponds to the change in the volume along the ray tube. We can express this change using the Jacobian of the change of coordinates, from the Cartesian coordinates to the coordinates given by the characteristics. We write expression (4.52) as

$$\nabla \cdot \dot{x}(x(t)) = \frac{\mathrm{d}}{\mathrm{d}t} \ln \Delta V_t, \tag{4.53}$$

where ΔV_t is the volume generated by surface S_t displaced by Δt, where Δt is constant; in other words, $\Delta V_t = S_t \Delta t$. To justify this step, we note that

$$\frac{\mathrm{d}}{\mathrm{d}t} \ln \Delta V_t = \frac{\mathrm{d}}{\mathrm{d}t} \ln (S_t \Delta t) = \frac{\mathrm{d}}{\mathrm{d}t} (\ln S_t + \ln \Delta t) = \frac{\mathrm{d}}{\mathrm{d}t} \ln S_t .$$

Thus, divergence $\nabla \cdot \dot{x}(x(t))$ can be expressed as the change of the logarithm of the volume of a raytube with respect to the arclength.[5] Using n ray-centred coordinates, $\xi_1, \xi_2, \ldots, \xi_{n-1}, t$, we can write this volume as

$$\Delta V_t = \det \begin{bmatrix} \frac{\partial x_1}{\partial \xi_1} & \frac{\partial x_1}{\partial \xi_2} & \cdots & \frac{\partial x_1}{\partial t} \\ \frac{\partial x_2}{\partial \xi_1} & \frac{\partial x_2}{\partial \xi_2} & \cdots & \frac{\partial x_2}{\partial t} \\ & & \ddots & \\ \frac{\partial x_n}{\partial \xi_1} & \frac{\partial x_n}{\partial \xi_2} & \cdots & \frac{\partial x_n}{\partial t} \end{bmatrix} \mathrm{d}\xi_1 \, \mathrm{d}\xi_2 \ldots \, \mathrm{d}t = J \, \mathrm{d}\xi_1 \, \mathrm{d}\xi_2 \ldots \, \mathrm{d}t,$$

where J stands for the determinant of the matrix. Using this expression, we restate equation (4.53) as

$$\nabla \cdot \dot{x}(x(t)) = \frac{\mathrm{d}}{\mathrm{d}t} \ln (J \, \mathrm{d}\xi_1 \, \mathrm{d}\xi_2 \ldots \, \mathrm{d}t) = \frac{\mathrm{d}}{\mathrm{d}t} (\ln J + \ln \, \mathrm{d}\xi_1 \, \mathrm{d}\xi_2 \ldots \, \mathrm{d}t) .$$

Since $\ln \, \mathrm{d}\xi_1 \, \mathrm{d}\xi_2 \ldots \, \mathrm{d}t$ is a constant, we obtain

$$\nabla \cdot \dot{x}(x(t)) = \frac{\mathrm{d}}{\mathrm{d}t} \ln J, \tag{4.54}$$

[5] Readers interested in an alternative proof might refer to Marsden and Tromba (1981, p. 184).

which relates the divergence of the tangents of the characteristics to the change of volume associated with these characteristics.

Let us examine solution (4.42) in the context of expression (4.52). In the second exponent, we have $\nabla\cdot((1/\zeta)\,\dot{x})$. Invoking the product rule, we write

$$\nabla\cdot\left(\frac{1}{\zeta}\dot{x}\right)=\left(\nabla\frac{1}{\zeta}\right)\cdot\dot{x}+\frac{1}{\zeta}\nabla\cdot\dot{x}. \tag{4.55}$$

Inserting expression (4.52), we get

$$\nabla\cdot\left(\frac{1}{\zeta}\dot{x}\right)=\frac{\mathrm{d}}{\mathrm{d}t}\frac{1}{\zeta}+\frac{1}{\zeta}\frac{\mathrm{d}}{\mathrm{d}t}\ln S_t,$$

where, using the chain rule, we rewrite also the first term on the right-hand side. To write the right-hand side of this equation as a total derivative, we multiply both sides by ζ to get

$$\zeta\nabla\cdot\left(\frac{1}{\zeta}\dot{x}\right)=\zeta\frac{\mathrm{d}}{\mathrm{d}t}\frac{1}{\zeta}+\frac{\mathrm{d}}{\mathrm{d}t}\ln S_t,$$

which can be written as

$$\zeta\nabla\cdot\left(\frac{1}{\zeta}\dot{x}\right)=\frac{\mathrm{d}}{\mathrm{d}t}\ln\frac{1}{\zeta}+\frac{\mathrm{d}}{\mathrm{d}t}\ln S_t\,.$$

Using the linearity of the differential operator, we write this expression as

$$\zeta\nabla\cdot\left(\frac{1}{\zeta}\dot{x}\right)=\frac{\mathrm{d}}{\mathrm{d}t}\left(-\ln\zeta+\ln S_t\right)=\frac{\mathrm{d}}{\mathrm{d}t}\ln\frac{S_t}{\zeta}. \tag{4.56}$$

Also, we can express this formula in terms of the Jacobian using expressions (4.55) and (4.54):

$$\zeta\nabla\cdot\left(\frac{1}{\zeta}\dot{x}\right)=\frac{\mathrm{d}}{\mathrm{d}t}\left(-\ln\zeta+\ln J\right)=\frac{\mathrm{d}}{\mathrm{d}t}\ln\frac{J}{\zeta}. \tag{4.57}$$

Substituting result (4.56) into solution (4.42), we write

$$\begin{aligned}A_0\left(x\left(t\right)\right)&=A_0\left(x\left(t_0\right)\right)\exp\left\{-\int_{t_0}^{t}\zeta\sum_{k\leqslant m}\sum_{|\alpha|=k-1}B_\alpha^k\left(x\right)\left(\frac{\partial\psi}{\partial x}\right)^{|\alpha|}\mathrm{d}s\right\}\\&\quad\exp\left\{-\frac{1}{2}\int_{t_0}^{t}\frac{\mathrm{d}}{\mathrm{d}s}\ln\frac{S_s}{\zeta}\,\mathrm{d}s\right\}\\&=A_0\left(x\left(t_0\right)\right)\exp\left\{-\int_{t_0}^{t}\zeta\sum_{k\leqslant m}\sum_{|\alpha|=k-1}B_\alpha^k\left(x\right)\left(\frac{\partial\psi}{\partial x}\right)^{|\alpha|}\mathrm{d}s\right\}\\&\quad\sqrt{\frac{S_0\zeta\left(x\left(t\right)\right)}{S_t\zeta\left(x\left(t_0\right)\right)}}.\end{aligned} \tag{4.58}$$

Substituting result (4.57) into solution (4.42), we write

$$A_0(x(t)) = A_0(x(t_0)) \exp\left\{-\int_{t_0}^{t} \zeta \sum_{k \leqslant m} \sum_{|\alpha|=k-1} B_\alpha^k(x) \left(\frac{\partial \psi}{\partial x}\right)^{|\alpha|} \mathrm{d}s\right\}$$
$$\sqrt{\frac{J_0 \zeta(x(t))}{J_t \zeta(x(t_0))}}. \tag{4.59}$$

Solution (4.58) requires an infinitesimally small surface S_0. Both solutions (4.58) and (4.59) require the arclength parametrization, which means that $\|\mathrm{d}x/\mathrm{d}s\| = 1$; hence, following equation (4.35),

$$\zeta = \left\|\frac{\partial \sigma(D)(p)}{\partial p}\right\|^{-1}. \tag{4.60}$$

To see the meaning of this expression, let us consider the following example.

Example 4.8. Let us consider the wave equation in an isotropic medium and its Hamiltonian, $F = p^2(x) - 1/v^2(x)$. The choice of parametrization fixes the expression for ζ. Following equation (4.60), where $\sigma(D)(p) = F$, we see that herein

$$\zeta = \frac{1}{\left|\frac{\partial F}{\partial p_i}\right|} = \frac{1}{2|p|}.$$

Along the characteristics, $F = 0$, and, hence, $|p| = 1/v(x)$, where v is the velocity of propagation. Thus, $\zeta = v(x)/2$. Using the reduced wave equation (4.31), we see that the exponent in expression (4.58) is zero, as shown in Exercise 4.8. Thus, we write expression (4.58) as

$$A_0(x(t)) = A_0(t_0)\sqrt{\frac{S_0 v(x(t))}{S_t v(x(0))}}. \tag{4.61}$$

Solutions (4.58) and (4.59) describe the evolution of the amplitude along the characteristics. This amplitude depends only on the radicand in expressions (4.58) and (4.59); the exponential does not depend on the characteristics since they are given only by symbol $\sigma(D)(p)$. This solution demonstrates that the square of the amplitude depends on the geometrical spreading of the characteristics; the greater the increase in the raytube cross-section, the greater the decrease of the square of the amplitude, provided that ζ remains the same. This spreading is illustrated in Figure 4.2. The case of the vanishing denominator implies that the amplitude tends to infinity; it is associated with caustics, which are discussed in Chapter 5.

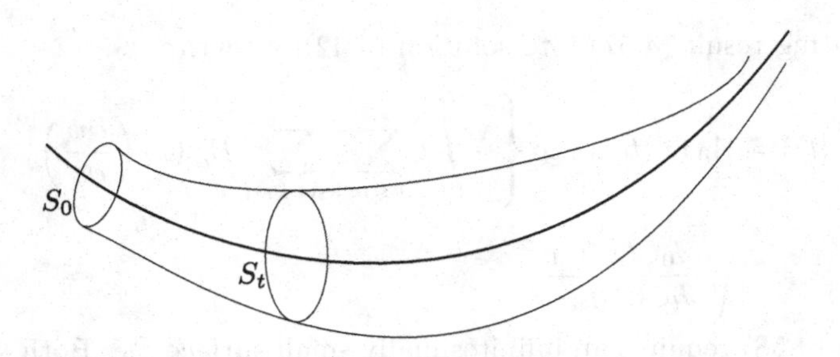

Fig. 4.2 Geometrical spreading of rays.

4.6 Higher-order transport equations

To obtain the entire asymptotic solution of a given differential equation, we need to compute all coefficients $A_j(x)$. In the preceding section we formulated the way of looking for the first of these coefficients, $A_0(x)$. Herein, we exemplify the procedure for finding these coefficients using the reduced wave equation. Example 4.9 can be viewed as a continuation of Example 4.7.

Example 4.9. Let us consider the reduced wave equation given in expression (4.31) and write it as

$$\frac{v^2}{(\iota\omega)^2}\nabla^2\hat{u}(x,\omega)=\hat{u}(x,\omega).$$

The asymptotic solution is of the form

$$\hat{u}(x,\omega)\sim\sum_{j=0}^{\infty}e^{\iota\omega\psi(x)}\frac{A_j(x)}{(\iota\omega)^j}.$$

Substituting this expression into the reduced wave equation and considering the three-dimensional case, we get

$$\frac{v^2}{(\iota\omega)^2}\sum_{i=1}^{3}\frac{\partial}{\partial x_i}\left(e^{\iota\omega\psi(x)}\left(\iota\omega\frac{\partial\psi}{\partial x_i}\sum_{j=0}^{\infty}\frac{A_j(x)}{(\iota\omega)^j}+\sum_{j=0}^{\infty}\frac{\frac{\partial A_j(x)}{\partial x_i}}{(\iota\omega)^j}\right)\right)$$
$$=\sum_{j=0}^{\infty}e^{\iota\omega\psi(x)}\frac{A_j(x)}{(\iota\omega)^{j-2}}.\quad(4.62)$$

Continuing the differentiation, we write the left-hand side of this equation as

$$v^2 \sum_{i=1}^{3} e^{\iota\omega\psi(x)} \sum_{j=0}^{\infty} \left(\frac{\partial^2 \psi}{\partial x_i^2} \frac{A_j(x)}{(\iota\omega)^{j+1}} + \frac{\partial \psi}{\partial x_i} \frac{\frac{\partial A_j(x)}{\partial x_i}}{(\iota\omega)^{j+1}} + \frac{\frac{\partial^2 A_j(x)}{\partial x_i^2}}{(\iota\omega)^{j+2}} \right.$$
$$\left. + \frac{\partial \psi}{\partial x_i} \left(\frac{\partial \psi}{\partial x_i} \frac{A_j(x)}{(\iota\omega)^j} + \frac{\frac{\partial A_j(x)}{\partial x_i}}{(\iota\omega)^{j+1}} \right) \right).$$

The procedure for finding the coefficients consists of equating the corresponding terms of $\iota\omega$ on both sides of equation (4.62).

Considering the terms with $(\iota\omega)^0$, we obtain

$$v^2 \sum_{i=1}^{3} \left(\frac{\partial \psi}{\partial x_i} \right)^2 A_0(x) = A_0(x),$$

which, for nonzero scalar $A_0(x)$, reduces to equation (4.32) as expected.

Considering the terms with $(\iota\omega)^{-1}$, we obtain

$$v^2 \sum_{i=1}^{3} e^{\iota\omega\psi(x)} \left(\frac{\partial^2 \psi}{\partial x_i^2} A_0(x) + \frac{\partial \psi}{\partial x_i} \frac{\partial A_0(x)}{\partial x_i} + \frac{\partial \psi}{\partial x_i} \left(\frac{\partial \psi}{\partial x_i} A_1(x) + \frac{\partial A_0(x)}{\partial x_i} \right) \right)$$
$$= e^{\iota\omega\psi(x)} A_1(x),$$

which simplifies to

$$\left(\sum_{i=1}^{3} \left(\frac{\partial \psi}{\partial x_i} \right)^2 - \frac{1}{v^2} \right) A_1(x) + \sum_{i=1}^{3} \left(\frac{\partial^2 \psi}{\partial x_i^2} + 2 \frac{\partial \psi}{\partial x_i} \frac{\partial}{\partial x_i} \right) A_0(x) = 0, \quad (4.63)$$

which is equation (4.38) as expected.

Considering terms with $(\iota\omega)^{-k}$, where $k \geqslant 2$, we obtain

$$v^2 \sum_{i=1}^{3} e^{\iota\omega\psi(x)} \left(\frac{\partial^2 \psi}{\partial x_i^2} A_{k-1}(x) + \frac{\partial \psi}{\partial x_i} \frac{\partial A_{k-1}(x)}{\partial x_i} + \frac{\partial^2 A_{k-2}(x)}{\partial x_i^2} \right.$$
$$\left. + \frac{\partial \psi}{\partial x_i} \left(\frac{\partial \psi}{\partial x_i} A_k(x) + \frac{\partial A_{k-1}(x)}{\partial x_i} \right) \right) = e^{\iota\omega\psi(x)} A_k(x),$$

which simplifies to

$$\left(\sum_{i=1}^{3} \left(\frac{\partial \psi}{\partial x_i} \right)^2 - \frac{1}{v^2} \right) A_k(x) + \sum_{i=1}^{3} \left(\frac{\partial^2 \psi}{\partial x_i^2} + 2 \frac{\partial \psi}{\partial x_i} \frac{\partial}{\partial x_i} \right) A_{k-1}(x)$$
$$+ \sum_{i=1}^{3} \frac{\partial^2 A_{k-2}(x)}{\partial x_i^2} = 0.$$

This equation allows us to compute $A_k(x)$, provided we know all $A_l(x)$ for $l < k$. Hence, we can compute iteratively any term in the asymptotic series of the solution. This equation is referred to as the kth-order transport equation.

Since ψ satisfies the eikonal equation, the first term of the above equation is zero, and the kth-order transport equation reduces to

$$\sum_{i=1}^{3}\left(\frac{\partial^2\psi}{\partial x_i^2}+2\frac{\partial\psi}{\partial x_i}\frac{\partial}{\partial x_i}\right)A_{k-1}(x)+\sum_{i=1}^{3}\frac{\partial^2 A_{k-2}(x)}{\partial x_i^2}=0.$$

Having solved the lower-order transport equations, we know $A_{k-2}(x)$. Thus, A_{k-1} is a solution of the following first-order linear equation:

$$\sum_{i=1}^{3}\left(\frac{\partial^2\psi}{\partial x_i^2}+2\frac{\partial\psi}{\partial x_i}\frac{\partial}{\partial x_i}\right)A_{k-1}(x)=-\sum_{i=1}^{3}\frac{\partial^2 A_{k-2}(x)}{\partial x_i^2}. \tag{4.64}$$

Using the chain rule, we write equation (4.64) as an ordinary differential equation along the characteristics $x(t)$, namely,

$$\frac{\mathrm{d}}{\mathrm{d}t}A_{k-1}(x(t))=-\frac{1}{2}\left(\sum_{i=1}^{3}\frac{\partial^2 A_{k-2}(x)}{\partial x_i^2}+\frac{\partial^2\psi}{\partial x_i^2}A_{k-1}(x)\right).$$

The particular form of the solution of this equation depends on functions $A_{k-2}(x)$ and $\psi(x)$ obtained in previous steps:

$$\begin{aligned}A_{k-1}(x(t))=\exp&\left\{-\int_0^t\frac{1}{2}\sum_{i=1}^{3}\frac{\partial^2 A_{k-2}(x(s_1))}{\partial x_i^2}\,\mathrm{d}s_1\right\}\\&\left(A_{k-1}(x(0))-\frac{1}{2}\int_0^t\exp\left\{\int_0^{s_2}\frac{1}{2}\sum_{i=1}^{3}\frac{\partial^2 A_{k-2}(x(s_3))}{\partial x_i^2}\,\mathrm{d}s_3\right\}\right.\\&\left.\frac{\partial^2\psi}{\partial x_i^2}A_{k-1}(x(s_2))\,\mathrm{d}s_2\right).\end{aligned}$$

This completes the construction of the asymptotic series of solutions of a general differential equation (4.21). This construction of the solution is often referred to as construction using asymptotic ray theory. In the following section, we focus our attention on the elastodynamic and Maxwell equations.

4.7 Physical applications

4.7.1 *Elastodynamic equations*

We examine solutions of the elastodynamic equations (B.24) derived in Appendix B,

$$\rho(x)\frac{\partial^2 u_i(x,t)}{\partial t^2}=\sum_{j=1}^{3}\sum_{k=1}^{3}\sum_{l=1}^{3}\left(\frac{\partial c_{ijkl}(x)}{\partial x_j}\frac{\partial u_k}{\partial x_l}+c_{ijkl}(x)\frac{\partial^2 u_k}{\partial x_j\partial x_l}\right),\quad(4.65)$$

where $i\in\{1,2,3\}$, considering that they might contain discontinuities. Following the reasoning of Section 4.2, we express equations (4.65) in terms of x and ω to consider the limit of ω tending to infinity, which is associated with discontinuities. We perform the Fourier transform from t to ω, as discussed in Appendix D.3, to get

$$(\iota\omega)^2\rho(x)\hat{u}_i(x,\omega)=\sum_{j=1}^{3}\sum_{k=1}^{3}\sum_{l=1}^{3}\left(\frac{\partial c_{ijkl}(x)}{\partial x_j}\frac{\partial \hat{u}_k}{\partial x_l}+c_{ijkl}(x)\frac{\partial^2 \hat{u}_k}{\partial x_j\partial x_l}\right),\tag{4.66}$$

where, following Appendix D.3,

$$\hat{u}_i(x,\omega):=\frac{1}{\sqrt{2\pi}}\int_{-\infty}^{\infty}u_i(x,t)\exp(-\iota\omega t)\,\mathrm{d}t,$$

and $i\in\{1,2,3\}$.

Following results of Sections 4.4.2 and 4.5.2, we would like to write the solutions of the eikonal and transport equations for the elastodynamic equations. First we express equations (4.66) as

$$\left(\frac{1}{(\iota\omega)^2}D_2+D_0\right)\hat{u}=0,$$

where

$$\sum_{l=1}^{3}(D_2)_{il}\hat{u}_l=\sum_{l=1}^{3}\left(\sum_{j=1}^{3}\sum_{k=1}^{3}\frac{\partial c_{ijkl}(x)}{\partial x_j}\frac{\partial}{\partial x_k}+c_{ijkl}(x)\left(\frac{\partial^2}{\partial x_j\partial x_k}\right)\right)\hat{u}_l\,,$$

$$\sum_{l=1}^{3}(D_0)_{il}\hat{u}_l=-\rho\sum_{l=1}^{3}\delta_{il}\hat{u}_l\,.$$

Using the general form of the equation that leads to the eikonal equation, namely equation (4.29), we write it for the Cauchy equations of motion as

$$\sum_{l=1}^{3}\left(\sum_{j=1}^{3}\sum_{k=1}^{3}c_{ijkl}(x)\frac{\partial\psi}{\partial x_j}\frac{\partial\psi}{\partial x_k}-\rho\delta_{il}\right)A_{0l}=0\,,\tag{4.67}$$

which we refer to as the Christoffel equation. The solutions of this equation are discussed in Section 3.4.1.

Similarly, using the general form of the zero-order transport equation, given in expression (4.36), we write it in this case as

$$\sum_{l=1}^{3}\left(\sum_{j=1}^{3}\sum_{k=1}^{3}c_{ijkl}\left(x\right)\frac{\partial\psi}{\partial x_j}\frac{\partial\psi}{\partial x_k}-\rho\delta_{il}\right)A_{1l}$$
$$+\sum_{l=1}^{3}\left(\sum_{j=1}^{3}\sum_{k=1}^{3}\frac{\partial c_{ijkl}\left(x\right)}{\partial x_j}\frac{\partial\psi}{\partial x_k}+\frac{1}{2}\sum_{j=1}^{3}\sum_{k=1}^{3}c_{ijkl}\left(x\right)\frac{\partial^2\psi}{\partial x_j\partial x_k}\right.$$
$$\left.+\sum_{j=1}^{3}\sum_{k=1}^{3}c_{ijkl}\left(x\right)\left(\frac{\partial\psi}{\mathrm{d}x_j}\right)\frac{\partial}{\partial x_k}\right)A_{0l}\left(x\right)=0\,.$$

To find the solutions of the transport equation, we assume that A_1 satisfies the Christoffel equation (4.67) and we follow the general solution given by expression (4.59), namely,

$$A_0\left(x\left(t\right)\right)=A_0\left(x\left(t_0\right)\right)\exp\left\{-\int_{t_0}^{t}\zeta\sum_{k\leqslant m}\sum_{|\alpha|=k-1}B_\alpha^k\left(x\right)\left(\frac{\partial\psi}{\partial x}\right)^{|\alpha|}\mathrm{d}s\right\}$$
$$\sqrt{\frac{J_0\zeta\left(x\left(t\right)\right)}{J_t\zeta\left(x\left(t_0\right)\right)}}\,.$$

We rewrite this expression using the particular B of the elastodynamic equations as

$$A_{0i}(x\left(t\right))=\sqrt{\frac{J_0\zeta\left(x\left(t\right)\right)}{J_t\zeta\left(x\left(t_0\right)\right)}}\sum_{l=1}^{3}\exp\left\{-\int_{t_0}^{t}\zeta\sum_{j=1}^{3}\sum_{k=1}^{3}\frac{\partial c_{ijkl}}{\partial x_j}\frac{\partial\psi}{\partial x_k}\,\mathrm{d}s\right\}A_{0l}\left(x\left(t_0\right)\right),$$

where A_{0i} is the ith component of vector A_0.

4.7.2 *Maxwell equations*

Following the derivation of the Maxwell equations in Appendix C, we use their potential formulation,

$$\nabla^2 A\left(x,t\right)-\frac{1}{c^2}\frac{\partial^2 A\left(x,t\right)}{\partial t^2}=-\frac{J\left(x,t\right)}{c^2\epsilon_0}\tag{4.68}$$

and

$$\nabla^2\phi\left(x,t\right)-\frac{1}{c^2}\frac{\partial^2\phi\left(x,t\right)}{\partial t^2}=-\frac{\rho\left(x,t\right)}{\epsilon_0},\tag{4.69}$$

to study the propagation of discontinuities. To examine discontinuities, which are associated with ω tending to infinity, we perform the Fourier transform of equations (4.68) and (4.69) in variable t, in a manner analogous to the one discussed in Section 4.7.1:

$$\nabla^2 \hat{A}(x,\omega) - \frac{1}{c^2}(\iota\omega)^2 \hat{A}(x,\omega) = -\frac{\hat{J}(x,\omega)}{c^2\epsilon_0} \tag{4.70}$$

and

$$\nabla^2 \hat{\phi}(x,\omega) - \frac{1}{c^2}(\iota\omega)^2 \hat{\phi}(x,\omega) = -\frac{\hat{\rho}(x,\omega)}{\epsilon_0}. \tag{4.71}$$

We express the vector equation in terms of components as

$$\sum_{j=1}^{3} \frac{\partial^2 \hat{A}_i}{\partial x_j^2}(x,\omega) - \frac{1}{c^2}(\iota\omega)^2 \hat{A}_i(x,\omega) = -\frac{\hat{J}_i(x,\omega)}{c^2\epsilon_0},$$

which, following definition (4.22), can be written as

$$\left(D_0 + \frac{1}{(\iota\omega)^2} D_2\right) \hat{A} = -\frac{\hat{J}(x,\omega)}{(\iota\omega)^2 \epsilon_0},$$

where

$$D_0 \hat{A}_i = -\hat{A}_i$$

and

$$D_2 \hat{A}_i = c^2 \sum_{j=1}^{3} \frac{\partial^2}{\partial x_j^2} \hat{A}_i.$$

Similarly, the scalar equation can be written as

$$\left(D_0 + \frac{1}{(\iota\omega)^2} D_2\right) \hat{\phi} = -\frac{c^2 \hat{\rho}}{(\iota\omega)^2 \epsilon_0},$$

where

$$D_0 \hat{\phi} = -\hat{\phi}$$

and

$$D_2 \hat{\phi} = c^2 \sum_{j=1}^{3} \frac{\partial^2}{\partial x_j^2} \hat{\phi}.$$

Since the right-hand sides of equations (4.70) and (4.71) are nonzero, they might contain terms with powers of $\iota\omega$. If we assume that the right-hand sides can be expressed as asymptotic series of the form

$$e^{\iota\omega\psi(x)}\left(f_0(x) + f_1(x)\frac{1}{(\iota\omega)} + \cdots\right), \tag{4.72}$$

then we can write the terms with given powers of $\iota\omega$ as shown below. For the vector equation, the $(\iota\omega)^0$ term is

$$\left(\sum_{j=1}^{3} c^2 \left(\frac{\partial \psi}{\partial x_j}\right)^2 - 1\right) \hat{A}_i(x) = 0,$$

where $i \in \{1,2,3\}$. For the scalar equation, it is

$$\left(\sum_{j=1}^{3} c^2 \left(\frac{\partial \psi}{\partial x_j}\right)^2 - 1\right) \hat{\phi}(x) = 0.$$

Nontriviality of solutions $\hat{A}(x)$ and $\hat{\phi}(x)$ requires $\psi(x)$ to satisfy the same equation for both cases—the eikonal equation (2.68), whose solutions are discussed in Section 3.4.2. This means that both vector and scalar potentials admit discontinuities along the same surfaces—the wavefronts.

Using the general form of the zero-order transport equation, (4.36), we write it for the vector equation as

$$\sum_{j=1}^{n} \left(\frac{\partial^2 \psi}{\partial x_j^2} + 2\frac{\partial \psi}{\partial x_j}\frac{\partial}{\partial x_j}\right) A_{Ai}(x) = f_0(x), \qquad i \in \{1,2,3\},$$

where A_{Ai} is the zero-order term in the asymptotic series for the ith component of the vector potential A. For the scalar equation, it is

$$\sum_{j=1}^{n} \left(\frac{\partial^2 \psi}{\partial x_j^2} + 2\frac{\partial \psi}{\partial x_j}\frac{\partial}{\partial x_j}\right) A_{\phi}(x) = g_0(x),$$

where A_ϕ is the zero-order term in the asymptotic series for the scalar potential. In view of series (4.72), the ith-order transport equations have the right-hand side equal to $f_i(x)$ or $g_i(x)$.

If the right-hand sides of equations (4.70) and (4.71) are not composed only of discontinuities given by series (4.72) but also contain continuous terms, we cannot write the right-hand side using the asymptotic series. However, we can try to solve the equation by finding a particular solution, which we would add to the homogeneous one. Since we are dealing with linear equations, a sum of solutions is also a solution.

Closing remarks

In this chapter, we consider solutions of differential equations as asymptotic series; in particular, series given by expression (4.24) as $\omega \to \infty$, namely

$$e^{\iota\omega\psi(x)}\left(A_0(x) + \frac{1}{\iota\omega}A_1(x) + \cdots + \frac{1}{(\iota\omega)^n}A_n(x) + \cdots\right).$$

Its Fourier transform with respect to ω is

$$A_0(x)\sqrt{2\pi}\delta(\psi(x)-t)+\sum_{j=1}^{\infty}A_j(x)\sqrt{2\pi}V_j(\psi(x)-t),$$

where

$$V_{j+1}(\eta)=\begin{cases}0 & \eta<0\\ & \\ \dfrac{1}{j!}\eta^j & \eta\geqslant 0\end{cases},\qquad j\in\{0,1,2,\ldots\}.$$

As $\omega\to\infty$, this expression describes the high-frequency content of the solution, and, hence, the behaviour of discontinuities of the solution. The coefficients A_i describe amplitudes of different types of discontinuities within the solution: the Dirac delta, the Heaviside step, and continuous functions with discontinuous derivatives of progressively higher order.

The asymptotic series of a solution of differential equations leads us to the eikonal equation and the transport equations. The eikonal equation describes the possible location of discontinuities, and the transport equations their magnitudes. If a solution has discontinuities, they happen along the characteristics only. However, characteristics are not associated only with discontinuities as is shown in Chapters 1, 2 and 3.

4.8 Exercises

Exercise 4.1. Show that conditions (4.13) and (4.14) give equivalent definitions of the existence of infinite asymptotic series.

Solution. Condition (4.13), namely,

$$\lim_{x\to x_0}\frac{f(x)-S_n(x)}{(x-x_0)^n}=0,\qquad \forall n\in\mathbb{N}$$

implies that

$$\left|\frac{f(x)-S_n(x)}{(x-x_0)^n}\right|<K|x-x_0|,\qquad \forall n\in\mathbb{N},$$

for any positive constant K. Thus condition (4.13) implies condition (4.14), namely,

$$\left|\frac{f(x)-S_n(x)}{(x-x_0)^{n+1}}\right|<K, \qquad \forall n\in\mathbb{N}.$$

Condition (4.14) implies that

$$f(x)-S_n(x)=O\left((x-x_0)^{n+1}\right)\Rightarrow$$

$$\lim_{x\to x_0}\frac{f(x)-S_n(x)-a_{n+1}(x-x_0)^{n+1}}{(x-x_0)^{n+1}}=\lim_{x\to x_0}\frac{f(x)-S_n(x)}{(x-x_0)^{n+1}}-a_{n+1}=0,$$

which means that

$$\lim_{x\to x_0}\frac{f(x)-S_n(x)}{(x-x_0)^n}=\lim_{x\to x_0}a_{n+1}(x-x_0)=0,$$

as desired.

Exercise 4.2. Evaluate

$$\lim_{x\to\infty}x\int_x^\infty t^{-1}e^{-t}\,\mathrm{d}t.$$

Solution. We rewrite this limit as

$$\lim_{x\to\infty}\frac{\int_x^\infty t^{-1}e^{-t}\,\mathrm{d}t}{x^{-1}}. \tag{4.73}$$

To evaluate this limit, we use the de l'Hôpital rule, which requires that $\lim_{x\to\infty}\int_x^\infty t^{-1}e^{-t}\,\mathrm{d}t=0$. This is true since the integral is bounded for $x>0$ by $\int_x^\infty e^{-t}\,\mathrm{d}t$, which approaches zero as $x\to\infty$. Returning to limit (4.73), we write

$$\lim_{x\to\infty}\frac{\int_x^\infty t^{-1}e^{-t}\,\mathrm{d}t}{x^{-1}}=\lim_{x\to\infty}\frac{-x^{-1}e^{-x}}{-x^{-2}}=0.$$

Exercise 4.3. Verify that the fourth derivative of a composed function satisfies the general chain rule:

$$\frac{\mathrm{d}^k}{\mathrm{d}x^k}f\circ g\equiv\frac{\mathrm{d}^k}{\mathrm{d}x^k}f(g(x)) \tag{4.74}$$

$$=\sum\frac{k!}{l_1!l_2!\cdots l_k!}f^{(l_1+l_2+\cdots+l_k)}\circ g\left(\frac{g'}{1!}\right)^{l_1}\left(\frac{g''}{2!}\right)^{l_2}\cdots\left(\frac{g^{(k)}}{k!}\right)^{l_k},$$

where the sum is over all nonnegative integer solutions of

$$l_1+2l_2+\cdots+kl_k=k.$$

Solution. The fourth derivative can be written as

$$\begin{aligned}
\frac{\mathrm{d}^4}{\mathrm{d}x^4} f \circ g &= \frac{\mathrm{d}^3}{\mathrm{d}x^3} f' \circ g g' = \frac{\mathrm{d}^2}{\mathrm{d}x^2} \left(f'' \circ g g'^2 + f' \circ g g'' \right) \qquad (4.75)\\
&= \frac{\mathrm{d}}{\mathrm{d}x} \left(f''' \circ g g'^3 + f'' \circ g 2g' g'' + f'' \circ g g' g'' + f' \circ g g''' \right)\\
&= f'''' \circ g g'^4 + 6 f''' \circ g g'^2 g'' + 3 f'' \circ g g''^2 + 4 f'' \circ g g' g''' + f' \circ g g''''.
\end{aligned}$$

To verify that this derivative satisfies the general formula, we find the solutions for

$$l_1 + 2l_2 + 3l_3 + 4l_4 = 4.$$

We see that l_4 can be either one or zero.

If $l_4 = 1$, then $l_1 = l_2 = l_3 = 0$.

If $l_4 = 0$, then either $l_3 = 0$ or $l_3 = 1$. If $l_3 = 1$, then $l_1 = 1$. If $l_3 = 0$, then $l_2 = 2$, $l_2 = 1$ or $l_2 = 0$. If $l_2 = 2$ then $l_1 = 0$. If $l_2 = 1$, then $l_1 = 2$. If $l_2 = 0$, then $l_1 = 4$.

Thus, quadruple (l_1, l_2, l_3, l_4) can have values $(0,0,0,1)$, $(1,0,1,0)$, $(0,2,0,0)$, $(2,1,0,0)$, $(4,0,0,0)$. Substituting these quadruples into the general formula, we obtain

$$\begin{aligned}
&\frac{4!}{1!} f' \circ g \left(\frac{g''''}{4!} \right) + \frac{4!}{1!1!} f'' \circ g \left(\frac{g'}{1!} \right) \left(\frac{g'''}{3!} \right) + \frac{4!}{2!} f'' \circ g \left(\frac{g''}{2!} \right)^2\\
&\qquad + \frac{4!}{2!1!} f''' \circ g \left(\frac{g'}{1!} \right)^2 \left(\frac{g''}{2!} \right) + \frac{4!}{4!} f'''' \circ g \left(\frac{g'}{1!} \right)^4,
\end{aligned}$$

which, upon simplification, becomes the last line of expression (4.75), as required.

Exercise 4.4. Find the term containing the highest power of $\iota\omega$ in

$$\frac{\partial^{|\alpha|}}{\partial x^{\alpha}} \exp\{\iota\omega\psi(x)\},$$

where α is a multiindex.

Solution. This derivative is a short-hand expression for

$$\frac{\partial^{\alpha_1}}{\partial x_1^{\alpha_1}} \frac{\partial^{\alpha_2}}{\partial x_2^{\alpha_2}} \cdots \frac{\partial^{\alpha_n}}{\partial x_n^{\alpha_n}} \exp\{\iota\omega\psi(x)\}.$$

Using expression (4.75) in Exercise 4.3, we write this derivative as

$$\frac{\partial^{\alpha_1}}{\partial x_1^{\alpha_1}} \cdots \frac{\partial^{\alpha_{n-1}}}{\partial x_{n-1}^{\alpha_{n-1}}} \sum \frac{\alpha_n!}{l_1! l_2! \cdots l_{\alpha_n}!} (\iota\omega)^{l_1+l_2+\cdots+l_{\alpha_n}}$$

$$\exp\{\iota\omega\psi\} \left(\frac{\frac{\partial\psi}{\partial x_n}}{1!} \right)^{l_1} \left(\frac{\frac{\partial^2\psi}{\partial x_n^2}}{2!} \right)^{l_2} \cdots \left(\frac{\frac{\partial^{\alpha_n}\psi}{\partial x_n^{\alpha_n}}}{\alpha_n!} \right)^{l_{\alpha_n}} .$$

Herein, the highest power of $\iota\omega$ results from

$$(l_1, l_2, \ldots, l_{\alpha_n}) = (\alpha_n, 0, \ldots, 0) .$$

Considering only this term of the sum, we write

$$\frac{\partial^{\alpha_1}}{\partial x_1^{\alpha_1}} \cdots \frac{\partial^{\alpha_{n-1}}}{\partial x_{n-1}^{\alpha_{n-1}}} (\iota\omega)^{\alpha_n} \exp\{\iota\omega\psi\} \left(\frac{\partial\psi}{\partial x_n} \right)^{\alpha_n} .$$

Differentiating this term with respect to x_{n-1}, we would get mixed terms resulting from the product rule. Since we are interested in the highest power of $\iota\omega$ only, we consider the part of the product rule in which we differentiate the exponential α_{n-1} times. Using the chain rule, we write the above expression as

$$\frac{\partial^{\alpha_1}}{\partial x_1^{\alpha_1}} \cdots \frac{\partial^{\alpha_{n-2}}}{\partial x_{n-2}^{\alpha_{n-2}}} \sum \frac{\alpha_{n-1}!}{l_1! l_2! \cdots l_{\alpha_{n-1}}!} (\iota\omega)^{l_1+l_2+\cdots+l_{\alpha_{n-1}}}$$

$$\exp\{\iota\omega\psi\} \left(\frac{\frac{\partial\psi}{\partial x_{n-1}}}{1!} \right)^{l_1} \left(\frac{\frac{\partial^2\psi}{\partial x_{n-1}^2}}{2!} \right)^{l_2} \cdots \left(\frac{\frac{\partial^{\alpha_{n-1}}\psi}{\partial x_{n-1}^{\alpha_{n-1}}}}{\alpha_{n-1}!} \right)^{l_{\alpha_n}} (\iota\omega)^{\alpha_n} \left(\frac{\partial\psi}{\partial x_n} \right)^{\alpha_n} .$$

Again, taking into account the highest power of $i\omega$, we consider

$$(l_1, l_2, \ldots, l_{\alpha_{n-1}}) = (\alpha_{n-1}, 0, \ldots, 0) ,$$

which results in

$$\frac{\partial^{\alpha_1}}{\partial x_1^{\alpha_1}} \cdots \frac{\partial^{\alpha_{n-2}}}{\partial x_{n-2}^{\alpha_{n-2}}} (\iota\omega)^{\alpha_{n-1}} (\iota\omega)^{\alpha_n} \exp\{\iota\omega\psi\} \left(\frac{\partial\psi}{\partial x_{n-1}} \right)^{\alpha_{n-1}} \left(\frac{\partial\psi}{\partial x_n} \right)^{\alpha_n} .$$

Following this pattern, we obtain the term with the highest power of $i\omega$, namely,

$$(\iota\omega)^{\alpha_1+\alpha_2+\cdots+\alpha_n} \exp\{\iota\omega\psi\} \left(\frac{\partial\psi}{\partial x_1} \right)^{\alpha_1} \left(\frac{\partial\psi}{\partial x_2} \right)^{\alpha_2} \cdots \left(\frac{\partial\psi}{\partial x_n} \right)^{\alpha_n} .$$

Exercise 4.5. Prove the generalized Leibniz rule for multiindex α, namely,

$$\frac{\partial^{|\alpha|}}{\partial x^\alpha} (fg) = \sum_{\beta+\gamma=\alpha} \frac{\alpha!}{\beta!\gamma!} \frac{\partial^{|\beta|} f}{\partial x^\beta} \frac{\partial^{|\gamma|} g}{\partial x^\gamma} ,$$

where $\alpha! = \alpha_1! \alpha_2! \ldots \alpha_n!$.

Solution. The left-hand side of the generalized Leibniz rule can be written as

$$\frac{\partial^{\alpha_1}}{\partial x_1^{\alpha_1}}\frac{\partial^{\alpha_2}}{\partial x_2^{\alpha_2}}\cdots\frac{\partial^{\alpha_n}}{\partial x_n^{\alpha_n}}(fg)=\frac{\partial^{\alpha_1}}{\partial x_1^{\alpha_1}}\frac{\partial^{\alpha_2}}{\partial x_2^{\alpha_2}}\cdots\frac{\partial^{\alpha_{n-1}}}{\partial x_{n-1}^{\alpha_{n-1}}}\sum_{\beta_n+\gamma_n=\alpha_n}\frac{\alpha_n!}{\beta_n!\gamma_n!}\frac{\partial^{\beta_n}f}{\partial x_n^{\beta_n}}\frac{\partial^{\gamma_n}g}{\partial x_n^{\gamma_n}},$$

where we use the standard Leibniz rule, namely,

$$\frac{\mathrm{d}^k}{\mathrm{d}x^k}(f\cdot g)=\sum_{l+m=k}\frac{k!}{l!m!}\frac{\mathrm{d}^l f}{\mathrm{d}x^l}\frac{\mathrm{d}^m g}{\mathrm{d}x^m}.$$

Continuing the differentiation, we obtain

$$\sum_{\beta_1+\gamma_1=\alpha_1}\sum_{\beta_2+\gamma_2=\alpha_2}\cdots\sum_{\beta_n+\gamma_n=\alpha_n}\frac{\alpha_1!}{\beta_1!\gamma_1!}\frac{\alpha_2!}{\beta_2!\gamma_2!}\cdots\frac{\alpha_n!}{\beta_n!\gamma_n!}\frac{\partial^{\beta_1}\partial^{\beta_2}\cdots\partial^{\beta_n}f}{\partial x_1^{\beta_1}\partial x_2^{\beta_2}\cdots\partial x_n^{\beta_n}}\frac{\partial^{\gamma_1}\partial^{\gamma_2}\cdots\partial^{\gamma_n}g}{\partial x_1^{\gamma_1}\partial x_2^{\gamma_2}\cdots\partial x_n^{\gamma_n}},$$

which can be written using the multiindex notation as the right-hand side of the generalized Leibniz rule,

$$\sum_{\beta+\gamma=\alpha}\frac{\alpha!}{\beta!\gamma!}\frac{\partial^{|\beta|}f}{\partial x^\beta}\frac{\partial^{|\gamma|}g}{\partial x^\gamma}.$$

Exercise 4.6. Following the general formulation on page 147, derive eikonal equation (4.32) from the reduced wave equation (4.31) written as

$$\frac{1}{(\iota\omega)^2}\sum_{i=1}^{3}\frac{\partial^2}{\partial x_i^2}\hat{u}(x,\omega)-\frac{1}{v^2}\hat{u}(x,\omega)=0.$$

Solution. Following the discussion in Section 4.3.2, we consider an asymptotic solution of the form

$$\hat{u}(x,\omega)=e^{\iota\omega\psi(x)}\sum_{j=0}^{\infty}A_j(x)\frac{1}{(\iota\omega)^j}.$$

Using this expression in the reduced wave equation, we write

$$\left(\frac{1}{(\iota\omega)^2}\sum_{i=1}^{3}\frac{\partial^2}{\partial x_i^2}-\frac{1}{v^2}\right)\left(e^{\iota\omega\psi(x)}\sum_{j=0}^{\infty}A_j(x)\frac{1}{(\iota\omega)^j}\right)=0.$$

Applying the product rule, we get

$$\frac{1}{(\iota\omega)^2}\left(\sum_{i=1}^{3}\left(\frac{\partial^2}{\partial x_i^2}e^{\iota\omega\psi(x)}\right)\sum_{j=0}^{\infty}A_j(x)\frac{1}{(\iota\omega)^j}\right.$$
$$+2\sum_{i=1}^{3}\frac{\partial}{\partial x_i}e^{\iota\omega\psi(x)}\frac{\partial}{\partial x_i}\sum_{j=0}^{\infty}A_j(x)\frac{1}{(\iota\omega)^j}$$
$$\left.+e^{\iota\omega\psi(x)}\sum_{i=1}^{3}\frac{\partial^2}{\partial x_i^2}\sum_{j=0}^{\infty}A_j(x)\frac{1}{(\iota\omega)^j}\right)-\frac{1}{v^2}e^{\iota\omega\psi(x)}\sum_{j=0}^{\infty}A_j(x)\frac{1}{(\iota\omega)^j}=0.$$

Differentiating the exponentials, we get

$$\frac{e^{\iota\omega\psi(x)}}{(\iota\omega)^2}\left(\sum_{i=1}^{3}\left((\iota\omega)^2\left(\frac{\partial\psi}{\partial x_i}\right)^2+(\iota\omega)\frac{\partial^2\psi}{\partial x_i^2}\right)\sum_{j=0}^{\infty}A_j\frac{1}{(\iota\omega)^j}\right.$$
$$\left.+2\sum_{i=1}^{3}(\iota\omega)\frac{\partial\psi}{\partial x_i}\sum_{j=0}^{\infty}\frac{\partial A_j}{\partial x_i}\frac{1}{(\iota\omega)^j}+\sum_{i=1}^{3}\sum_{j=0}^{\infty}\frac{\partial^2 A_j}{\partial x_i^2}\frac{1}{(\iota\omega)^j}\right)$$
$$-\frac{1}{v^2}e^{\iota\omega\psi(x)}\sum_{j=0}^{\infty}A_j\frac{1}{(\iota\omega)^j}=0. \tag{4.76}$$

The terms with the zeroth power of $\iota\omega$ in this asymptotic series are

$$e^{\iota\omega\psi(x)}A_0(x)\left(\sum_{i=1}^{3}\left(\frac{\partial\psi}{\partial x_i}\right)^2-\frac{1}{v^2}\right).$$

Viewing the right-hand side of equation (4.76) as an asymptotic series and equating the coefficients of the zeroth power of $\iota\omega$, we obtain

$$A_0(x)\left(\sum_{i=1}^{3}\left(\frac{\partial\psi}{\partial x_i}\right)^2-\frac{1}{v^2}\right)=0.$$

If we consider only nonzero functions A_0, then this equation reduces to

$$\sum_{i=1}^{3}\left(\frac{\partial\psi}{\partial x_i}\right)^2=\frac{1}{v^2},$$

which is equation (4.32), as required.

Exercise 4.7. Following equation (4.30), show that the zero-order approximation of the solution of the three-dimensional reduced wave equation (4.31), namely,

$$\frac{1}{(\iota\omega)^2}\nabla^2\hat{u}(x,\omega)-\frac{1}{v^2}\hat{u}(x,\omega)=0,$$

results in eikonal equation (4.32), namely,

$$\sum_{i=1}^{3}\left(\frac{\partial\psi}{\partial x_i}\right)^2=\frac{1}{v^2}.$$

Solution. Considering the general form of an asymptotic differential equation given by expression (4.25), we see that the coefficients $B_\alpha(x)$ of the reduced wave equation are

$$\begin{aligned}B_{(2,0,0)}&=1,\\B_{(0,2,0)}&=1,\\B_{(0,0,2)}&=1,\\B_{(0,0,0)}&=-\frac{1}{v^2}.\end{aligned}$$

Substituting these coefficients into equation (4.30), we obtain

$$\sum_i\left(\frac{\partial\psi}{\partial x_i}\right)^2-\frac{1}{v^2}=0,$$

which is equation (4.32), as required.

Exercise 4.8. Show that in the case of the reduced wave equation (4.31), namely,

$$\frac{v^2}{(i\omega)^2}\nabla^2\hat{u}(x,\omega)=\hat{u}(x,\omega), \tag{4.77}$$

equation (4.36) reduces to equation (4.38).

Solution. As stated in Exercise 4.7, coefficients $B_\alpha(x)$ for the reduced wave equation are $B_{(2,0,0)}=1$, $B_{(0,2,0)}=1$, $B_{(0,0,2)}=1$ and $B_{(0,0,0)}=-1/v^2$. Substituting these coefficients into equation (4.36), we obtain

$$\left(\sum_i\left(\frac{\partial\psi}{\partial x_i}\right)^2-\frac{1}{v^2}\right)A_1(x)+\left(\sum_i\frac{\partial^2\psi}{\partial x_i^2}+\sum_i 2\frac{\partial\psi}{\partial x_i}\frac{\partial}{\partial x_i}\right)A_0(x)=0, \tag{4.78}$$

which is equation (4.38), as required.

Exercise 4.9. Using the symbols of the differential operators discussed in Appendix E.3, rewrite equation (4.36).

Solution. The symbol of the differential operator D is

$$\sigma(D)(p) = \sum_{k \leqslant m} \sigma(D_k)(p),$$

where

$$\sigma(D_k)(p) = \sum_{|\alpha|=k} B_\alpha(x) p^\alpha.$$

Letting $p = \mathrm{d}\psi$, which we can write in components as

$$(p_1, p_2, \ldots, p_n) = \left(\frac{\partial \psi}{\partial x_1}, \frac{\partial \psi}{\partial x_2}, \ldots, \frac{\partial \psi}{\partial x_n}\right),$$

we write

$$\sigma(D)(\mathrm{d}\psi) = \sum_{k \leqslant m} \sum_{|\alpha|=k} B_\alpha(x) \left(\frac{\partial \psi}{\partial x}\right)^{|\alpha|}.$$

Similarly,

$$\sum_{k \leqslant m} \sigma(D_{k-1})(\mathrm{d}\psi) = \sum_{k \leqslant m} \sum_{|\alpha|=k-1} B_\alpha(x) \left(\frac{\partial \psi}{\partial x}\right)^{|\alpha|}.$$

Differentiation of the symbol with respect to its argument results in

$$\sum_{k \leqslant m} \left.\frac{\partial \sigma(D_k)(p)}{\partial p_i}\right|_{p=\mathrm{d}\psi} = \sum_{k \leqslant m} \sum_{|\alpha|=m} B_\alpha \alpha_i \left(\frac{\partial \psi}{\partial x}\right)^{\alpha-(0,\ldots,0,1,0,\ldots,0)},$$

where 1 in $(0, \ldots, 0, 1, 0, \ldots, 0)$ appears in the ith location. Similarly, considering the second derivatives of the symbol, we obtain

$$\sum_{k \leqslant m} \left.\frac{\partial^2 \sigma(D_k)(p)}{\partial p_i \partial p_j}\right|_{p=\mathrm{d}\psi}$$
$$= \sum_{k \leqslant m} \sum_{|\alpha|=k} B_\alpha (\alpha_i \alpha_j (1 - \delta_{ij}) + \alpha_i (\alpha_j - 1) \delta_{ij})$$
$$\frac{\partial^2 \psi}{\partial x_i \partial x_j} \left(\frac{\partial \psi}{\partial x}\right)^{\alpha-(0,\ldots,0,1,0,\ldots,0,1,0,\ldots,0)},$$

where 1 in multiindex $(0, \ldots, 0, 1, 0, \ldots, 0, 1, 0, \ldots, 0)$ appears in the ith and jth locations. Summing these four results, we obtain the desired expression, namely,

$$\sigma(D)(\mathrm{d}\psi) A_1(x) + \left(\sum_{k \leqslant m} \sigma(D_{k-1})(\mathrm{d}\psi)\right.$$
$$\left. + \sum_{k \leqslant m} \frac{1}{2} \sum_{i,j=1}^{n} \frac{\partial^2 \sigma(D_k)(p)}{\partial p_i \partial p_j} \frac{\partial^2 \psi}{\partial x_i \partial x_j} + \sum_{k \leqslant m} \sum_{i=1}^{n} \frac{\partial \sigma(D_k)(p)}{\partial p_i} \frac{\partial}{\partial x_i}\right) A_0(x) = 0,$$

where $p = \mathrm{d}\psi$.

Exercise 4.10. Solve

$$\frac{\mathrm{d}\ln A_0\left(x\left(t\right)\right)}{\mathrm{d}t}=-\zeta\left(\sum_{k\leqslant m}\sigma\left(D_{k-1}\right)(\,\mathrm{d}\psi)+\sum_{k\leqslant m}\frac{1}{2}\sum_{i,j=1}^{n}\frac{\partial^2\sigma\left(D_k\right)(p)}{\partial p_i\partial p_j}\frac{\partial^2\psi}{\partial x_i\partial x_j}\right)$$

for A_0.

Solution. Integrating both sides with respect to t between 0 and t, we get

$$\ln A_0\left(x\left(t\right)\right)-\ln A_0\left(x\left(0\right)\right)=\ln\frac{A_0\left(x\left(t\right)\right)}{A_0\left(x\left(0\right)\right)}$$

$$=\int_0^t-\zeta\left(\sum_{k\leqslant m}\sigma\left(D_{k-1}\right)(\,\mathrm{d}\psi)+\sum_{k\leqslant m}\frac{1}{2}\sum_{i,j=1}^{n}\frac{\partial^2\sigma\left(D_k\right)(p)}{\partial p_i\partial p_j}\frac{\partial^2\psi}{\partial x_i\partial x_j}\right)\mathrm{d}s$$

$$=\int_0^t-\zeta\left(\sum_{k\leqslant m}\sigma\left(D_{k-1}\right)(\,\mathrm{d}\psi)+\sum_{k\leqslant m}\frac{1}{2}\sum_{i,j=1}^{n}\frac{\partial^2\sigma\left(D_k\right)(p)}{\partial p_i\partial p_j}\frac{\partial^2\psi}{\partial x_i\partial x_j}\right)\mathrm{d}s\,.$$

Exponentiating and multiplying by $A_0\left(x\left(0\right)\right)$ we obtain

$$A_0\left(x\left(t\right)\right)=A_0\left(x\left(0\right)\right)$$

$$\exp\left\{\int_0^t-\zeta\left(\sum_{k\leqslant m}\sigma\left(D_{k-1}\right)(\,\mathrm{d}\psi)+\sum_{k\leqslant m}\frac{1}{2}\sum_{i,j=1}^{n}\frac{\partial^2\sigma\left(D_k\right)(p)}{\partial p_i\partial p_j}\frac{\partial^2\psi}{\partial x_i\partial x_j}\right)\mathrm{d}s\right\},$$

as required.

Exercise 4.11. Applying Definition 4.4 to expression (4.20), we obtain

$$\lim_{\omega\to\infty}\left(\frac{(\iota\omega)^N}{\exp\left\{\iota\omega\psi\left(x\right)\right\}}\left(\hat{u}\left(x,\omega\right)-\exp\left\{\iota\omega\psi\left(x\right)\right\}\sum_{n=0}^{N}\frac{A_n\left(x\right)}{\left(\iota\omega\right)^n}\right)\right)=0,\tag{4.79}$$

which allows us to determine uniquely all A_n. Determine A_0, A_1 and A_2.

Solution. For $N=0$ also $n=0$, and we write expression (4.79) as

$$\lim_{\omega\to\infty}\left(\frac{1}{\exp\left\{\iota\omega\psi\left(x\right)\right\}}\left(\hat{u}\left(x,\omega\right)-\exp\left\{\iota\omega\psi\left(x\right)\right\}A_0\left(x\right)\right)\right)=0,$$

which means that

$$A_0\left(x\right)=\lim_{\omega\to\infty}\frac{\hat{u}\left(x,\omega\right)}{\exp\left\{\iota\omega\psi\left(x\right)\right\}}.\tag{4.80}$$

For $N=1$, $n=0,1$, and we write expression (4.79) as

$$\lim_{\omega\to\infty}\left(\frac{\iota\omega}{\exp\left\{\iota\omega\psi\left(x\right)\right\}}\left(\hat{u}\left(x,\omega\right)-\exp\left\{\iota\omega\psi\left(x\right)\right\}\left(A_0\left(x\right)+\frac{A_1\left(x\right)}{\iota\omega}\right)\right)\right)=0.$$

We can rewrite it as

$$\lim_{\omega\to\infty}\left(i\omega\left(\frac{\hat{u}(x,\omega)}{\exp\{\iota\omega\psi(x)\}}-\left(A_0(x)+\frac{A_1(x)}{\iota\omega}\right)\right)\right)=0$$

to get

$$\lim_{\omega\to\infty}\left(\iota\omega\left(\frac{\hat{u}(x,\omega)}{\exp\{\iota\omega\psi(x)\}}-A_0(x)\right)-A_1(x)\right)=0,$$

which means that

$$A_1(x)=\lim_{\omega\to\infty}\left(\iota\omega\left(\frac{\hat{u}(x,\omega)}{\exp\{\iota\omega\psi(x)\}}-A_0(x)\right)\right), \tag{4.81}$$

where A_0 is known from equation (4.80). For $N=2$, $n=1,2,3$, and we write expression (4.79) as

$$\lim_{\omega\to\infty}\left(\frac{(\iota\omega)^2}{e^{\iota\omega\psi(x)}}\left(\hat{u}(x,\omega)-e^{\iota\omega\psi(x)}\left(A_0(x)+\frac{A_1(x)}{\iota\omega}+\frac{A_2(x)}{(\iota\omega)^2}\right)\right)\right)=0.$$

We can rewrite it as

$$\lim_{\omega\to\infty}\left((\iota\omega)^2\left(\frac{\hat{u}(x,\omega)}{\exp\{\iota\omega\psi(x)\}}-A_0(x)\right)-\iota\omega A_1(x)-A_2(x)\right),$$

which means that

$$A_2(x)=\lim_{\omega\to\infty}\left((\iota\omega)^2\left(\frac{\hat{u}(x,\omega)}{\exp\{\iota\omega\psi(x)\}}-A_0(x)\right)-\iota\omega A_1(x)\right), \tag{4.82}$$

where A_0 and A_1 are known from equations (4.80) and (4.81), respectively. Continuing this process we could obtain all A_n uniquely.

Exercise 4.12. Verify that if

$$\phi_n=\exp\{\iota\omega\psi(x)\}\frac{1}{(\iota\omega)^n},$$

then

$$\lim_{\omega\to\infty}\frac{\phi_{n+1}(x,\omega)}{\phi_n(x,\omega)}=0$$

and $\partial^m\phi_n/\partial x^m$ and $(\iota\omega)^{-k}\phi_n$ are a finite linear combination of ϕ_n, where the coefficients can depend on x.

Solution. The first condition is

$$\lim_{\omega\to\infty}\frac{\phi_{n+1}(x,\omega)}{\phi_n(x,\omega)}=\lim_{\omega\to\infty}\frac{(\iota\omega)^n}{(\iota\omega)^{n+1}}=\lim_{\omega\to\infty}\frac{1}{\iota\omega}=0,$$

as required. The expression $(\iota\omega)^{-k}\phi_n$ is

$$(\iota\omega)^{-k}\exp\{\iota\omega\psi(x)\}\frac{1}{(\iota\omega)^n}=\exp\{\iota\omega\psi(x)\}\frac{1}{(\iota\omega)^{n+k}}=\phi_{n+k},$$

which is a finite sum of terms $a_k(x)\phi_k$. The mth derivative of ϕ_n with respect to x is

$$\frac{\partial^m\psi(x)}{\partial x^m}\exp\left((\iota\omega)^l\psi(x)\right)\frac{1}{(\iota\omega)^{n-l}}=\frac{\partial^m\psi(x)}{\partial x^m}(\iota\omega)^l\phi_n=\frac{\partial^m\psi(x)}{\partial x^m}\phi_{n-l},$$

which is $a_{n-l}(x)\phi_{n-l}$ for $a_{n-l}(x)=\partial^m\psi(x)/\partial x^m$ and hence a finite sum of terms $a_k(x)\phi_k$.

Exercise 4.13. Consider the wave equation given by

$$\frac{\partial^2 u}{\partial x_1^2}+\frac{\partial^2 u}{\partial x_2^2}=\frac{1}{v^2}\frac{\partial^2 u}{\partial t^2},\tag{4.83}$$

with its characteristic equation given by

$$\left(\frac{\partial f}{\partial x_1}\right)^2+\left(\frac{\partial f}{\partial x_2}\right)^2=\frac{1}{v^2}\left(\frac{\partial f}{\partial t}\right)^2.\tag{4.84}$$

Show that the relation between these two equations is preserved if they are both transformed into polar coordinates. In other words, show that characteristics are invariant under coordinate transformations.

Solution. To express the wave equation and its characteristic equation in polar coordinates given by

$$x_1=r\cos\varphi\tag{4.85}$$

$$x_2=r\sin\varphi,\tag{4.86}$$

or, equivalently, by

$$r=\sqrt{x_1^2+x_2^2}\tag{4.87}$$

$$\varphi=\arctan\frac{x_2}{x_1},\tag{4.88}$$

we express differential operators in terms of r and φ. Thus, we write

$$\frac{\partial}{\partial x_1}=\frac{\partial r}{\partial x_1}\frac{\partial}{\partial r}+\frac{\partial\varphi}{\partial x_1}\frac{\partial}{\partial\varphi}$$

and

$$\frac{\partial}{\partial x_2}=\frac{\partial r}{\partial x_2}\frac{\partial}{\partial r}+\frac{\partial\varphi}{\partial x_2}\frac{\partial}{\partial\varphi}.$$

Considering $\partial/\partial x_1$ and using expressions (4.87) and (4.88), we get

$$\frac{\partial}{\partial x_1} = \frac{\partial\sqrt{x_1^2+x_2^2}}{\partial x_1}\frac{\partial}{\partial r} + \frac{\partial \arctan\dfrac{x_2}{x_1}}{\partial x_1}\frac{\partial}{\partial \varphi}$$
$$= \frac{x_1}{\sqrt{x_1^2+x_2^2}}\frac{\partial}{\partial r} - \frac{x_2}{x_1^2+x_2^2}\frac{\partial}{\partial \varphi} = \frac{x_1}{r}\frac{\partial}{\partial r} - \frac{x_2}{r^2}\frac{\partial}{\partial \varphi}.$$

Using expressions (4.85) and (4.86), we obtain

$$\frac{\partial}{\partial x_1} = \cos\varphi\frac{\partial}{\partial r} - \frac{\sin\varphi}{r}\frac{\partial}{\partial \varphi}. \tag{4.89}$$

Following a similar procedure for $\partial/\partial x_2$, we obtain

$$\frac{\partial}{\partial x_2} = \sin\varphi\frac{\partial}{\partial r} + \frac{\cos\varphi}{r}\frac{\partial}{\partial \varphi}. \tag{4.90}$$

Repeating similar procedures for the second derivatives, $\partial^2/\partial x_1^2$ and $\partial^2/\partial x_2^2$, we obtain the Laplace operator, $\nabla^2 := \left[\partial^2/\partial x_1^2 + \partial^2/\partial x_2^2\right]$, in polar coordinates, namely,

$$\nabla^2 := \frac{\partial^2}{\partial r^2} + \frac{2}{r}\frac{\partial}{\partial r} + \frac{1}{r^2}\left(\frac{\partial^2}{\partial \varphi^2} + \cot\varphi\frac{\partial}{\partial \varphi}\right).$$

Thus, we can write wave equation (4.83) in polar coordinates as

$$\nabla^2 u = \frac{\partial^2 u}{\partial r^2} + \frac{2}{r}\frac{\partial u}{\partial r} + \frac{1}{r^2}\left(\frac{\partial^2 u}{\partial \varphi^2} + \cot\varphi\frac{\partial u}{\partial \varphi}\right) = \frac{1}{v^2}\frac{\partial^2 u}{\partial t^2}.$$

Hence, we write the corresponding characteristic equation, which involves only the highest-order derivatives, as

$$\left(\frac{\partial f}{\partial r}\right)^2 + \frac{1}{r^2}\left(\frac{\partial f}{\partial \varphi}\right)^2 = \frac{1}{v^2}\left(\frac{\partial f}{\partial t}\right)^2. \tag{4.91}$$

Also, using expressions (4.89) and (4.90), we can write equation (4.84) in polar coordinates as

$$\left(\cos\varphi\frac{\partial f}{\partial r} - \frac{\sin\varphi}{r}\frac{\partial f}{\partial \varphi}\right)^2 + \left(\sin\varphi\frac{\partial f}{\partial r} + \frac{\cos\varphi}{r}\frac{\partial f}{\partial \varphi}\right)^2 = \frac{1}{v^2}\left(\frac{\partial f}{\partial t}\right)^2.$$

Squaring and simplifying, we obtain

$$\left(\frac{\partial f}{\partial r}\right)^2 + \frac{1}{r^2}\left(\frac{\partial f}{\partial \varphi}\right)^2 = \frac{1}{v^2}\left(\frac{\partial f}{\partial t}\right)^2,$$

which is equation (4.91), as expected.

Exercise 4.14. Show that the Fourier transform of asymptotic series (4.24), namely

$$e^{\iota\omega\psi(x)}\sum_{j=0}^{\infty}A_j(x)\frac{1}{(\iota\omega)^j},$$

is

$$A_0(x)\sqrt{2\pi}\delta(\psi(x)-t)+\sum_{j=1}^{\infty}A_j(x)\sqrt{2\pi}V_j(\psi(x)-t),$$

where

$$V_{j+1}(\eta)=\begin{cases}0 & \eta<0\\ \dfrac{1}{j!}\eta^j & \eta\geqslant 0\end{cases},$$

for $j\in\{0,1,2,\ldots\}$.

Solution. Following Appendix D.3, the Fourier transform of the jth term of series (4.24) is

$$A_j(x)\frac{1}{\sqrt{2\pi}}\int e^{-\iota\omega t}e^{\iota\omega\psi(x)}\frac{1}{(\iota\omega)^j}\,\mathrm{d}\omega=A_j(x)\frac{1}{\sqrt{2\pi}}\int e^{\iota\omega(\psi(x)-t)}\frac{1}{(\iota\omega)^j}\,\mathrm{d}\omega.$$

Let us examine the integral in the above expression. For simplicity, we write it as

$$V_j(\eta):=\int e^{\iota\omega\eta}\frac{1}{(\iota\omega)^j}\,\mathrm{d}\omega.$$

The derivative of this expression is

$$\frac{\mathrm{d}}{\mathrm{d}\eta}V_j(\eta)=\int e^{\iota\omega\eta}\frac{1}{(\iota\omega)^{j-1}}\,\mathrm{d}\omega=V_{j-1}(\eta).$$

If we find $V_0(\eta)$, we are able to find all V_j. The expression for V_0 is

$$V_0(\eta)=\int e^{\iota\omega\eta}\,\mathrm{d}\omega,$$

which is, as shown in expression (E.7), $V_0(\eta)=2\pi\delta(\eta)$. Consequently,

$$V_1(\eta)=2\pi H(\eta)+C,$$

where C is the integration constant, and H is the Heavyside function. We can choose this constant to be such that $V_j(\eta)=0$ for $\eta<0$. Then,

$$V_1(\eta)=2\pi\begin{cases}0 & \eta<0\\ 1 & \eta\geqslant 0\end{cases}.$$

Similarly,

$$V_2(\eta) = 2\pi \begin{cases} 0 & \eta < 0 \\ \eta & \eta \geqslant 0 \end{cases}$$

and

$$V_{j+1} = 2\pi \begin{cases} 0 & \eta < 0 \\ \dfrac{1}{j!}\eta^j & \eta \geqslant 0 \end{cases},$$

as desired. We set the integration constants to zero to get $V_j(\eta) = 0$ for $\eta < 0$; otherwise we obtain additional terms in the series that include powers of η.

Chapter 5

Caustics

Intuitively, if an expression oscillates wildly, then we expect its average value to be relatively small in magnitude, since the positive and negative parts, or in the complex case the parts with a wide range of different arguments, will cancel out. [...] *Generalizations of this phenomenon include the so-called principle of stationary phase.*

Terence Tao (2008)

Preliminary remarks

In the previous chapter we describe amplitudes of waves propagating along rays. In this chapter, we look at points at which this description fails. Such points belong to caustics, which are intrinsic entities of ray theory, and can be viewed as its limitation. Even though, at these points, the amplitude of the displacement is infinite—a nonphysical result—caustics provide an insight into the physical phenomenon of focusing, which is associated with high amplitudes. The Greek term for caustics, $\kappa\alpha\upsilon\sigma\tau o\sigma$, means combustible; herein, it refers to burning of an object due to light focussed upon it.

We begin this chapter by relating the singularities of transport equations to caustics. Subsequently, we formulate caustics as the envelopes of characteristics in a manner similar to the one used in the context of the Monge cone. We conclude the chapter by discussing the phase shift of a wave that has either passed through or touched the caustic.

Readers might find it useful to study this chapter together with Appendix D, where we discuss the Fourier transform.

5.1 Singularities of transport equations

In Section 4.5.2, we obtained solution (4.59) of the transport equation. This solution is

$$A_0(x(t)) = A_0(x(0)) \exp\left\{-\int_{t_0}^{t} \zeta \sum_{k\leqslant m} \sigma(D_{k-1})(\mathrm{d}\psi)\,\mathrm{d}s\right\}\sqrt{\frac{J_0\zeta(x(t))}{J_t\zeta(x(0))}}, \tag{5.1}$$

where A_0 denotes the amplitude, $\zeta = |\partial\sigma(D)(p)/\partial p_i|$, and J denotes the Jacobian of the change from the Cartesian to the ray-centred coordinates described by Červený (2001). In this chapter we study the singularities of this solution, namely the points for which the Jacobian vanishes. Referring to the alternative expression for solution (5.1) given by formula (4.58), we see that the vanishing of the Jacobian is equivalent to the vanishing of the area of the raytube cross-section. Intuitively, we expect such vanishing to happen at points where neighbouring characteristics converge. Such a convergence is called a caustic.

5.2 Caustics as envelopes of characteristics

In this section, we show that caustics are the envelopes of the characteristics by constructing such an envelope. We follow the construction analogous to the one discussed in Section 3.2 on page 104, where we described the Monge cone as the envelope of a family of planes.

Consider a hypersurface that is transverse to the characteristic curves, and on which we define the side conditions. Let us parametrize it by $\xi_1, \xi_2, \ldots, \xi_{n-1}$. The characteristics passing through this surface are denoted by $\gamma(\xi, t)$, where t is a parameter along the characteristic. The envelope is formed by intersections of characteristics that are infinitesimally close to each other. In other words, for point x to be on the envelope, there must be characteristics, $\gamma(\xi, t)$, that pass through that point such that

$$x = \gamma(\xi, t) = \gamma(\xi + \mathrm{d}\xi, t),$$

where $\mathrm{d}\xi$ is an infinitesimally small vector in the $\xi_1\xi_2\ldots\xi_{n-1}$-space. This is a stationarity condition with respect to a direction on the hypersurface. This direction is given by a vector in the kernel of the derivative of map $\gamma : \xi_1, \xi_2, \ldots, \xi_{n-1}, t \to \mathbb{R}^n$. In other words, this vector corresponds to the zero eigenvalue of the derivative of the map γ. This map is the coordinate change from the ray-centred coordinates to the Cartesian ones. Thus, a point with coordinates (ξ, t) is on the envelope if the Jacobian matrix of γ is degenerate: the Jacobian of γ is zero. As shown in Section 5.1, such points correspond to caustics.

To illustrate caustics as ray envelopes, let us consider the following example.

Example 5.1. To find the envelope of rays in a two-dimensional anisotropic homogeneous medium with wavefront velocity $v(N)$, we consider an initial wavefront S parametrized by its arclength ξ. Following Huygens's principle, wavefront $S(\xi)$ propagates according to

$$\gamma(\xi, t) = S(\xi) + N(\xi)\, v(N(\xi))\, t, \tag{5.2}$$

where $N(\xi)$ is the unit normal vector to the wavefront S at point $S(\xi)$. For subsequent times t, we view curves $\gamma(\xi, t)$ as wavefronts at those times. These wavefronts are parametrized by ξ, but not with the arclength parametrization, which means that the length of the tangent vector $\partial\gamma/\partial\xi$ is not one, as can be seen from the wavefronts "shrinking" or "expanding". The points where this vector vanishes are called the singularities of the wavefront. At such points the wavefront folds into itself as illustrated in Exercise 5.4. These points form the envelope of rays and hence satisfy

$$\frac{\partial\gamma(\xi, t)}{\partial\xi} = 0,$$

which, using expression (5.2), we write as

$$\frac{\partial S}{\partial\xi} + \frac{\partial N}{\partial\xi} v(N(\xi))\, t + N(\xi)\frac{\partial v}{\partial N}\frac{\partial N}{\partial\xi} t = 0.$$

Since $v(N)$ is a zero-degree homogeneous function in N—it does not depend on the magnitude of N— it follows from the Euler homogeneous-function theorem[1] that $N(\xi)\,\partial v/\partial N = 0$. Hence, the above equation becomes

$$\frac{\partial S}{\partial\xi} + \frac{\partial N}{\partial\xi} v(N(\xi))\, t = 0.$$

In view of the arclength parametrization of S by ξ, this equation becomes

$$l - \kappa l v t = 0\,,$$

[1]Readers interested in this theorem might refer to Courant and Hilbert (1989, Volume 2, p. 11).

where l is the unit tangent to S given by $\partial S/\partial\xi$ and κ is the curvature: the rate of change of the normal vector to the curve. Herein, $\kappa = -\partial N/\partial\xi$. Thus, we conclude that the points on the envelope are given by

$$t = \frac{1}{\kappa v}.$$

In other words, the envelope is given by substituting this expression for t in equation (5.2), namely,

$$\epsilon(\xi) = S(\xi) + \frac{1}{\kappa}N(\xi). \tag{5.3}$$

The singular points of wavefronts at different times, illustrated in Figure 5.6 on page 205, belong to the envelope of rays emanating from the original wavefront.

We express the above results in the language of differential geometry. Since the curvature is the reciprocal of the radius of curvature, $\kappa = 1/\rho$, we could write expression (5.3) as $\epsilon(\xi) = S(\xi) + \rho(\xi)N(\xi)$, which is the locus of the centres of curvatures of the initial wavefront, and is called the evolute. In other words, the evolute is the locus of the wavefront-osculating circles with radius ρ. Furthermore, the envelope of the normals to a curve is called an involute. Thus, a wavefront is the involute of the caustic, the rays are the wavefront normals, and the envelope of these rays is the evolute of the wavefront, as shown in Exercise 5.5.

An example of constructing the envelopes in isotropic homogeneous media is given in Exercise 5.2. A particular example of such an envelope is discussed in Exercise 5.4.

5.3 Phase change on caustics

5.3.1 *Formulation*

As indicated in Section 4.5.2, the solution of the transport equation might have a singularity, which is a caustic. It is observed that the phase of a signal travelling along a characteristic changes as the signal touches a caustic. To describe this phenomenon, we begin with a description of spherical waves in homogeneous media, which allows us, with the help of Huygens's principle,[2] to find an expression for the solution of the wave equation. We

[2]Readers interested in Huygens's principle might refer to Baker and Copson (1939).

investigate the behaviour of this expression at caustics using the method of stationary phase.

To use Huygens's principle to investigate the behaviour of amplitudes, we need to use its version that Augustin-Jean Fresnel modified to agree with his experiments. The resulting principle can be expressed as

$$\hat{u}(x,\omega) = \frac{\iota\omega}{v\sqrt{2\pi}} \int_S \frac{e^{-\iota\omega\frac{\|x-\xi\|}{v}}}{\|x-\xi\|} A(\xi) \frac{1+\cos\theta}{2}\,\mathrm{d}\xi\,, \tag{5.4}$$

where A denotes the amplitude of the wave originating at point ξ, and θ is the angle between the normal to surface S at ξ—which is the direction of the phase-slowness vector, p, at that point—and vector $x-\xi$. The first of the two terms introduced by Fresnel is $\iota\omega$; it is needed to change the phase of the secondary wavefronts compared to the original wavefront. The second term, $(1+\cos\theta)/2$, ensures that the secondary wavefronts radiate only in the propagation direction of the original wave.

5.3.2 *Waves in isotropic homogeneous media*

Let us consider the wave equation in three spatial dimensions,

$$\frac{\partial^2 u(x,t)}{\partial x_1^2} + \frac{\partial^2 u(x,t)}{\partial x_2^2} + \frac{\partial^2 u(x,t)}{\partial x_3^2} - \frac{1}{v^2}\frac{\partial^2 u(x,t)}{\partial t^2} = 0\,. \tag{5.5}$$

Expressing this equation in spherical coordinates, namely,

$$x_1 = r\sin\phi\cos\vartheta,$$
$$x_2 = r\sin\phi\sin\vartheta,$$
$$x_3 = r\cos\vartheta,$$

we write the wave equation as

$$\frac{\partial^2 u}{\partial r^2} + \frac{2}{r}\frac{\partial u}{\partial r} + \frac{1}{r^2}\left(\frac{\partial^2 u}{\partial \phi^2} + \cot\phi\frac{\partial u}{\partial \phi} + \frac{1}{\sin^2\phi}\frac{\partial^2 u}{\partial \vartheta^2}\right) - \frac{1}{v^2}\frac{\partial^2 u}{\partial t^2} = 0.$$

Since v is a constant, the wave equation is spherically symmetric. If we consider side conditions that are symmetric as well, the solutions depend on neither ϕ nor ϑ, and the wave equation becomes

$$\frac{\partial^2 u(r,t)}{\partial r^2} + \frac{2}{r}\frac{\partial u(r,t)}{\partial r} - \frac{1}{v^2}\frac{\partial^2 u(r,t)}{\partial t^2} = 0\,. \tag{5.6}$$

We let $u(r,t) = w(r,t)/r$, which, after taking the derivatives, results in the following wave equation:

$$\frac{\partial^2 w}{\partial r^2} - \frac{1}{v^2}\frac{\partial^2 w}{\partial t^2} = 0\,.$$

Formally, in a manner analogous to the d'Alembert solution of the one dimensional wave equation (2.29),[3,4] we write

$$w(r,t) = F(r+vt) + G(r-vt) \, ,$$

where functions F and G are constrained by the side conditions. Since $w(r,t) = r u(r,t)$, we write the solution of equation (5.6) as

$$u(r,t) = \frac{1}{r}\left(F(r+vt) + G(r-vt)\right).$$

Let us express this solution as its Fourier transform with t and ω being the transformation variables. We write

$$\hat{u}(r,\omega) = \frac{1}{r}\frac{1}{\sqrt{2\pi}}\int\limits_{-\infty}^{\infty}\left(F(r+vt) + G(r-vt)\right)e^{-\iota\omega t}\,\mathrm{d}t.$$

Substituting $r \pm vt = s$, which implies that $\mathrm{d}t = \mp\mathrm{d}s/v$, we rewrite it as

$$\hat{u}(r,\omega) = \frac{1}{vr}\frac{1}{\sqrt{2\pi}}\int\limits_{-\infty}^{\infty}\left[F(s)\exp\left\{-\iota\omega\frac{s-r}{v}\right\} + G(s)\exp\left\{-\iota\omega\frac{r-s}{v}\,\mathrm{d}s\right\}\right].$$

We split the integral into the sum of two integrals and factor out the terms that do not depend on s to get

$$\hat{u}(r,\omega) = \frac{1}{vr}\frac{1}{\sqrt{2\pi}}\left(\exp\left\{\iota\omega\frac{r}{v}\right\}\int\limits_{-\infty}^{\infty}F(s)\exp\left\{-\iota\frac{\omega}{v}s\right\}\mathrm{d}s\right.$$
$$\left. + \exp\left\{-\iota\omega\frac{r}{v}\right\}\int\limits_{-\infty}^{\infty}G(s)\exp\left\{-\iota\left(-\frac{\omega}{v}\right)s\right\}\mathrm{d}s\right).$$

The two integrals are Fourier transforms. For the first one, the transformation changes variable s to ω/v, and for the second one, s to $-\omega/v$. Thus, we write

$$\hat{u}(r,\omega) = \frac{1}{vr}\left(\exp\left\{\iota\omega\frac{r}{v}\right\}\hat{F}\left(\frac{\omega}{v}\right) + \exp\left\{-\iota\omega\frac{r}{v}\right\}\hat{G}\left(-\frac{\omega}{v}\right)\right).$$

Let us consider the solution with $\hat{F} = 0$ and $\hat{G}$ constant equal to $A/\sqrt{2\pi}$, namely,

$$\hat{u}(r,\omega) = \frac{1}{vr}\frac{1}{\sqrt{2\pi}}\exp\left\{-\iota\omega\frac{r}{v}\right\}A.$$

[3]Readers interested in d'Alembert solution might refer to many books, including McOwen (2003, p. 74).

[4]*See also*: Slawinski (2015, Section 6.5.1).

We take the inverse Fourier transform to write

$$u(r,t) = \frac{A}{r}\delta\left(t - \frac{r}{v}\right),$$

which describes a spherical wavefront propagating from the point source at the origin. The amplitude, which is proportional to the square root of energy, decreases proportionally to the distance, r, as expected in three spatial dimensions.

If we consider a surface, S, from which the wave is propagating, we construct the corresponding solution following the Huygens-Fresnel principle by adding the point sources from each location, ξ, of S, namely equation (5.4),

$$\hat{u}(x,\omega) = \frac{\iota\omega}{v\sqrt{2\pi}}\int_S \frac{e^{-\iota\omega\frac{\|x-\xi\|}{v}}}{\|x-\xi\|} A(\xi)\,\frac{1+\cos\theta}{2}\,\mathrm{d}\xi\,, \tag{5.7}$$

where A stands for the amplitude of the wave originating at point ξ. To solve this integral, we introduce the method of stationary phase.

5.3.3 *Method of stationary phase*[5]

To evaluate integral (5.7) in the context of ray theory, we study its behaviour for large ω by introducing the method of stationary phase. Let us investigate

$$I = \int_S a(\xi)\,e^{\iota\omega\psi(\xi)}\,\mathrm{d}\xi. \tag{5.8}$$

We consider the directional derivative given by the vector field

$$X = \sum_{i=1}^{m}\frac{\partial\psi}{\partial\xi_i}\frac{\partial}{\partial\xi_i},$$

where m is the dimension of surface S. The action of X on the exponential term in expression (5.8) is

$$Xe^{\iota\omega\psi} = \sum_{i=1}^{m}\frac{\partial\psi}{\partial\xi_i}\frac{\partial}{\partial\xi_i}e^{\iota\omega\psi} = \iota\omega p^2 e^{\iota\omega\psi},$$

where $p = \mathrm{d}\psi$. If $p^2 \neq 0$, we solve for the exponential term to get

$$e^{\iota\omega\psi} = \frac{Xe^{\iota\omega\psi}}{\iota\omega p^2}.$$

[5] In this section, we follow Guillemin and Sternberg (1990, Chapter I).

Inserting this result into equation (5.8), we get

$$I = \frac{1}{\iota\omega} \int_S \frac{a(\xi)}{p^2} X e^{\iota\omega\psi} \, d\xi \, .$$

Using the product rule, we obtain

$$\begin{aligned} I &= \frac{1}{\iota\omega} \int_S X \left(\frac{a(\xi)}{p^2} e^{\iota\omega\psi} \right) - X \left(\frac{a(\xi)}{p^2} \right) e^{\iota\omega\psi} \, d\xi \\ &= \frac{1}{\iota\omega} \left(\int_S X \left(\frac{a(\xi)}{p^2} e^{\iota\omega\psi} \right) d\xi - \int_S X \left(\frac{a(\xi)}{p^2} \right) e^{\iota\omega\psi} \, d\xi \right) . \end{aligned}$$

Let us consider the first integral. If we change the order of integration of this multidimensional integral to first the integration over the integral lines of X and then the integration over the quotient of S and these integral lines; then the integration over the integral lines of X vanishes. This vanishing is due to the fundamental theorem of calculus: the integral of the derivative, $X(f)$, of a function f that vanishes at both ends of the integral lines of X is zero. Thus,

$$I = -\frac{1}{\iota\omega} \int_S X \left(\frac{a(\xi)}{p^2} \right) e^{\iota\omega\psi} \, d\xi \, .$$

Letting $b_1(\xi) := X\left(a(\xi)/p^2\right)$, we write

$$I = -\frac{1}{\iota\omega} \int_S b_1(\xi) \, e^{\iota\omega\psi} \, d\xi \, .$$

Repeating the process of making the substitution for the exponential term and integrating by parts n times, we obtain

$$I = \left(-\frac{1}{\iota\omega} \right)^n \int_S b_n(\xi) \, e^{\iota\omega\psi} \, d\xi \, .$$

If all b_i are compactly supported and continuous, all of the integrals are bounded.[6] Thus, we can say that

$$I = O\left(\frac{1}{(-\iota\omega)^n} \right), \tag{5.9}$$

which means that if $d\psi \neq 0$ then I vanishes as fast as any negative power of ω, as $\omega \to \infty$. Thus, for large ω, the only nonzero contribution to I can

[6] For the definition of compactly supported functions in $\mathbb{R}^n$ see page 267.

come from points where $d\psi = 0$. This property justifies the name of the discussed method for evaluation of such integrals for large ω as the method of stationary phase. In the next section, we evaluate I for the stationary values of $\psi(\xi)$.

To evaluate for stationary values of ψ, we invoke the Morse Lemma,[7] which deals with a change of coordinates in which a symmetric quadratic form is diagonal with eigenvalues ± 1.

Lemma 5.1. *If* $d\psi(\xi_0) = 0$ *and* $\partial^2\psi/\partial\xi_i\partial\xi_j$ *is nondegenerate, then there exist coordinates* $z_1, \ldots, z_m$ *in a neighbourhood of* ξ_0 *such that* $\psi = \psi(\xi_0) + \frac{1}{2}\left(z_1^2 + z_2^2 + \ldots + z_l^2 - z_{l+1}^2 - \ldots - z_m^2\right)$, *where* l *is the number of positive eigenvalues of* $\partial^2\psi/\partial\xi_i\partial\xi_j$ *and* m *is the number of variables* ξ.

If view of this lemma, we express integral (5.8) in terms of coordinates z_i as

$$I = \int_S a(z) \exp\left\{i\omega\left(\psi(\xi_0) + \frac{1}{2}\left(z_1^2 + z_2^2 + \ldots + z_l^2 - z_{l+1}^2 - \ldots - z_m^2\right)\right)\right\} \left|\det\left[\frac{\partial\xi}{\partial z}\right]\right| dz,$$

where $\det[\partial\xi/\partial z]$ is the Jacobian. Factoring out the term independent of z_i, we write

$$I = e^{i\omega\psi(\xi_0)} \int_S a(z) \exp\left\{\frac{i\omega}{2}\left(z_1^2 + z_2^2 + \ldots + z_l^2 - z_{l+1}^2 - \ldots - z_m^2\right)\right\} \left|\det\left[\frac{\partial\xi}{\partial z}\right]\right| dz.$$

We express $a(z)\det[\partial\xi/\partial z]$ as

$$b_0 + b_1(z) z_1 + b_2(z) z_2 + \ldots + b_l(z) z_l - b_{l+1}(z) z_{l+1} - \ldots - b_m(z) z_m,$$

for functions $b_i(z)$. Denoting $Q(z) := z_1^2 + z_2^2 + \ldots + z_l^2 - z_{l+1}^2 - \ldots - z_m^2$, we write the integral as

$$I = e^{i\omega\psi(y_0)}\left[b_0 \int_S e^{\frac{i\omega}{2}Q(z)} dz + \int_S \left(\sum_{i=1}^{l} b_i(z) z_i - \sum_{i=l+1}^{m} b_i(z) z_i\right) e^{\frac{i\omega}{2}Q(z)} dz\right],$$

[7] Readers interested in the proof might refer to Guillemin and Sternberg (1990, Appendix I).

which can be written as

$$I = e^{i\omega\psi(\xi_0)}\left[b_0\int\limits_S e^{\frac{\iota\omega}{2}Q(z)}\,\mathrm{d}z + \frac{1}{i\omega}\int\limits_S\sum_{i=1}^m b_i(z)\frac{\partial}{\partial z_i}e^{\frac{\iota\omega}{2}Q(z)}\,\mathrm{d}z\right].$$

Following the method used to arrive at equation (5.9), we obtain

$$I = e^{i\omega\psi(\xi_0)}b_0\int\limits_S e^{\frac{i\omega}{2}Q(z)}\,\mathrm{d}z + O\left(\frac{1}{(i\omega)^n}\right),$$

which, in view of the definition of asymptotic equivalence given in expression (4.12), means that

$$I \sim e^{i\omega\psi(\xi_0)}b_0\int\limits_S e^{\frac{i\omega}{2}Q(z)}\,\mathrm{d}z, \tag{5.10}$$

as $\omega \to \infty$. In other words, I is asymptotically equivalent to the right-hand side of expression (5.10).

To evaluate the integral on the right-hand side of expression (5.10), we consider the following lemma.

Lemma 5.2. *It is true that*

$$\int\limits_S e^{\frac{\iota\omega}{2}Q(z)}\,\mathrm{d}z = \left(\frac{2\pi}{\iota\omega}\right)^{\frac{m}{2}}e^{\iota\frac{\pi}{4}(2(l+1)-m)},$$

where $Q(z) := z_1^2 + z_2^2 + \ldots + z_l^2 - z_{l+1}^2 - \ldots - z_m^2$.

Proof. We evaluate the left-hand side integral by writing it as

$$\left(\int e^{\frac{\iota\omega}{2}z_1^2}\,\mathrm{d}z_1\right)\left(\int e^{\frac{\iota\omega}{2}z_2^2}\,\mathrm{d}z_2\right)\cdots\left(\int e^{\frac{\iota\omega}{2}z_l^2}\,\mathrm{d}z_l\right)\left(\int e^{-\frac{\iota\omega}{2}z_{l+1}^2}\,\mathrm{d}z_{l+1}\right)$$
$$\cdots\left(\int e^{-\frac{\iota\omega}{2}z_m^2}\,\mathrm{d}z_m\right).$$

Each of these integrals is

$$\int\limits_S e^{\pm\frac{\iota\omega}{2}z^2}\,\mathrm{d}z = \sqrt{\frac{2\pi}{\mp\iota\omega}} = \sqrt{\frac{2\pi}{\omega}}e^{\mp\iota\frac{\pi}{4}} = \sqrt{\frac{2\pi}{\iota\omega}}e^{\mp\iota\frac{\pi}{4}+\iota\frac{\pi}{4}},$$

where we use the result of Exercise 5.6 with $a \to \mp\iota\omega/2$; for example, by considering sequence $a_n = \mp\iota\omega/2 + 1/n$, which ensures that integral I in this exercise is well-defined since $\Re(a_n) > 0$. □

In view of this lemma and expression (5.10), we sum over all ξ such that $\mathrm{d}\psi(\xi) = 0$ to obtain

$$I = \int_S a(\xi)\, e^{\iota\omega\psi(\xi)}\, \mathrm{d}\xi \sim \left(\frac{2\pi}{\iota\omega}\right)^{\frac{m}{2}} \sum_{\xi \mid \mathrm{d}\psi = 0} \frac{e^{\iota\omega\psi(\xi)} a(\xi)}{\sqrt{\left|\det\left[\dfrac{\partial^2\psi}{\partial\xi_i \partial\xi_j}\right]\right|}} e^{\iota\frac{\pi}{2}(m-l)}, \tag{5.11}$$

as $\omega \to \infty$. We proceed to evaluate integral (5.7) using the above result.

5.3.4 *Phase change*

In this section we study the asymptotic behavior of integral (5.7) by using the method of stationary phase on

$$\frac{\hat{u}(x,\omega)}{\iota\omega} = \frac{1}{v\sqrt{2\pi}} \int_S \frac{e^{-\iota\omega\frac{\|x-\xi\|}{v}}}{\|x-\xi\|} A(\xi)\, \frac{1+\cos\theta}{2}\, \mathrm{d}\xi. \tag{5.12}$$

Comparing the right-hand sides of expressions (5.12) and (5.8), we see that

$$\psi(\xi) = -\frac{\|x-\xi\|}{v} \tag{5.13}$$

and

$$a(\xi) = \frac{A(\xi)}{v\sqrt{2\pi}\,\|x-\xi\|} \cdot \frac{1+\cos\theta}{2}.$$

Thus, using expression (5.11) on (5.12), we write expression (5.7) as

$$\begin{aligned}
\hat{u}(x,\omega) &= \frac{\iota\omega}{v\sqrt{2\pi}} \int_S \frac{e^{-\iota\omega\frac{\|x-\xi\|}{v}}}{\|x-\xi\|} A(\xi)\, \frac{1+\cos\theta}{2}\, \mathrm{d}\xi \\
&\sim \left(\frac{2\pi}{\iota\omega}\right)^{\frac{m}{2}-1} \sum_{\xi \mid \mathrm{d}\frac{\|x-\xi\|}{v} = 0} \frac{\sqrt{2\pi} A(\xi)\, e^{-\iota\omega\frac{\|x-\xi\|}{v}} e^{\iota\frac{\pi}{2}(m-l)} \dfrac{1+\cos\theta}{2}}{v\,\|x-\xi\| \sqrt{\left|\det\left[\dfrac{-1}{v}\dfrac{\partial^2 \|x-\xi\|}{\partial\xi_i \partial\xi_j}\right]\right|}},
\end{aligned} \tag{5.14}$$

as $\omega \to \infty$. For $m = 2$, which is the dimension of the initial wavefront, the right-hand side of this asymptotic equivalence contains only the zero power of $1/(\iota\omega)$; all other terms in asymptotic series (4.20) are zero. In other words, $A_i = 0$, except for $i = 0$. The term $(1+\cos\theta)/2$ for the stationary points $\mathrm{d}\dfrac{\|\mathrm{x}-\xi\|}{\mathrm{v}} = 0$ equals to one, since this corresponds to $\theta = 0$, the case of the normal direction to the surface S.

At points where the Hessian, $\det\left[\partial^2\psi/\partial\xi_i\partial\xi_j\right]$ with ψ given by expression (5.13), is zero, the solution is infinite; these points belong to a caustic.

We can compare this asymptotic description of $\hat{u}$ to the three-dimensional case of expression (4.61), which is the solution of transport equation (4.63) derived using the asymptotic series. In particular, we compare the behaviour of

$$\hat{u}(x,\omega) \sim e^{\iota\omega\psi(\xi)} \frac{\sqrt{2\pi}A(\xi)\, e^{\iota\frac{\pi}{4}(2-l)}}{v\,\|x-\xi\|\sqrt{\left|\det\left[\dfrac{-1}{v}\dfrac{\partial^2\|x-\xi\|}{\partial\xi_i\partial\xi_j}\right]\right|}} \tag{5.15}$$

to the behaviour of

$$\hat{u}(x,\omega) \sim e^{\iota\omega\psi(\xi)} A_0(t_0)\sqrt{\frac{S_0 v(x(t))}{S_t v(x(0))}}. \tag{5.16}$$

In the following example, we consider an isotropic and homogeneous medium.

Example 5.2. Consider an isotropic and homogeneous three-dimensional medium. The asymptotic-series solution that corresponds to the solution of the transport equation stated in expression (4.61) is given by expression (5.15). Herein, we compare solution (5.15) to solution (5.16) obtained by the method of stationary phase. To emphasize the usefulness of the stationary-phase solution, we consider scenarios resulting in caustics.

Consider two wavefronts, $S_1(\xi)$ and $S_2(\xi)$. $S_1(\xi)$ is given locally by a sphere whose centre is located in the propagation direction, as shown in Figure 5.1. $S_2(\xi)$ is given locally by a cylinder whose centre is located in the propagation direction, as shown in Figure 5.2.

These wavefronts are given in the Cartesian coordinates by

$$S_1(\xi): \quad \left((R-x)^2 + y^2 + z^2 = R^2\right),$$

$$S_2(\xi): \quad ((R-x)^2 + y^2 = R),$$

and are parametrized by $\xi_1 = y$ and $\xi_2 = z$.

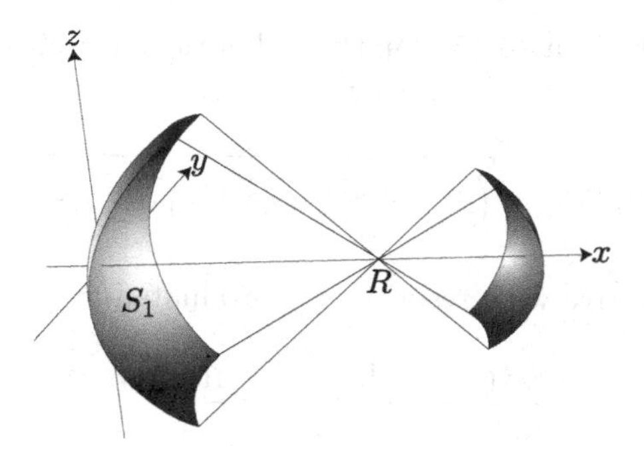

Fig. 5.1 Caustic generated by a spherical initial wavefront.

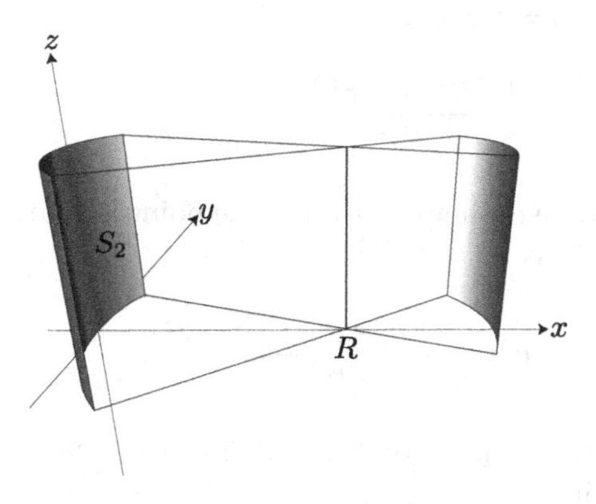

Fig. 5.2 Caustic generated by a cylindrical initial wavefront.

To evaluate expression (5.15) for the spherical wavefront at point $x = (x_0, 0, 0)$, where $x_0 > 0$, shown in Figure 5.1, we write

$$\|x - \xi\| = \sqrt{\left(R - \sqrt{R^2 - y^2 - z^2} - x_0\right)^2 + y^2 + z^2}\,.$$

The second derivative with respect to y or z, evaluated at $y = z = 0$ is

$$\left.\frac{\partial^2 \|x - \xi\|}{\partial y^2}\right|_{y=z=0} = \left.\frac{\partial^2 \|x - \xi\|}{\partial z^2}\right|_{y=z=0} = \frac{R - x_0}{R x_0}\,. \tag{5.17}$$

The mixed derivative evaluated at the same point is zero. The determinant in expression (5.15) is

$$\det\left[\frac{-1}{v}\frac{\partial^2 \|x - \xi\|}{\partial \xi^2}\right] = \left(\frac{R - x_0}{vRx_0}\right)^2\,.$$

To evaluate expression (5.15) for the cylindrical wavefront at point $x = (x_0, 0, 0)$, shown in Figure 5.2, we write

$$\|x - \xi\| = \sqrt{\left(R - \sqrt{R^2 - y^2} - x_0\right)^2 + y^2 + z^2}\,.$$

The second derivative with respect to y, evaluated at $y = z = 0$ is

$$\left.\frac{\partial^2 \|x - \xi\|}{\partial y^2}\right|_{y=z=0} = \frac{R - x_0}{R x_0}\,,$$

which is the same as expression (5.17). The second derivative with respect to z, evaluated at $y = z = 0$ is

$$\left.\frac{\partial^2 \|x - \xi\|}{\partial z^2}\right|_{y=z=0} = \frac{1}{x_0}\,.$$

The mixed derivative evaluated at the same point is zero. The determinant in expression (5.15) is

$$\det\left[\frac{-1}{v}\frac{\partial^2 \|x - \xi\|}{\partial \xi^2}\right] = \frac{R - x_0}{v^2 R x_0^2}\,.$$

Using these results in expression (5.15) and considering the spherical initial wavefront, we write

$$\hat{u}(x, \omega) \sim \frac{\sqrt{2\pi} A(\xi)\, e^{\iota\frac{\pi}{4}(m-l)}}{v\,\|x - \xi\| \sqrt{\left|\det\left[\frac{-1}{v}\frac{\partial^2 \|x - \xi\|}{\partial \xi_i \partial \xi_j}\right]\right|}} = \frac{\sqrt{2\pi} R A(0, 0)\, e^{\iota\frac{\pi}{4}(m-l)}}{|R - x_0|}\,, \tag{5.18}$$

where $m = 2$ is the dimension of the initial wavefront, l is the number of positive eigenvalues of $\left[{}^{-1}/_{v}\partial^2\|x-\xi\|/\partial\xi_i\partial\xi_j\right]$, and $\sqrt{2\pi}c = A$. This expression describes the change of amplitude with the distance from the focal point, $R - x_0$. The amplitude increases as it approaches the focal point, where it tends to infinity. Beyond the focal point, the amplitude decreases proportionally to the inverse of distance from the focal point. Also, since l changes from 2 to 0, the exponential term changes its value at the focal point from 1 to -1, as illustrated in Figure 5.4. The total change of amplitude around the focal point is illustrated in Figure 5.3.

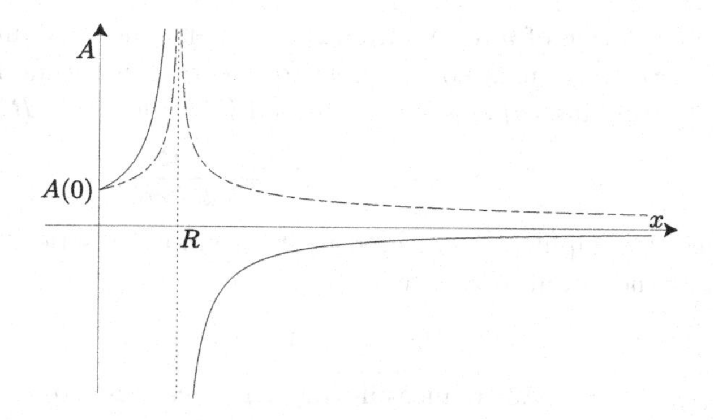

Fig. 5.3 Amplitude change around the focal point located at $x = R$ for a spherical initial wavefront, shown as a solid line, and for a cylindrical initial wavefront, shown as a dashed line.

Using the results in expression (5.15) and considering the cylindrical initial wavefront, we write

$$\hat{u}(x,\omega) \sim e^{\iota\omega\psi(\xi)} \frac{\sqrt{2\pi} A(\xi)\, e^{\iota\frac{\pi}{4}(2(l+1)-m)}}{v\,\|x-\xi\| \sqrt{\left|\det\left[\dfrac{-1}{v}\dfrac{\partial^2 \|x-\xi\|}{\partial\xi_i \partial\xi_j}\right]\right|}}$$

$$= e^{\iota\omega\psi(\xi)} \frac{\sqrt{2\pi} A(0,0) \sqrt{R} e^{\iota\frac{\pi}{2}l}}{\sqrt{|R-x_0|}}. \tag{5.19}$$

This expression describes the change of amplitude with the distance from the focal point, $R - x_0$. The amplitude increases as it approaches the focal line, where it tends to infinity. Beyond the focal point, the amplitude decreases proportionally to the inverse of the square-root of the distance from the focal point. Since l changes from 2 to 1, the exponent term changes its value at the focal point from 1 to ι. The total change of amplitude around the focal line is due only to the distance from the focal point and does not change sign, as illustrated in Figure 5.3; the phase changes by $\pi/2$.

For $x_0 < R$, expression (5.16) reduces in the spherical case to

$$\hat{u}(x,\omega) \sim e^{\iota\omega\psi} A_0(t_0) \sqrt{\frac{S_0 v(x(t))}{S_t v(x(0))}}$$

$$= e^{\iota\omega\psi} A_0(t_0) \sqrt{\frac{S_0}{S_t}} = e^{\iota\omega\psi} A_0(t_0) \frac{R}{R - r_0}, \tag{5.20}$$

where the cancellation of v results from the fact that the medium is homogeneous. For $x_0 > R$, it is not possible to find the amplitude from that expression. Comparing expressions (5.18) and (5.20) for $x_0 < R$, we write

$$\sqrt{2\pi}A(0,0)\frac{R}{R-x_0} = A_0(t_0)\frac{R}{R-r_0},$$

which relates the amplitude, A_0, used in the asymptotic series to the amplitude, A, at the initial wavefront by

$$A_0 = \sqrt{2\pi}A.$$

Similarly, expression (5.16) reduces in the cylindrical case to

$$\hat{u}(x,\omega) \sim e^{\iota\omega\psi}A_0(t_0)\sqrt{\frac{S_0 v(x(t))}{S_t v(x(0))}}$$

$$= e^{\iota\omega\psi}A_0(t_0)\sqrt{\frac{S_0}{S_t}} = e^{\iota\omega\psi}A_0(t_0)\sqrt{\frac{R}{R-r_0}}. \tag{5.21}$$

For $x_0 > R$, it is not possible to find the amplitude from that expression. Comparing expressions (5.19) and (5.21) for $x_0 < R$, we conclude again that $A_0 = \sqrt{2\pi}A$. The factor $\sqrt{2\pi}$ is a consequence of our choice of the asymptotic sequence stated in expression (4.20); there is no physical interpretation of this factor.

The value of the exponent is given by the number of positive and negative eigenvalues of matrix

$$\frac{-1}{v}\frac{\partial^2\|x-\xi\|}{\partial\xi_i\partial\xi_j}.$$

In particular, l is the number of positive eigenvalues. This value depends on the location of x.

If x is between the focal point, $r = 0$, and wavefront S, then the negative of the traveltime function, $-\|x-\xi\|/v$, has its maximum with respect to y at $r = R$. If x is beyond the focal point, then this function has its minimum with respect to y at $r = R$, as illustrated in Figure 5.4. Thus, the second derivatives of this function with respect to y change sign at the focal point from negative to positive as the distance from the initial wavefront increases.

For the spherical initial wavefront, the second derivative with respect to z changes sign the same way as the second derivative with respect to y. Thus, both eigenvalues change sign; the value of l changes from 2 to 0. Examining the exponential in expression (5.18), we see that the exponent

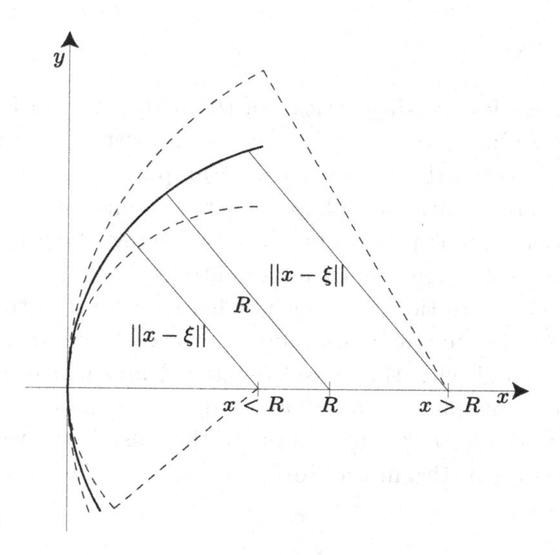

Fig. 5.4 Illustration of the derivatives of the traveltime function $\|x-\xi\|/v$, with respect to y, if one of ξ is equal to y. The curves correspond to the constant values of the function for different values of x. The solid curve represents the wavefront.

changes from 0 to $\iota\pi$. In other words, we have a polarity change—a π shift in phase.

For the cylindrical initial wavefront, the second derivative of the negative traveltime function with respect to z does not change sign; it remains negative. Thus, the value of l changes from 2 to 1, as the distance from the initial wavefront increases. Examining expression (5.19), we see that the exponent changes from 0 to $\iota\pi/2$; in other words, there is a $\pi/2$ shift in phase.

There are many different types of caustics, as illustrated in Exercise 5.4. However, for all types, the phase shifts are multiples of $\pi/2$, as illustrated in Example 5.2 and as can be seen from the exponential in expression (5.15), where l is the only variable that might change at a caustic. A propagating signal might traverse several caustics, and the progressive phase shift is accounted for by the so-called Maslov index.[8]

[8]Readers interested in caustics might refer to Arnold (1989, Appendix 11.B).

Closing remarks

Even though caustics are singularities of the mathematical formulation of ray theory, they illustrate physical phenomena observable in experimental studies.[9] To describe the propagation of amplitude across caustics, we invoke the method of stationary phase, which accounts for the phase change of waves as observed in experiments, thus confirming a physical meaning of the formulation. Although the actual amplitude is not infinite, in contrast to the ray-theory description, it is much higher near caustics than elsewhere. An attempt to compute true amplitudes is provided by the Airy functions, which are associated with the Bessel functions, and which are defined as linearly independent solutions of $u''(x)+xu(x)=0$; these functions do not tend to infinity at $x=0$.[10] An example of their use is the work of Harold Jeffreys (1939) on caustics in the Earth's core.

[9]Readers interested in physical meaning of caustics might refer to Arnold (1989, Appendix 16); Arnold (1991, 1992); Batterman (2002); Dahlen and Tromp (1998); Duistermaat (1996); Kravtsov and Orlov (1999); Landau and Lifschitz (2006); Nye (1999), and Porteous (1994).

[10]Readers interested in these functions might refer to Kravtsov and Orlov (1999).

5.4 Exercises

Exercise 5.1. Find the envelope of the family of straight lines in the xz-plane given by

$$a(t)\,x + b(t)\,z = c(t)\,. \tag{5.22}$$

Solution. Let us write a line within this family that neighbours line (5.22) as

$$a(t+\Delta t)\,x + b(t+\Delta t)\,z = c(t+\Delta t)\,. \tag{5.23}$$

We wish to write expression (5.23) in a manner that allows us for a convenient comparison with expression (5.22). In view of the Mean-value Theorem, there exists a real number, $\xi_a \in (t, t+\Delta t)$, such that

$$\left.\frac{\mathrm{d}a(t)}{\mathrm{d}t}\right|_{t=\xi_a} = \frac{a(t+\Delta t) - a(t)}{\Delta t}\,.$$

Solving for $a(t+\Delta t)$, we obtain

$$a(t+\Delta t) = a(t) + \left.\frac{\mathrm{d}a(t)}{\mathrm{d}t}\right|_{t=\xi_a}\Delta t\,.$$

Analogous expressions hold for b and c. Consequently, we can write expression (5.23) as

$$\left(a(t) + \left.\frac{\mathrm{d}a(t)}{\mathrm{d}t}\right|_{t=\xi_a}\Delta t\right)x + \left(b(t) + \left.\frac{\mathrm{d}b(t)}{\mathrm{d}t}\right|_{t=\xi_b}\Delta t\right)z$$
$$= c(t) + \left.\frac{\mathrm{d}c(t)}{\mathrm{d}t}\right|_{t=\xi_c}\Delta t.$$

Rearranging, we rewrite it as

$$(a(t)\,x + b(t)\,x - c(t)) + \left.\frac{\mathrm{d}a(t)}{\mathrm{d}t}\right|_{t=\xi_a}\Delta t x + \left.\frac{\mathrm{d}b(t)}{\mathrm{d}t}\right|_{t=\xi_b}\Delta t z = \left.\frac{\mathrm{d}c(t)}{\mathrm{d}t}\right|_{t=\xi_c}\Delta t\,.$$

In view of equation (5.22), the term in parentheses is zero. Also, since $\Delta t \neq 0$, we divide the remaining part by Δt to get

$$a'(\xi_a)\,x + b'(\xi_b)\,z = c'(\xi_c)\,,$$

where the prime stands for the derivative with respect to the argument. Since we are interested in the intersection of the neighbouring lines, we set $\Delta t \to 0$, which implies that $\xi_a \to t$, $\xi_b \to t$ and $\xi_c \to t$. Thus we write

$$a'(t)\,x + b'(t)\,z = c'(t)\,. \tag{5.24}$$

The condition for the two lines to intersect one another is that they both contain point (x, z). To formulate this condition, we solve a system consisting of equations (5.22) and (5.24), namely,

$$\begin{bmatrix} a(t) & b(t) \\ a'(t) & b'(t) \end{bmatrix} \begin{bmatrix} x \\ z \end{bmatrix} = \begin{bmatrix} c(t) \\ c'(t) \end{bmatrix}. \tag{5.25}$$

Following Cramer's rule, we write the solution as

$$x(t) = \frac{\det \begin{bmatrix} c(t) & b(t) \\ c'(t) & b'(t) \end{bmatrix}}{W}$$

and

$$z(t) = \frac{\det \begin{bmatrix} a(t) & c(t) \\ a'(t) & c'(t) \end{bmatrix}}{W},$$

where

$$W = \det \begin{bmatrix} a(t) & b(t) \\ a'(t) & b'(t) \end{bmatrix}.$$

Thus, the envelope of the family of lines given by expression (5.22) is solution $[x(t), z(t)]$. A particular example of such an envelope is discussed in Exercise 5.3 and shown in Figure 5.5.

Exercise 5.2. Use the results of Exercise 5.1 to show that we can view a caustic as the envelope of rays.

Solution. Let us consider wavefront $S(\xi) = [S_x(\xi), S_z(\xi)]$, where ξ is the arclength parameter. To write the equations for rays corresponding to this wavefront, we express them by

$$a(\xi)\, x + b(\xi)\, z = c(\xi).$$

The normals to the wavefront, $N(\xi)$, are tangent to the rays. Hence, vectors $[a(\xi), b(\xi)]$ and $[N_1(\xi), N_2(\xi)]$ are orthogonal to one another, namely,

$$[a(\xi), b(\xi)] = [-N_2(\xi), N_1(\xi)], \tag{5.26}$$

which allows us to write the equations for rays as

$$a(\xi)\, x + b(\xi)\, z \equiv \det \begin{bmatrix} x & z \\ -b(\xi) & a(\xi) \end{bmatrix} = -\det \begin{bmatrix} x & z \\ N_1(\xi) & N_2(\xi) \end{bmatrix} = c(\xi).$$

Each ray passes through the wavefront $S(\xi) = [S_x(\xi), S_z(\xi)]$, and hence the ray satisfies

$$-\det\begin{bmatrix} S_x(\xi) & S_z(\xi) \\ N_1(\xi) & N_2(\xi) \end{bmatrix} = c(\xi) .$$

Combining the last two expressions, we obtain the equation for rays:

$$a(\xi)x + b(\xi)z = -\det\begin{bmatrix} S_x(\xi) & S_z(\xi) \\ N_1(\xi) & N_2(\xi) \end{bmatrix} \equiv -\det[S(\xi), N(\xi)] . \tag{5.27}$$

To obtain the equation for the envelope of rays, following equation (5.25) and in view of expressions (5.26) and (5.27), we write

$$\begin{bmatrix} N_2(\xi) & -N_1(\xi) \\ \frac{\mathrm{d}}{\mathrm{d}\xi}N_2(\xi) & -\frac{\mathrm{d}}{\mathrm{d}\xi}N_1(\xi) \end{bmatrix}\begin{bmatrix} x \\ z \end{bmatrix} = \begin{bmatrix} \det[S(\xi), N(\xi)] \\ \frac{\mathrm{d}}{\mathrm{d}\xi}\det[S(\xi), N(\xi)] \end{bmatrix} . \tag{5.28}$$

To use Cramer's rule for solving system (5.28), we write the main determinant as

$$W = \det\begin{bmatrix} N_2(\xi) & -N_1(\xi) \\ N_2'(\xi) & -N_1'(\xi) \end{bmatrix} = \det\begin{bmatrix} N_1(\xi) & N_2(\xi) \\ N_1'(\xi) & N_2'(\xi) \end{bmatrix} \equiv \det[N(\xi), N'(\xi)] .$$

Invoking one of the Serret-Frenet equations,[11] namely,

$$N'(\xi) = -\kappa(\xi)T(\xi) , \tag{5.29}$$

where κ denotes curvature of the wavefront and T denotes its tangent, we rewrite this determinant as

$$W = \det[N(\xi), -\kappa(\xi)T(\xi)] = -\kappa(\xi)\det[N(\xi), T(\xi)] . \tag{5.30}$$

We can show that the determinant on the right-hand side is equal to one. We write

$$\det[N(\xi), T(\xi)] = N_1(\xi)T_2(\xi) - N_2(\xi)T_1(\xi) = [N_1, N_2]\cdot[T_2, -T_1] .$$

Recognizing that N and T are unit length vectors orthogonal to one another and that $[T_2, -T_1] = N$, we write

$$\det[N(\xi), T(\xi)] = N\cdot N = 1, \tag{5.31}$$

as required. Thus, returning to equation (5.30) and using result (5.31), we see that the main determinant is

$$W = -\kappa(\xi) .$$

[11]Readers interested in the Serret-Frenet equations might refer to a standard book on differential geometry, such as Spivak (1999, Volume 2, p. 34).

Continuing the search for the solution of system (5.28), we write

$$W_x = \det \begin{bmatrix} \det\left[S(\xi), N(\xi)\right], & -N_1(\xi) \\ \dfrac{\mathrm{d}}{\mathrm{d}\xi}\det\left[S(\xi), N(\xi)\right] & -N_1'(\xi) \end{bmatrix}.$$

First, let us consider

$$\frac{\mathrm{d}}{\mathrm{d}\xi}\det\left[S(\xi), N(\xi)\right] = \det\left[S'(\xi), N(\xi)\right] + \det\left[S(\xi), N'(\xi)\right].$$

Since $S'(\xi)$ is tangent to $S(\xi)$, we write

$$\frac{\mathrm{d}}{\mathrm{d}\xi}\det\left[S(\xi), N(\xi)\right] = \det\left[T(\xi), N(\xi)\right] + \det\left[S(\xi), N'(\xi)\right].$$

Using equation (5.31), we write

$$\frac{\mathrm{d}}{\mathrm{d}\xi}\det\left[S(\xi), N(\xi)\right] = -1 + \det\left[S(\xi), N'(\xi)\right]. \tag{5.32}$$

Consequently,

$$W_x = \det \begin{bmatrix} \det\left[S(\xi), N(\xi)\right] & -N_1(\xi) \\ -1 + \det\left[S(\xi), N'(\xi)\right] & -N_1'(\xi) \end{bmatrix}.$$

Proceeding to compute this determinant, we write

$$\begin{aligned} W_x &= -\det\left[S(\xi), N(\xi)\right] N_1'(\xi) + \left(-1 + \det\left[S(\xi), N'(\xi)\right]\right) N_1(\xi) \\ &= -\det \begin{bmatrix} S_x(\xi) & S_z(\xi) \\ N_1(\xi) & N_2(\xi) \end{bmatrix} N_1'(\xi) - N_1(\xi) + \det \begin{bmatrix} S_x(\xi) & S_z(\xi) \\ N_1'(\xi) & N_2'(\xi) \end{bmatrix} N_1(\xi). \end{aligned}$$

Performing algebraic manipulations and gathering terms with $S_x(\xi)$ and with $S_z(\xi)$, we get

$$\begin{aligned} W_x = &-S_x(\xi)\left(N_2(\xi) N_1'(\xi) - N_2'(\xi) N_1(\xi)\right) \\ &+ S_z(\xi)\left(N_1'(\xi) N_1(\xi) - N_1(\xi) N_1'(\xi)\right) - N_1(\xi). \end{aligned}$$

Since the term associated with $S_z(\xi)$ is equal to zero, we write

$$W_x = -N_1(\xi) - S_x(\xi)\left(N_1'(\xi) N_2(\xi) - N_1(\xi) N_2'(\xi)\right).$$

Let us write the term in parentheses as a scalar product of vectors, namely,

$$W_x = -N_1(\xi) - S_x(\xi)\left[N_1'(\xi), N_2'(\xi)\right] \cdot \left[N_2(\xi), -N_1(\xi)\right].$$

Since, in view of expression (5.31), the second vector in the product is $-T(\xi)$, we write

$$W_x = -N_1(\xi) + S_x(\xi)\, N'(\xi) \cdot T(\xi).$$

Invoking Serret-Frenet equation (5.29), we get

$$W_x = -N_1(\xi) + S_x(\xi)\,\kappa(\xi)\,T(\xi) \cdot T(\xi)\,.$$

Since T has unit length, we conclude that

$$W_x = -N_1(\xi) - \kappa(\xi)\,S_x(\xi)\,.$$

Similarly,

$$W_z = -N_2(\xi) - \kappa(\xi)\,S_z(\xi)\,.$$

Having obtained W, W_x and W_z, we write the solution of system (5.28) as

$$\begin{aligned}\epsilon(\xi) &= \left[\frac{W_x}{W}, \frac{W_z}{W}\right] \\ &= \left[\frac{-N_1(\xi) - \kappa(\xi)\,S_x(\xi)}{-\kappa(\xi)}, \frac{-N_2(\xi) - \kappa(\xi)\,S_z(\xi)}{-\kappa(\xi)}\right] \\ &= \left[S_x(\xi) + \frac{1}{\kappa(\xi)}N_1(\xi), S_z(\xi) + \frac{1}{\kappa(\xi)}N_2(\xi)\right].\end{aligned}$$

Using properties of vector addition, we write

$$\epsilon(\xi) = [S_x(\xi), S_z(\xi)] + \frac{1}{\kappa(\xi)}[N_1(\xi), N_2(\xi)]\,.$$

Recognizing that the first term is the wavefront, we write

$$\epsilon(\xi) = S(\xi) + \frac{1}{\kappa(\xi)}N(\xi)\,, \tag{5.33}$$

which is a special case of expression (5.3) for isotropic media with $v = 1$. This is an illustration of caustics being envelopes of rays.

Exercise 5.3. Consider a family of straight lines given by

$$\xi^2 x + \xi z = 1\,.$$

Find the equation of their envelope, and plot it.

Solution. Following equation (5.25) with $a = \xi^2$ and $b = \xi$, we write the equations of the envelope as

$$\begin{bmatrix}\xi^2 & \xi \\ 2\xi & 1\end{bmatrix}\begin{bmatrix}x \\ z\end{bmatrix} = \begin{bmatrix}1 \\ 0\end{bmatrix}.$$

The solution of this system is

$$[x(\xi), z(\xi)] = \left(-\frac{1}{\xi^2}, \frac{2}{\xi}\right),$$

which is a curve shown in Figure 5.5.

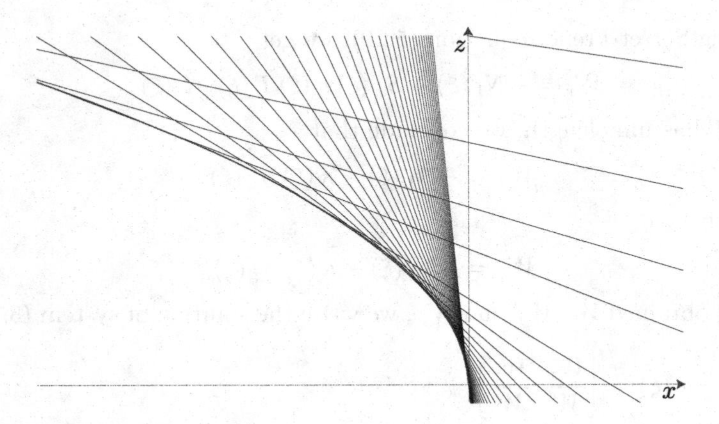

Fig. 5.5 Lines $z = -\xi x + 1/\xi$ and their envelope given by $\left[-1/\xi^2, 2/\xi\right]$

Exercise 5.4. Consider the parametric representation of a wavefront, $S(\xi) = \left[\xi, \xi^2\right]$, in an isotropic homogeneous medium. Find the expression of the caustic generated by this wavefront, and plot it.

Solution. According to Huygens's principle, each point on the wavefront is a source. In an isotropic homogeneous medium, the rays are normal to the wavefront and their length is directly proportional to traveltime. For simplicity, we set $v = 1$ in equation (5.3) to write

$$\epsilon(\xi) = S(\xi) + \frac{1}{\kappa(\xi)} N(\xi). \tag{5.34}$$

To find N, we use the expression for the unit normal to S, namely,

$$[N_1, N_2] = \left[\frac{-\left(S_z'\right)^2}{\sqrt{\left(S_x'\right)^2 + \left(S_z'\right)^2}}, \frac{\left(S_x'\right)^2}{\sqrt{\left(S_x'\right)^2 + \left(S_z'\right)^2}}\right],$$

and write the normal to the given wavefront as

$$N(\xi) = \left[\frac{-2\xi}{\sqrt{1+4\xi^2}}, \frac{1}{\sqrt{1+4\xi^2}}\right].$$

To find κ, we invoke the standard expression for the curvature to write

$$\kappa = \frac{S_x' S_z'' - S_x'' S_z'}{\left(\left(S'\right)^2 + \left(S_z'\right)^2\right)^{\frac{3}{2}}} = \frac{2}{\left(1+4\xi^2\right)^{\frac{3}{2}}}.$$

Thus, inserting expressions for N and κ into equation (5.34), we get

$$\epsilon(\xi) = [\xi, \xi^2] + \frac{(1+4\xi^2)^{\frac{3}{2}}}{2}\left[\frac{-2\xi}{\sqrt{1+4\xi^2}}, \frac{1}{\sqrt{1+4\xi^2}}\right]$$
$$= [\xi, \xi^2] + \left[-\xi(1+4\xi^2), \frac{1}{2}(1+4\xi^2)\right].$$

Adding the two vectors and simplifying, we obtain

$$\epsilon(\xi) = \left[-4\xi^3, 3\xi^2 + \frac{1}{2}\right]. \tag{5.35}$$

The plots of the original wavefront, S, and the caustic, ϵ, which is the locus of singularities of the subsequent wavefronts, are shown in Figure 5.6. Also, in view of the comment on page 184 and as shown in Exercise 5.5, the caustic is the evolute of the original wavefront.

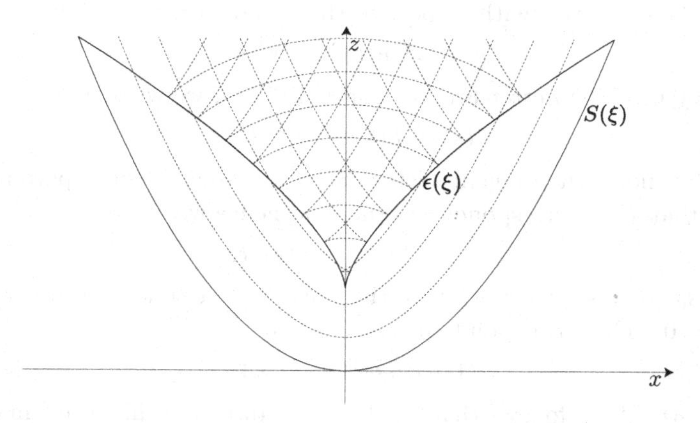

Fig. 5.6 Propagation of a parabolic wavefront, $S(\xi) = [\xi, \xi^2]$. The locus of singularities, $\epsilon(\xi)$, of the wavefronts is the caustic.

Remark 5.1. Curve (5.35), which can be also written as

$$z(x) = \frac{3}{4}\sqrt[3]{4x^2} + \frac{1}{2},$$

is Neile's semicubical parabola. It was discovered by an English mathematician William Neile in 1659. It was the first nontrivial algebraic curve whose arclength was calculated.[12]

[12]Readers interested in geometrical—including the global, local and spurious extrema—as well as algebraic properties of Neile's parabola might refer to Kot (2014, Section 9.4).

Exercise 5.5. Prove the following theorem.

Theorem 5.1. *In an isotropic homogeneous medium, the caustic curve, which is the locus of wavefront singularities, is the evolute of the original wavefront:*

$$F_t(s_0) = F(s_0) + \frac{1}{\kappa(s_0)} N(s_0), \tag{5.36}$$

where t is the traveltime, s is the arclength parameter whose value at the original wavefront is s_0, κ is the curvature and N is a vector normal to the wavefront.

Solution.

Proof. According to Huygens's principle, the subsequent wavefronts are

$$F_t(s) = F(s) + t\,N(s). \tag{5.37}$$

Taking the derivative with respect to the parameter, we obtain

$$F_t'(s) = F'(s) + tN'(s),$$

which, by invoking Serret-Frenet equation (5.29), we write as

$$F_t'(s) = T(s) + t\kappa(s)T(s),$$

where T denotes the tangent vector. Since s is the arclength parameter, it follows that $T(s)$ corresponds to time, and hence we write

$$F_t'(s) = (1 + t\kappa(s))\,t.$$

Let $F_t'(s_0)$ be a singular point on the wavefront at time t; in other words, $F_t'(s_0) = 0$. Thus, this point corresponds to

$$(1 + t\kappa(s_0))\,t(s_0) = 0.$$

Since $t(s_0) \neq 0$, it follows that $t = 1/\kappa(s_0)$. Inserting this result in expression (5.37) and considering the original wavefront, we write

$$F_t(s_0) = F(s_0) + \frac{1}{\kappa(s_0)} N(s_0),$$

which is expression (5.36), as required. The fact that the singular points of all wavefronts are the evolute of the original wavefront is illustrated in Figure 5.6. □

Exercise 5.6. Evaluate

$$I = \int_{-\infty}^{\infty} e^{-az^2}\,\mathrm{d}z,$$

where $\Re(a) > 0$.

Solution. Let us consider

$$M := \int_{-\infty}^{\infty}\int_{-\infty}^{\infty} e^{-a\left(x^2+y^2\right)}\,\mathrm{d}x\,\mathrm{d}y = \int_{-\infty}^{\infty}\int_{-\infty}^{\infty} e^{-ax^2}e^{-ay^2}\,\mathrm{d}x\,\mathrm{d}y.$$

Viewing x and y as integration variables, we write

$$M = \left(\int_{-\infty}^{\infty} e^{-ax^2}\,\mathrm{d}x\right)\left(\int_{-\infty}^{\infty} e^{-ay^2}\,\mathrm{d}y\right)$$

as $\left(\int_{-\infty}^{\infty} e^{-az^2}\,\mathrm{d}z\right)^2$, which we recognize to be I^2. Upon changing from the Cartesian to polar coordinates,

$$x = r\cos\theta\,, \qquad y = r\sin\theta\,,$$

where $r^2 = x^2 + y^2$, we write integral M as

$$M = \int_0^{2\pi}\int_0^{\infty} re^{-ar^2}\,\mathrm{d}r\,\mathrm{d}\theta = 2\pi\int_0^{\infty}\frac{e^{-u}}{2a}\,\mathrm{d}u = \frac{\pi}{a}\,,$$

where $u = ar^2$. Since $I = \sqrt{M}$, it follows that

$$I = \int_{-\infty}^{\infty} e^{-az^2}\,\mathrm{d}z = \sqrt{\frac{\pi}{a}}\,.$$

Afterword

On vous a sans doute souvant demandé à quoi servent les mathématiques [...] le langage ordinaire est trop pauvre, il est d'ailleur trop vague, pour exprimer des rapports delicats, si riches et si précis. Voilà donc une première raison pour laquelle le physicien ne peut se passer des mathématiques; elles lui fournissent la seule langue qu'il puisse parler.[1]

Henri Poincaré (1999)

A quantitative study of wave phenomena requires the study of the equations of motion, which are partial differential equations. These equations can be examined and solved by a variety of methods. We choose to employ the method of characteristics to examine these equations because for equations of motion, characteristics have physical meanings: they are wavefronts and rays.

Wavefronts and rays represent two different types of characteristics, those that pertain to linear equations and those that pertain to first-order equations, respectively, as shown in Figure 1. For linear equations, characteristics are surfaces along which we cannot construct a solution by a convergent Taylor series; in other words, we are not free to specify side conditions to obtain a unique solution. For first-order equations, charac-

[1] *We often ask what purpose mathematics serves. [...] the ordinary language is too limited; besides, it is too loose to express subtle, rich and precise relations. Hence, the primary reason for which physicists cannot work without mathematics: it is the only language in which they can communicate.*

Feynman (1967) states a similar view.

> You might say, '[...] Why not tell me in words instead of in symbols? Mathematics is just a language, and I want to be able to translate the language.' But I do not think it is possible, because mathematics is *not* just another language. [...] Mathematics is a tool for reasoning.

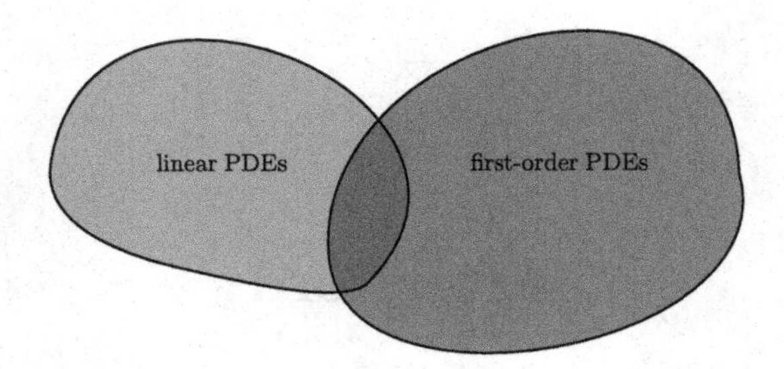

Fig. 1 Characteristics for linear partial differential equations and for first-order partial differential equations. For the former, we are not free to set the side conditions along characteristics; for the latter, we can propagate side conditions along the characteristics to construct solutions. Both properties coincide for first-order linear equations. Equations belonging to the intersection are discussed in Chapter 1; equations belonging to the complement, in Chapters 2 and 3.

teristics are trajectories along which we can propagate the side conditions to construct solutions. These side conditions must be differentiable along these trajectories but may be nondifferentiable in an oblique direction. In other words, we can propagate side conditions that contain discontinuities. In addition, along the characteristics of first-order equations, partial differential equations become ordinary differential equations.

The investigation of characteristics presented in this book allows us to clarify common views about ray theory. Wavefronts and rays are intrinsic entities of fundamental equations of mathematical physics, such as the Cauchy equations and the Maxwell equations. In general, wavefronts and rays are neither approximations nor are they introduced *a priori* as in the case of geometrical optics.

Ray theory is described commonly as a high-frequency approximation in view of its derivation using the asymptotic series, as discussed in Chapter 4. This statement has to be considered critically;[2] we obtain rays as characteristics of the eikonal equation, which is the characteristic equation of the equations of motion. Since the eikonal equation is the solution

[2]Readers interested in foundational issues might refer to Batterman (2002, Sections 6.5 and 7.4), where the author writes

> [...] understanding of various observed and observable universal behaviours "contained in" the finer theories (wave optics and quantum mechanics) requires reference to structures present only in the coarser theories [ray theory, classical mechanics] to which they are related by asymptotic limits [...]

of the characteristics of the equations of motion, the rays are referred to as bicharacteristics. Finding these bicharacteristics does not require any assumptions about high frequency or discontinuities. Also, the transport equations can be obtained without such approximations. However, to use characteristics to find the solution, we need to invoke the high-frequency assumption.

Characteristics allow us to describe the traveltime of wavefronts, their shapes, and the shape of rays. They do not suffice to obtain information about the amplitude or phase of the wave. These require the use of asymptotic series, which are high-frequency approximations. Thus, it is important to note that although rays and wavefronts are not approximations, a complete description that includes amplitude and phase requires approximations; hence the justification for the name, asymptotic ray theory.

In the context of geometrical optics, rays are postulated. As such, they are not viewed as an intrinsic entity of the study of wave phenomena. However, any method that examines these phenomena in the context of differential equations must acknowledge the intrinsic existence of rays, which are bicharacteristics of these equations.

A study of the method of characteristics illustrates the reciprocal relationship between physics and mathematics. Mathematics is the language capable of revealing the intrinsic physical meaning of wave phenomena, while also providing an effective tool for reasoning within this physical framework.

Appendix A

Integral theorems

The equations we shall study are really mathematical theorems. They will be useful not only for interpreting the meaning and the content of the divergence and the curl, but also in working out general physical theories. These mathematical theorems are, for the theory of fields, what the theorem of conservation of energy is to the mechanics of particles.

Feynman *et al.* (1964, Vol. 2)

Preliminary remarks

In this appendix, we state and justify two integral theorems that play an important role in mathematical physics: the Divergence Theorem and the Curl Theorem, which are three-dimensional and two-dimensional cases of Stokes's Theorem, named in honour of George Gabriel Stokes. Also, in honour of Carl Friedrich Gauss, it is common to refer to the Divergence Theorem as Gauss's Theorem.

Both the Divergence Theorem and the Curl Theorem relate the values of a function on the boundary of a geometrical object to the values of the derivatives of that function in the interior of that object; this relation is the essence of Stokes's Theorem.

Remark A.1. The general form of Stokes's Theorem—written in modern notation—is

$$\int_{\Omega} \mathrm{d}\omega = \int_{\partial\Omega} \omega \,, \tag{A.1}$$

where Ω is a piecewise smooth n-dimensional manifold with $\partial\Omega$ being its boundary, and ω is an $n-1$ compactly supported continuously differentiable differential form on Ω. Notably, the Fundamental Theorem of Calculus,

$$\int_a^b \frac{\mathrm{d}f}{\mathrm{d}x}\mathrm{d}x = f(b) - f(a) . \tag{A.2}$$

is a one-dimensional case of Stokes's Theorem; it relates the values of a function, $f(a)$ and $f(b)$, on the boundary of an interval, $[a,b]$, to the value of the derivative of that function within that interval.

Commenting on the nomenclature, Hutchinson (1962) writes

> Most names are soon forgotten. Some become impersonal units—watts, ohms, and amperes. A few achieve immortality with major generalizations such as gravitation or evolution. Others have become attached to statements of quantitative laws which may represent only a minute part of the achievement of their authors. It is this somewhat epigrammatic survival which has been the lot of Sir George Gabriel Stokes [...]

Stokes's name is attached to Theorem A.2, stated on page 223, below, and to its generalization stated by expression (A.1).

To appreciate the essence of expression (A.1) and Remark A.1, it is necessary to engage in the theory of differential forms, which is not discussed in this book.[1] However, the following comment of Garrity (2001, Chapter 5) can serve as an intuitive explanation, whose meaning should become clear upon studying this appendix.

> Stokes's Theorem, in all of its many manifestations, comes down to equating the average of a function on the boundary of some geometric object with the average of its derivative (in a suitable sense) on the interior of the object. Of course, a correct statement about averages must be put into the language of integrals.

A *suitable sense* refers to the fact that, in this appendix, the derivatives are expressed by divergence, curl or gradient, as required by the context.

We begin this appendix by stating and discussing the Divergence Theorem. Our discussion contains a plausibility argument of its validity, a couple of examples, and a corollary. Subsequently, we examine the Curl Theorem in a similar manner. We close this appendix with a comment on the Gradient Theorem.

[1]Readers interested in the theory of differential forms might refer to many books on that subject, including an insightful introduction of Garrity (2001, Chapter 6).

A.1 Divergence Theorem

A.1.1 *Statement*

Herein, we state and justify the Divergence Theorem, which relates a surface integral of a vector field to a volume integral of a derivative of this field. We use this theorem to derive differential equations for studying physical systems. In Appendix B, we formulate the elastodynamic equations, and in Appendix C, Coulomb's law and no-monopole law.

Theorem A.1. *Divergence theorem: The surface integral of a continuously differentiable vector field along a closed surface can be expressed as the integral of the divergence of this field over the volume enclosed by this surface; specifically,*

$$\iint_S F \cdot N \, \mathrm{d}S = \iiint_V \nabla \cdot F \, \mathrm{d}V \,, \tag{A.3}$$

where N is the unit outward normal vector on surface S.

Since the flux of a vector field across a surface is defined as the surface integral of the normal component of this field, we see that the left-hand side of equation (A.3) is the flux of field F across surface S. Hence, the theorem states that the flux of F across S can be expressed as the volume integral of the divergence of F.

A.1.2 *Plausibility argument*

A general proof of the Divergence Theorem is quite involved and might be formulated in the vocabulary of expression (A.1). Herein, we provide an argument that makes the general statement of the theorem plausible.[2]

Let us consider vector field $F = [F_1, F_2, F_3]$ and a rectangular box within it. For convenience, we set the edges of the box to be parallel with the three coordinate axes. As shown in Figure A.1, each edge extends between a_1 and a_2, along the x_1-axis, between b_1 and b_2, along the x_2-axis, and between c_1 and c_2, along the x_3-axis.

Let us proceed from the left-hand side of equation (A.3). We parametrize side S_1 in Figure A.1 by

$$S_1 : (x_2, x_3) \mapsto (a_2, x_2, x_3) \,,$$

[2]Readers interested in a more general proof might refer to Marsden and Tromba (1981, pp. 441–443).

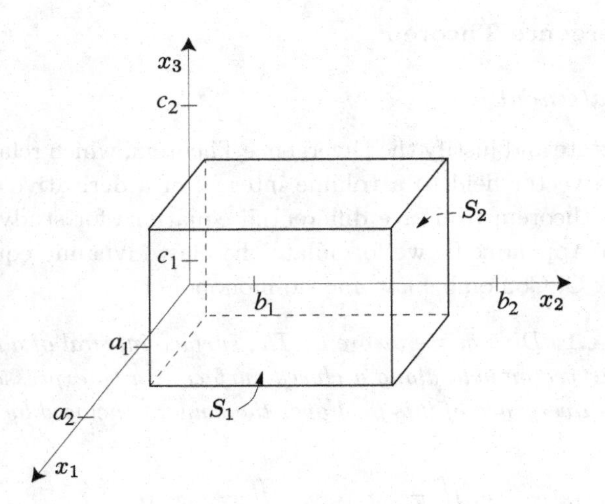

Fig. A.1 Rectangular box used to formulate Divergence Theorem.

with $x_2 \in [b_1, b_2]$ and $x_3 \in [c_1, c_2]$. The unit outward normal on this side is $N_1 = [1, 0, 0]$. Hence, the flux across surface S_1 is

$$\iint_{S_1} F \cdot N_1 \, \mathrm{d}S = \int_{c_1}^{c_2} \int_{b_1}^{b_2} [F_1, F_2, F_3] \cdot [1, 0, 0] \, \mathrm{d}x_2 \, \mathrm{d}x_3$$

$$= \int_{c_1}^{c_2} \int_{b_1}^{b_2} F_1 (a_2, x_2, x_3) \, \mathrm{d}x_2 \, \mathrm{d}x_3 .$$

We parametrize the side opposite to S_1 as

$$S_2 : (x_2, x_3) \mapsto (a_1, x_2, x_3) ,$$

where, again, $x_2 \in [b_1, b_2]$ and $x_3 \in [c_1, c_2]$, but the unit outward normal is $N_2 = [-1, 0, 0]$. In this case, the flux is

$$\iint_{S_2} F \cdot N_2 \, \mathrm{d}S = - \int_{c_1}^{c_2} \int_{b_1}^{b_2} F_1 (a_1, x_2, x_3) \, \mathrm{d}x_2 \, \mathrm{d}x_3 .$$

To consider the flux across these two faces, we sum the two integral expressions to get

$$\iint\limits_{S_1} F \cdot N_1 \, \mathrm{d}S + \iint\limits_{S_2} F \cdot N_2 \, \mathrm{d}S = \int\limits_{c_1}^{c_2} \int\limits_{b_1}^{b_2} F_1 \left(a_2, x_2, x_3\right) \mathrm{d}x_2 \, \mathrm{d}x_3 - \int\limits_{c_1}^{c_2} \int\limits_{b_1}^{b_2} F_1 \left(a_1, x_2, x_3\right) \mathrm{d}x_2 \, \mathrm{d}x_3$$

$$= \int\limits_{c_1}^{c_2} \int\limits_{b_1}^{b_2} \left[F_1 \left(a_2, x_2, x_3\right) - F_1 \left(a_1, x_2, x_3\right)\right] \mathrm{d}x_2 \, \mathrm{d}x_3 \, . \qquad \text{(A.4)}$$

In view of the Fundamental Theorem of Calculus, we can write the integrand of the right-hand side of equation (A.4) as

$$F_1 \left(a_2, x_2, x_3\right) - F_1 \left(a_1, x_2, x_3\right) = \int\limits_{a_1}^{a_2} \frac{\partial F_1 \left(x_1, x_2, x_3\right)}{\partial x_1} \, \mathrm{d}x_1 \, .$$

Inserting the right-hand side of this equation into the right-hand side of equation (A.4), we get

$$\int\limits_{c_1}^{c_2} \int\limits_{b_1}^{b_2} \int\limits_{a_1}^{a_2} \frac{\partial F_1 \left(x_1, x_2, x_3\right)}{\partial x_1} \, \mathrm{d}x_1 \, \mathrm{d}x_2 \, \mathrm{d}x_3 = \iiint\limits_{V} \frac{\partial F_1}{\partial x_1} \, \mathrm{d}V;$$

hence, equation (A.4) becomes

$$\iint\limits_{S_1} F \cdot N_1 \, \mathrm{d}S + \iint\limits_{S_2} F \cdot N_2 \, \mathrm{d}S = \iiint\limits_{V} \frac{\partial F_1}{\partial x_1} \, \mathrm{d}V \, . \qquad \text{(A.5)}$$

Considering the fluxes across the two faces perpendicular to the x_2-axis and the two faces perpendicular to the x_3-axis, we get expressions analogous to expression (A.5). Subsequently, adding the six surface and the three volume integrals, we write

$$\iint\limits_{S} F \cdot N \, \mathrm{d}S = \iiint\limits_{V} \left(\frac{\partial F_1}{\partial x_1} + \frac{\partial F_2}{\partial x_2} + \frac{\partial F_3}{\partial x_3}\right) \mathrm{d}V \, ,$$

where S denotes the surface of the box and V denotes its volume. Examining this equation, we recognize that the term in parentheses is the divergence of F. Thus, we write

$$\iint\limits_{S} F \cdot N \, \mathrm{d}S = \iiint\limits_{V} \nabla \cdot F \, \mathrm{d}V \, ,$$

which is equation (A.3), as required.

At this point, the Divergence Theorem is verified for a rectangular box. However, as argued below, we can extend this result to an arbitrary shape. The motivation for our approach is the fact that different-size boxes allow us to approximate a volume of arbitrary shape.

Let us consider a rectangular box whose portion of the bottom face coincides with a portion of the top face of the original box, as shown in Figure A.2. We assume that the second box is also entirely contained within field F. Since we could demonstrate the validity of equation (A.3) for the new box by following the argument identical to the one described above, let us focus our attention on the portion of the plane that is common to both boxes, and let us denote this portion by S_c. We can write the flux through S_c that is associated with the lower box as

$$\iint\limits_{S_c} [F_1, F_2, F_3] \cdot [0, 0, 1] \, \mathrm{d}S = \iint\limits_{S_c} F_3 \, \mathrm{d}S$$

and the flux through S_c that is associated with the upper box as

$$\iint\limits_{S_c} [F_1, F_2, F_3] \cdot [0, 0, -1] \, \mathrm{d}S = -\iint\limits_{S_c} F_3 \, \mathrm{d}S \,,$$

where the only difference consists of the opposite directions of the normal vectors. If we consider the flux associated with both boxes, we see, by examining the right-hand sides of the above expressions, that the effects across the coinciding portions cancel one another. In view of the cancelling of the coinciding portions, we can conclude that the flux through the outer surface is equal to the sum of the fluxes out of all interior pieces. Hence, if we build an arbitrary shape from many boxes, the result describes the flux across the outer surface.

Studying the Divergence Theorem in the context of mathematical physics, we learn about the physical properties of the entities involved. For instance, if surface S is contained in the domain of the divergence-free vector field, $\nabla \cdot F = 0$, then there is no flux across this surface. A physical example of such a vector field is given in expression (C.5). Furthermore, using vector-calculus identities, it is possible to show that $\nabla \cdot F = 0$ implies that $F = \nabla \times A$, where A is a vector field.

Example A.1. Let us examine the meaning of differentiability of a vector field for Theorem A.1. To do so, we consider

$$F = \left[\frac{x}{\sqrt{(x^2 + y^2 + z^2)^3}}, \frac{y}{\sqrt{(x^2 + y^2 + z^2)^3}}, \frac{z}{\sqrt{(x^2 + y^2 + z^2)^3}} \right], \tag{A.6}$$

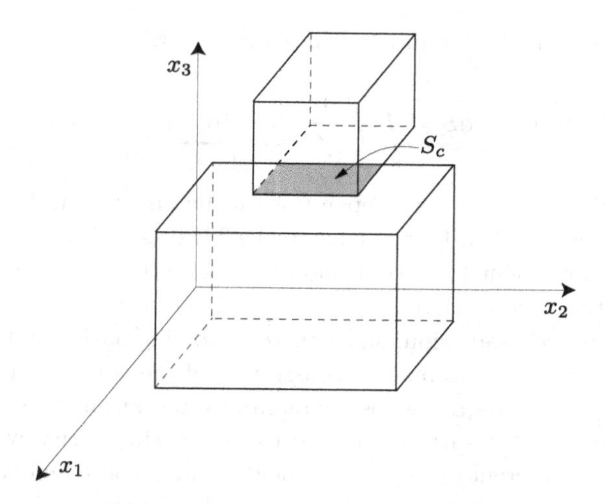

Fig. A.2 Two connected rectangular boxes used to formulate the Divergence Theorem.

which is a vector field not differentiable at the origin. Everywhere else, its divergence is zero,

$$\begin{aligned} \nabla \cdot F &= \frac{\left(-2x^2 + y^2 + z^2\right) + \left(x^2 - 2y^2 + z^2\right) + \left(x^2 + y^2 - 2z^2\right)}{\sqrt{\left(x^2 + y^2 + z^2\right)^5}} \\ &\equiv 0 \,. \end{aligned} \tag{A.7}$$

Hence, the right-hand side of equation (A.3) is zero, and—in accordance with the Divergence Theorem—a flux of this vector field through a surface that does not enclose the origin is zero. However, a flux through a surface that does enclose the origin is nonzero. To see it, we write F in spherical coordinates,

$$F = \left[\frac{r\sin\theta\cos\phi}{r^3}, \frac{r\sin\theta\sin\phi}{r^3}, \frac{r\cos\theta}{r^3}\right] = \left[\frac{\sin\theta\cos\phi}{r^2}, \frac{\sin\theta\sin\phi}{r^2}, \frac{\cos\theta}{r^2}\right] .$$

If we let S, in equation (A.3), be a sphere centred at the origin, its unit outward normal vector is

$$N = \left[\sin\theta\cos\phi, \sin\theta\sin\phi, \cos\theta\right] ,$$

and

$$F \cdot N = \frac{\sin^2\theta\cos^2\phi + \sin^2\theta\sin^2\phi + \cos^2\theta}{r^2} = \frac{1}{r^2} .$$

Thus, we write the double integral in expression (A.3) as

$$\iint_S F \cdot N \, \mathrm{d}S = \int_0^{2\pi}\int_0^{\pi} \frac{1}{r^2} \underbrace{r^2 \sin\theta \, \mathrm{d}\theta \, \mathrm{d}\phi}_{\mathrm{d}S} = 4\pi \neq 0\,.$$

The zero and nonzero fluxes—depending on the differentiability of a vector field, the former implied by the Divergence Theorem—illustrate the meaning of the requirement for a continuously differentiable vector field, in the statement of Theorem A.1.[3]

There is no contradiction, since there is no singularity of F, along the surface, S, which is the domain of integration of the double integral, in contrast to the triple integral, whose domain of integration is the volume, V, contained within this surface. F can exhibit a singularity within V but not along S. Differentiability of the field ensures the equality of the two integrals; it is not necessary for the integrability of the double integral.

Example A.2. Let us examine a possibility of applying the Divergence Theorem to a vector field with a singularity. To do so, we consider a vector field given by the gradient of $1/r$, where $1/r$ is a scalar field expressed in spherical coordinates. The vector field is $F = \nabla(1/r) = -\hat{r}/r^2$, where $\hat{r}$ is the radial unit vector, and where to obtain the gradient we use the fact that f is spherically symmetric.

We would like to use the Divergence Theorem to find the surface integral of the vector field over a unit sphere centred at the origin. To this end, we invoke the definition of $\nabla^2 := \nabla \cdot \nabla$, which in spherical coordinates for a spherically symmetric function is $\partial^2/\partial r^2 + (2/r)\partial/\partial r$.[4] We see that the integrand on the right-hand side of equation (A.3), namely, $\nabla \cdot F$, is zero, which means that the volume integral is zero. However, the left-hand side is

$$\int_0^{2\pi}\int_0^{\pi} F\, r^2 \sin\theta \, \mathrm{d}\theta \, \mathrm{d}\phi = -\int_0^{2\pi}\int_0^{\pi} \sin\theta \, \mathrm{d}\theta \, \mathrm{d}\phi = -4\pi\,.$$

The fact that the left- and right-hand sides are not equal to one another can be explained in view of the differentiability requirements in Theorem A.1, which are not fulfilled at the origin for the volume integral; therein lies the singularity illustrated in Figure A.3. Since we could isolate this singularity from the rest of the volume, for which the vector field is differentiable, we

[3] *See also*: Slawinski (2018, Exercise 2.3).
[4] *See also*: Slawinski (2018, expression (5.40)).

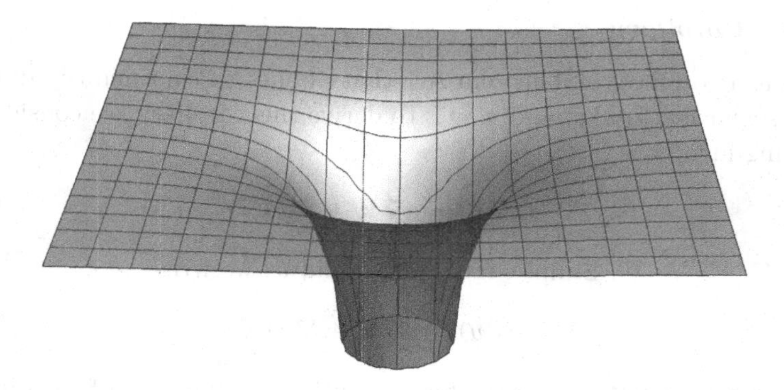

Fig. A.3 A coordinate-plane cross-section of the magnitude of $F = \nabla(1/r)$, in Example A.2, whose domain is $\mathbb{R}^3$, and where F tends to infinity at the origin of the coordinate system.

could argue that the value of the right-hand side would be also -4π, if we let this value to be the contribution of the singularity. We can formally write the integrand of $\iiint \nabla \cdot \nabla(1/r)\mathrm{d}V$ as $-4\pi\delta(r)$, where δ is the Dirac delta discussed in Appendix E, and in particular in Example E.4, where we present a distributional method to evaluate the volume integral whose vector field contains a singularity.

Invoking the rotational invariance of the problem in question, Arnold (2004, pp. 81-83) explains this result in the following manner.

> We compute the flux of the gradient field across two concentric spheres [...] The flux across a sphere equals the magnitude of the field multiplied by the surface area of the sphere. The area of a sphere of radius R is $\omega_{n-1}R^{n-1}$, where ω_{n-1} is the area of the unit sphere (4π for $n = 3$, 2π for $n = 2$). As a result, the flux is $(2 - n)\omega_{n-1}$. In particular, for $n = 3$ it is -4π.

Also, $\nabla^2(1/r) = -4\pi\delta(r)$ can be obtained using Corollary A.1, below, as exemplified by Boccara (1997, Chapter 3.2).

Remark A.2. Following the argument presented in Example A.2, we could conjecture that—for F given by expression (A.6), in Example A.1—the triple integral of equation (A.3), for any domain of integration that contains $x = y = z = 0$, is equal to 4π. If so, since—in accordance with expression (A.7)—this integral is zero for any domain that does not contain $x = y = z = 0$, its nonzero value is due solely to the contribution of the singularity.

A.1.3 *Corollary*

There is a corollary of Theorem A.1 that applies to continuously differentiable scalar fields, $u(x)$ and $v(x)$. To derive this corollary, we consider the following identity,

$$\nabla\cdot(u\nabla v)=u\nabla^2 v+\nabla u\cdot\nabla v\,, \tag{A.8}$$

where $\nabla^2:=\nabla\cdot\nabla$. Simply, renaming u and v, we write

$$\nabla\cdot(v\nabla u)=v\nabla^2 u+\nabla v\cdot\nabla u\,.$$

Subtracting the latter from the former—using the linearity of the differential operators and the commutativity of the scalar product—we obtain

$$\nabla\cdot(u\nabla v-v\nabla u)=u\nabla^2 v-v\nabla^2 u\,. \tag{A.9}$$

Integrating both sides over a volume, we write

$$\iiint\limits_V \nabla\cdot(u\nabla v-v\nabla u)\,\mathrm{d}V=\iiint\limits_V (u\nabla^2 v-v\nabla^2 u)\,\mathrm{d}V\,.$$

Applying Theorem A.1 to the integral on the left-hand side, we write the corollary of that theorem; this corollary is referred to as Green's Theorem[5] in recognition of George Green's contributions.

Corollary A.1. *Green's Theorem I: If $u(x)$ and $v(x)$ are continuously differentiable scalar fields, then*

$$\iint\limits_S (u\nabla v-v\nabla u)\cdot N\,\mathrm{d}S=\iiint\limits_V (u\nabla^2 v-v\nabla^2 u)\,\mathrm{d}V\,, \tag{A.10}$$

where N is the unit outward normal vector on surface S that encloses volume V.

Equation (A.10) can be verified directly by invoking Theorem A.1. Comparing the left-hand sides of equations (A.3) and (A.10), we let $F=u\nabla v-v\nabla u$. Taking the divergence of F, and using the linearity of the differential operator, we write

$$\nabla\cdot(u\nabla v-v\nabla u)=\nabla\cdot(u\nabla v)-\nabla\cdot(v\nabla u)\,. \tag{A.11}$$

[5] Another Green's Theorem is discussed in Appendix A.2.3; that theorem is more commonly associated with the name of George Green.

Considering the first term on the right-hand side, we obtain

$$\begin{aligned}\nabla\cdot(u\nabla v) &= \nabla\cdot\left(u\frac{\partial v}{\partial x_1},u\frac{\partial v}{\partial x_2},u\frac{\partial v}{\partial x_3}\right)\\ &= \frac{\partial}{\partial x_1}\left(u\frac{\partial v}{\partial x_1}\right)+\frac{\partial}{\partial x_2}\left(u\frac{\partial v}{\partial x_2}\right)+\frac{\partial}{\partial x_3}\left(u\frac{\partial v}{\partial x_3}\right)\\ &= \frac{\partial u}{\partial x_1}\frac{\partial v}{\partial x_1}+u\frac{\partial^2 v}{\partial x_1^2}+\frac{\partial u}{\partial x_2}\frac{\partial v}{\partial x_2}+u\frac{\partial^2 v}{\partial x_2^2}+\frac{\partial u}{\partial x_3}\frac{\partial v}{\partial x_3}+u\frac{\partial^2 v}{\partial x_3^2},\end{aligned}$$

which—gathering similar terms and recognizing the scalar product of the gradients and the definition of the Laplace operator—we write concisely as $\nabla u\cdot\nabla v+u\nabla^2 v$; this derivation is tantamount to demonstrating identity (A.8). Similarly, considering the second term on the right-hand side of equation (A.11), we obtain $\nabla v\cdot\nabla u+v\nabla^2 u$. Using the commutativity of the scalar product, we write the right-hand side of equation (A.11) as $\nabla\cdot F=u\nabla^2 v-v\nabla^2 u$, which—by comparison of the right-hand sides of equations (A.3) and (A.10)—is the required result.

A.2 Curl Theorem

A.2.1 *Statement*

Herein, we state and justify the Curl Theorem, which relates a line integral of a vector field to a surface integral of a derivative of this field. In Appendix C, we use this theorem to formulate Faraday's and Ampère's laws.

The theorem discussed in Appendix A.1 relates the volume integral of a derivative of a vector field to the surface integral of this field over the closed surface bounding this volume. Herein, we consider the theorem that relates the surface integral of a derivative of a vector field to the line integral of this field along the loop that bounds an area of this surface.

Theorem A.2. *Curl theorem: The line integral of a continuously differentiable vector field along a closed loop can be expressed as the surface integral of the curl of this field along a surface bounded by this loop; specifically,*

$$\int\limits_C F\cdot n\ \mathrm{d}s=\iint\limits_S\left[(\nabla\times F)\cdot N\right]\mathrm{d}S\,, \tag{A.12}$$

where $\mathrm{d}s$ *is the element of length along the curve,* n *is the unit vector tangent to the curve and* N *is the unit normal vector to* S *.*

Hence, the integral of the tangential component of F around boundary C is equal to the integral of the normal component of the curl of F on the surface enclosed by C.

Note that, in contrast to Theorem A.1, the surface is not closed. One could emphasize the requirement for an open or closed surface by writing $\iint_S$ or $\oiint_S$, respectively. Similarly, the requirement for a loop could be stated as $\oint_C$. In general, surfaces must be closed if they are boundaries of volumes and curves must be closed if they are boundaries of surfaces.

In Appendix A.1, we invoke the definition of the flux of a vector field as the surface integral of the normal component of this field. Herein, we invoke the definition of the circulation of a vector field as the integral around the loop of the component of the field that is tangent to this loop. Thus, Theorem A.2 states that the circulation of field F around loop C can be expressed as the surface integral of the component of the curl of this field that is normal to S.

Since the direction of the unit normal vector, N, changes the sign of the surface integral and the orientation of $\mathrm{d}s$ changes the sign of the curve integral, we must decide on the sign convention. We choose to orient the surface in such a way that vector N points in the direction of the thumb of the right hand with the fingers curled in the direction of $\mathrm{d}s$.

A.2.2 *Plausibility argument*

As in the case of the Divergence Theorem, a general proof of the Curl Theorem is quite involved. Herein, we provide an argument that makes the general statement of the theorem plausible.

Let us consider a rectangle within vector field $F = [F_1, F_2, F_3]$. For convenience, we position the rectangle in the x_1x_2-plane and set the edges of this rectangle to be parallel with two coordinate axes. As shown in Figure A.4, each edge extends between a_1 and a_2, along the x_1-axis, and b_1 and b_2, along the x_2-axis.

Let us proceed from the left-hand side of equation (A.12). We parametrize the far edge in Figure A.4 by

$$C_1 : x_2 \mapsto (a_1, x_2, 0) ,$$

with $x_2 \in [b_1, b_2]$. In all integrations, we consider the counterclockwise direction, which we could emphasize by writing $\circlearrowleft_C$, where the arrow indicates the direction. Hence, using the corresponding tangent vector, $[0, -1, 0]$, we see that the line integral of the component of the vector field that is tangent

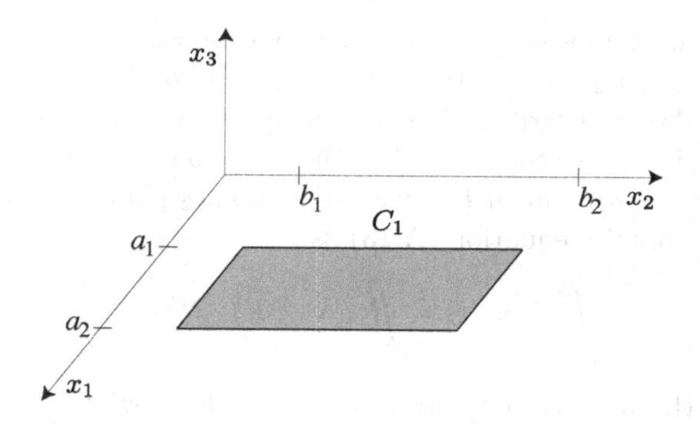

Fig. A.4 Rectangle used to formulate the Curl Theorem.

to C_1 is $\int_{b_2}^{b_1} F_2(a_1, x_2, 0)\, \mathrm{d}x_2$. Similarly, the line integral along the near edge is $\int_{b_1}^{b_2} F_2(a_2, x_2, 0)\, \mathrm{d}x_2$. Following the same approach for the other two edges of the rectangle and, then, summing all four segments, we get

$$\int_C F \cdot n \,\mathrm{d}s = \int_{b_2}^{b_1} F_2(a_1, x_2, 0)\, \mathrm{d}x_2 + \int_{b_1}^{b_2} F_2(a_2, x_2, 0)\, \mathrm{d}x_2 + \int_{a_1}^{a_2} F_1(x_1, b_1, 0)\, \mathrm{d}x_1 + \int_{a_2}^{a_1} F_1(x_1, b_2, 0)\, \mathrm{d}x_1 .$$

Changing the limits of integration, we can rewrite this equation as

$$\int_C F \cdot n \;\mathrm{d}s = \int_{b_1}^{b_2} \left[F_2(a_2, x_2, 0) - F_2(a_1, x_2, 0)\right] \mathrm{d}x_2 - \int_{a_1}^{a_2} \left[F_1(x_1, b_2, 0) - F_1(x_1, b_1, 0)\right] \mathrm{d}x_1 .$$

In view of the Fundamental Theorem of Calculus, we can write

$$\int_C F \cdot n \;\mathrm{d}s = \int_{b_1}^{b_2}\int_{a_1}^{a_2} \frac{\partial F_2(x_1, x_2, 0)}{\partial x_1}\, \mathrm{d}x_1\, \mathrm{d}x_2 - \int_{a_1}^{a_2}\int_{b_1}^{b_2} \frac{\partial F_1(x_1, x_2, 0)}{\partial x_2}\, \mathrm{d}x_2\, \mathrm{d}x_1 .$$

Changing the order of integration, we combine the two integrals to get

$$\int_C F \cdot n \;\mathrm{d}s = \int_{b_1}^{b_2}\int_{a_1}^{a_2} \left[\frac{\partial F_2(x_1, x_2, 0)}{\partial x_1} - \frac{\partial F_1(x_1, x_2, 0)}{\partial x_2}\right] \mathrm{d}x_1\, \mathrm{d}x_2 . \quad \text{(A.13)}$$

The integrand in brackets is the x_3-component of the curl of the vector field, $(\nabla \times F(x_1, x_2, x_3))_3$, evaluated at $x_3 = 0$, which corresponds to the x_1x_2-plane. We can write this component as $(\nabla \times F) \cdot N$, where $N = [0, 0, 1]$, which is a vector normal to the x_1x_2-plane. Since we can orient the coordinate system in such a way that its x_1x_2-plane coincides with the rectangle, we rewrite equation (A.13) as

$$\int\limits_C F \cdot n \, \mathrm{d}s = \iint\limits_A [(\nabla \times F) \cdot N] \, \mathrm{d}A,$$

where N is the unit vector normal to rectangle A whose area element is $\mathrm{d}A = \mathrm{d}x_1 \, \mathrm{d}x_2$. Now, we have the Curl Theorem for a plane segment bounded by a rectangle. We can extend its validity to an arbitrary area in the x_1x_2-plane or to a surface bounded by a loop in three dimensions. The former is discussed in Appendix A.2.3; herein, we proceed with the latter.

The motivation for our approach is the fact that small adjacent rectangles of different orientation allow us to approximate a surface of arbitrary shape. Let us consider two rectangles that touch one another along the

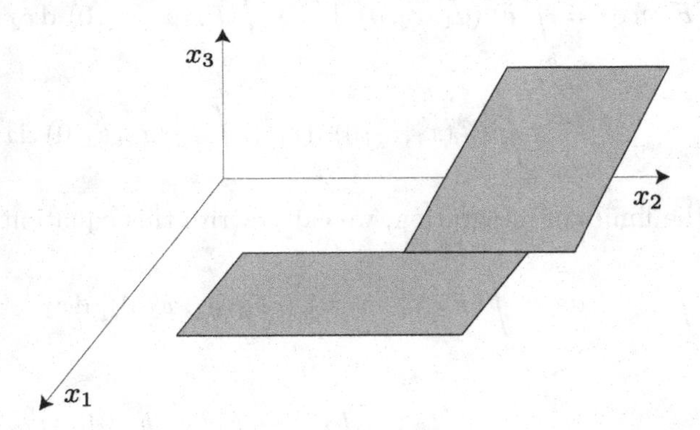

Fig. A.5 Two connected rectangles used to formulate the Curl Theorem.

portion of one edge, as shown in Figure A.5. Since we could demonstrate the validity of equation (A.12) for the new rectangle by following the argument identical to the one described above, let us focus our attention on the portion of the edge that is common to both rectangles. Since we choose the counterclockwise direction to describe the circulation along any curve, this means that for either rectangle the direction is opposite along the common portion. Consequently, along this portion, we get two integrals whose values differ by a sign. These integrals cancel one another in the summation. In view of this cancelling, we conclude that the circulation through the outer loop is equal to the sum of the circulations through all interior pieces. Furthermore, since the adjacent rectangles need not be coplanar, we can approximate an arbitrary surface bounded by a curve in three dimensions to obtain equation (A.12), as required.

Example A.3. Let us examine the meaning of differentiability of a vector field for Theorem A.2. To do so, we consider

$$F = \left[\frac{y}{x^2+y^2}, -\frac{x}{x^2+y^2}, 0\right], \tag{A.14}$$

which is a vector field not differentiable along the z-axis, where $x = y = 0$. Everywhere else, its curl is zero,

$$\begin{aligned}\nabla \times F &= \left[0, 0, -\frac{\partial}{\partial x}\frac{x}{x^2+y^2} - \frac{\partial}{\partial y}\frac{y}{x^2+y^2}\right] \\ &= \left[0, 0, \frac{2x^2}{(x^2+y^2)^2} + \frac{2y^2}{(x^2+y^2)^2} - \frac{2}{x^2+y^2}\right] \equiv [0,0,0]\,. \end{aligned} \tag{A.15}$$

Hence, everywhere else, the right-hand side of equation (A.12) is zero, and—in accordance with the Curl Theorem—circulation of this vector field along a loop that does not enclose the z-axis is zero. However, a circulation along a loop that does enclose the z-axis is nonzero. To see it, let us consider F in the xy-plane. In polar coordinates, we write

$$F = \left[\frac{r\sin\theta}{r^2}, -\frac{r\cos\theta}{r^2}, 0\right] = \left[\frac{\sin\theta}{r}, -\frac{\cos\theta}{r}, 0\right].$$

If we let C, in equation (A.12), be a circle, in the xy-plane, centred at the origin, its unit tangent vector is

$$n = [-\sin\theta, \cos\theta, 0]\,,$$

and

$$F \cdot n = -\frac{\sin^2\theta + \cos^2\theta}{r} = -\frac{1}{r}\,.$$

Thus, we write the single integral in equation (A.12) as

$$\int_C F \cdot n \, \mathrm{d}s = -\int_0^{2\pi} \frac{1}{r} \underbrace{r\,\mathrm{d}\theta}_{\mathrm{d}s} = -2\pi \neq 0\,.$$

The zero and nonzero circulations—depending on the differentiability of a vector field, the former implied by the Curl Theorem—illustrate the meaning of the requirement for a continuously differentiable vector field, in the statement of Theorem A.2.

There is no contradiction, since there is no singularity of F, along the curve, C, which is the domain of integration of the single integral, in contrast to the double integral, whose domain of integration is the surface, S, bounded by this curve. Differentiability of the field ensures the equality of the two integrals; it is not necessary for the integrability of the single integral.

Remark A.3. In Example A.3, following the argument of Example A.2, we could conjecture that—for F given in expression (A.14)—the double integral of equation (A.12), for any domain of integration that contains $x = y = 0$, is equal to -2π. If so, since—in accordance with expression (A.15)—this integral is zero for any domain that does not contain $x = y = 0$, its nonzero value is due solely to the contribution of the singularity.

A.2.3 *Corollary*

There is a corollary of Theorem A.2 that applies to vector fields with one Cartesian component that is zero, for instance, $F = [F_1, F_2, 0]$, and—in this case—the nonzero components do not depend on x_3. In such a case, the right-hand side of expression (A.12) becomes

$$\iint\limits_S \left[\frac{\partial 0}{\partial x_2} - \frac{\partial F_2(x_1,x_2)}{\partial x_3}, \frac{\partial F_1(x_1,x_2)}{\partial x_3} - \frac{\partial 0}{\partial x_1}, \frac{\partial F_2(x_1,x_2)}{\partial x_1} - \frac{\partial F_1(x_1,x_2)}{\partial x_2}\right] \cdot [0,0,1]\,\mathrm{d}S$$

$$= \iint\limits_S \left(\frac{\partial F_2}{\partial x_1} - \frac{\partial F_1}{\partial x_2}\right)\mathrm{d}S\,.$$

The left-hand side becomes

$$\int\limits_C F \cdot \mathrm{d}\ell = \int\limits_C [F_1, F_2, 0] \cdot [\mathrm{d}x_1, \mathrm{d}x_2, 0] = \int\limits_C F_1\,\mathrm{d}x_1 + F_2\,\mathrm{d}x_2\,,$$

where $\mathrm{d}\ell = n\,\mathrm{d}s$ stands for the element of a vector tangent to the curve. Thus, we have the following corollary of Theorem A.2.

Corollary A.2. *Green's Theorem II*: *If* $F = [F_1(x_1,x_2), F_2(x_1,x_2), 0]$ *is a continuously differentiable vector field and* S *is an area in the* x_1x_2*-plane, whose circumference,* C, *is a simply connected piecewise smooth curve with a counterclockwise orientation, then*

$$\int\limits_C F_1\,\mathrm{d}x_1 + F_2\,\mathrm{d}x_2 = \iint\limits_S \left(\frac{\partial F_2}{\partial x_1} - \frac{\partial F_1}{\partial x_2}\right)\mathrm{d}S\,. \tag{A.16}$$

Corollary A.2 can be viewed as a theorem of its own, which is commonly referred to as Green's Theorem. To see that, consider the fact that expression (A.13) is the Curl Theorem for a rectangular segment in the x_1x_2-plane, and proceed in a manner analogous to the one described in Appendix A.2.2. Small adjacent rectangles, in this plane, allow us to approximate an area of arbitrary shape. Since we choose the counterclockwise direction to describe the circulation along the circumference of each rectangle, the circulation is opposite along the common portion of the adjacent rectangles. Along this portion, the values of respective integrals differ by a sign, and they cancel one another upon summation. Hence, the circulation through the circumference is equal to the sum of circulations through all interior pieces. Thus, we obtain equation (A.16), where C and S are arbitrary.

Example A.4. Corollary A.2 and expression (A.16) underly the concept of a planimeter, which is a mechanical instrument to measure the area of a flat surface by tracing its boundary. Since the area, A, is $\iint_S \mathrm{d}S$ it follows—by the right-hand side of equation (A.16)—that the integrand is

$$\frac{\partial F_2}{\partial x_1} - \frac{\partial F_1}{\partial x_2} = 1\,, \tag{A.17}$$

which is satisfied by $F_1 = -x_2/2$ and $F_2 = x_1/2$. Thus, in accordance with the left-hand side of equation (A.16), we write

$$A = \frac{1}{2}\int\limits_C x_1\,\mathrm{d}x_2 - x_2\,\mathrm{d}x_1\,. \tag{A.18}$$

In general, C can be quite complicated, but to exemplify expression (A.16), let us consider an ellipse centred at the origin, whose width is $2\,a$ and height is $2\,b$. Using θ as a parameter, we write $x_1(\theta) = a\cos\theta$, $x_2(\theta) = b\sin\theta$, where $0 \leqslant \theta \leqslant 2\pi$. Inserting into expression (A.18), we obtain

$$A = \frac{1}{2}\int\limits_0^{2\pi} \underbrace{(a\cos\theta)}_{x_1}\underbrace{(b\cos\theta)\,\mathrm{d}\theta}_{\mathrm{d}x_2} - \underbrace{(b\sin\theta)}_{x_2}\underbrace{(-a\sin\theta)\,\mathrm{d}\theta}_{\mathrm{d}x_1} = a\,b\,\pi\,.$$

Remark A.4. In terms of vector fields, such an ellipse can be visualized as a wavefront within a medium—subject to an elliptical velocity dependence with direction—as discussed on page 111, in Example 3.1, where $a := v_x$ and $b := v_z$. Therein, the magnitude of the field is given by expressions (3.23), with $\theta := \vartheta$.

Equation (A.17) is also satisfied by $F_1 = 0$, $F_2 = x_1$ and by $F_1 = -x_2$, $F_2 = 0$. Hence, equation (A.18) can be extended to include

$$A = \int_C x_1 \, \mathrm{d}x_2 = -\int_C x_2 \, \mathrm{d}x_1 . \tag{A.19}$$

To exemplify its use, let us find the area of a square whose vertices are $(0,0)$, $(1,0)$, $(1,1)$, $(0,1)$. Since a square is a simply connected piecewise smooth curve, we parametrize it by t and proceed by starting from the origin in a counterclockwise direction along

$$C = \sum_{i=1}^{4} C_i \,, \quad \text{where} \quad \begin{cases} C_1 : x_1 = t, x_2 = 0 \,, & 0 \leqslant t \leqslant 1 \\ C_2 : x_1 = 1, x_2 = t \,, & 0 \leqslant t \leqslant 1 \\ C_3 : x_1 = 1 - t, x_2 = 1 \,, & 0 \leqslant t \leqslant 1 \\ C_4 : x_1 = 0, x_2 = 1 - t \,, & 0 \leqslant t \leqslant 1 \end{cases} .$$

Let us consider the first equality of equation (A.19). Since along C_1 and C_3, $\mathrm{d}x_2/\mathrm{d}t = 0$, we have

$$A = \int_C x_1 \, \mathrm{d}x_2 = \underbrace{\int_0^1 1 \, \mathrm{d}t}_{C_2} - \underbrace{\int_0^1 0 \, \mathrm{d}t}_{C_4} = \int_0^1 \mathrm{d}t = t|_0^1 = 1 \,,$$

as expected. For the second equality of equation (A.19), along C_2 and C_4, $\mathrm{d}x_1/\mathrm{d}t = 0$; hence,

$$A = -\int_C x_2 \, \mathrm{d}x_1 = -\left(\underbrace{\int_0^1 0 \, \mathrm{d}t}_{C_1} - \underbrace{\int_0^1 1 \, \mathrm{d}t}_{C_3} \right) = t|_0^1 = 1 \,,$$

as expected. The same result is obtained with expression (A.18), for which the four segments, C_1, C_2, C_3, C_4, have equal nonzero contributions.

Closing remarks

As suggested by Remark A.1, the Curl Theorem and the Divergence Theorem are important for mathematics itself and for the language of mathematical physics. Our discussion in this appendix is motivated, to a large extent, by the latter. Notably, physical laws are formulated commonly as integral equations. The aforementioned theorems allow us to express these laws as differential equations, which we find more convenient to examine, in general, and which allow us to study the concept of characteristics, in particular. Another importance of these theorems is the possibility of obtaining the values of double integrals by solving single integrals or of triple integrals by solving double integrals, as discussed in Example A.2, as well as in Remarks A.2 and A.3.

Green's Theorem, stated as Corollary A.2—which is a corollary of the Curl Theorem—is commonly presented as a theorem of its own. Being confined to a plane, it is convenient to use for many two-dimensional problems. It is commonly used to obtain the areas of planar surfaces or the values of single integrals by solving double integrals or *vice versa*. Be that as it may, the Divergence Theorem, the Curl Theorem and Green's Theorem are special cases of Stokes's Theorem, stated in expression (A.1).

To conclude, let us return to expression (A.1), which is a generalization of integral theorems, whose statement—in contrast to Theorems A.1 and A.2—is not limited to specific dimensions. In modern terminology, expression (A.1) relates the exterior derivative of the $k-1$-form, ω, which is a k-form, $\mathrm{d}\omega$, integrated over a k-dimensional manifold, Ω, to ω, integrated over the boundary of that manifold, $\partial\Omega$. Notably, beyond three dimensions, integral theorems cannot be stated in terms of functions and vector fields.[6]

Remaining within standard vector-calculus operations, however, we can generalize expression (A.2) from integrals along an interval in $\mathbb{R}$ to line integrals in $\mathbb{R}^n$,

$$\int_{\Omega}\mathrm{d}\omega=\int_{\partial\Omega}\omega\implies\int_{C}\nabla f=\int_{\partial C}f=\underbrace{f(b)-f(a)}_{\int_{\partial C} f},\qquad\text{(A.20)}$$

where C is a curve in $\mathbb{R}^n$, between points a and b, and

$$\nabla f:=\sum_{i=1}^{n}\frac{\partial f}{\partial x_i}\,\mathrm{d}x_i\,;$$

the last equality in expression (A.20) corresponds to, by definition, the boundary along C, namely, its endpoints. Since the value of $\int_C\nabla f$ depends only on the values of f at the endpoints, it is path-independent; hence, ∇f is a conservative field, as expected for any vector field that is the gradient of a scalar field. In particular, for a closed curve, $\oint_C\nabla f=0$. This generalization

[6]Readers interested in differential forms and the exterior derivative for integral theorems, in the context of an undergraduate textbook, might refer to Colley (2012, Section 8.3).

of the Fundamental Theorem of Calculus is referred to as the Fundamental Theorem of Line Integrals and, in the context of integral theorems, as the Gradient Theorem; it applies to vector fields that can be expressed as gradients of a scalar field.

Integral theorems are used explicitly in Appendices B, C and F, where Corollary A.1 is used in Appendix F.1.

Appendix B

Elastodynamic equations

Any science must stop at a certain level of description, and Continuum Mechanics stops at what might be regarded as a rather high level, whereby its description of material response loses the character of what the philosopher of science Mario Bunge would call an interpretive explanation.[1,2] *This is hardly a deficiency: it is simply a definition of the point of view already conveyed by the very name of the discipline. It is this particular point of view,* [...], *that has paved the way for the great success of Continuum Mechanics, both theoretical and practical.* [...] *The price to pay is that the representation of the material response is not universal but must be tailored to each material or class of materials.*

Marcelo Epstein (2010)

[1]Readers interested in interpretive explanations might refer to Bunge (1967, p. 79), where the author states that

> we shall call *interpretive explanation* the kind of explanation afforded by nonphenomenological theories. Interpretive theories are the deeper and the more challenging. They satisfy our wish to understand, a wish which classical positivists [...] considered sinful.

[2]In the same spirit, Lanczos (1949/1986, pp. 4–5) writes the following.

> It frequently happens that certain kinematical conditions exist between the particles of a moving system which can be stated *a priori*. For example, the particles of a solid body may move as if the body were "rigid". [...] Such kinematical conditions do not actually exist on *priori* grounds. They are maintained by strong forces. It is of great advantage, however, that the analytical treatment does not require the knowledge of these forces, but can take the given kinematical conditions for granted. We can develop the dynamical equations of a rigid body without knowing what forces produce the rigidity of the body.

Preliminary remarks

In this appendix, we derive equations that describe the motion within a linearly elastic solid. These equations are called the elastodynamic equations, and result from combining Cauchy's equations of motion with the constitutive equations of linear elasticity. Cauchy's equations of motion are rooted in the balance law of linear momentum, which states that the rate of change of momentum of a portion of continuum is equal to the sum of all forces acting upon it. The origin of the constitutive equations is in empirical studies, according to which we can commonly approximate the relation between forces and deformations as linear; in such a case we refer to the theory as linear. The constitutive equations used in this appendix are the stress-strain equations known as Hooke's law.

We begin this appendix by formulating Cauchy's equations of motion and the stress-strain equations. Subsequently, we combine these equations to derive the elastodynamic equations whose characteristic equations are the subject of our study in this book.

Readers might find it useful to study this appendix together with Appendix A, where we discuss the Curl Theorem and the Divergence Theorem.

B.1 Cauchy's equations of motion

In this section, we present a derivation of Cauchy's equations of motion, which is rooted in the balance of linear momentum. To study the momentum within a continuum, we have to formulate velocity therein. However, descriptions of motion, which—in particle mechanics—are tantamount to observing a displacement of a material object travelling through space, require within continuum mechanics other subtleties due to the absence of discrete points. We are led to a description in which the flow of continuum through a fixed position in space is also considered, as opposed to considering only—as in the case of particle mechanics—the displacement of an object travelling through a sequence of spatial positions.[3,4] As illustrated in this section, both descriptions are pertinent in examining concepts of continuum mechanics.

[3]Readers interested in examination of these two descriptions and their consequences might refer to Malvern (1969, Section 4.3) or Epstein (2010, Appendix A).
[4]*See also*: Slawinski (2015, Sections 1.3 and 2.2).

Let us label an infinitesimal portion of a continuum by coordinate X; we can view X as denoting a material point. In other words, this coordinate is attached to a point within the continuum, and is assumed to move with it. We can express the position, x, of a material point at time t as

$$x = x(X, t) , \tag{B.1}$$

where x is a coordinate system fixed in space. Equivalently, we can express point X at position x and time t as

$$X = X(x, t) .$$

Functions $x(X, t)$ and $X(x, t)$ are the inverses of one another, namely $x(X(x,t), t) = x$ and $X(x(X,t), t) = X$; their relationship is one-to-one, since two material points cannot occupy the same spatial location nor can one point occupy more than a single spatial coordinate at a given time. We refer to X and x as material and spatial coordinates, respectively.

Having expressed the position, we can discuss the velocity. The velocity of point X is

$$\tilde{v}(X, t) := \frac{\mathrm{d}}{\mathrm{d}t} x(X, t) , \tag{B.2}$$

which is given in material coordinates. This is a velocity that would be measured by an observer attached to, and travelling with, X. The velocity in spatial coordinates is

$$v(x, t) = \tilde{v}(X(x, t), t) . \tag{B.3}$$

This is a velocity that would be measured by an observer located at x and observing the motion of X. Relations (B.2) and (B.3) are true for any quantity expressed in the spatial and material coordinates, since if q is a quantity in the spatial coordinates and $\tilde{q}$ is the same quantity in the material coordinates, they are related by

$$q(x, t) = \tilde{q}(X(x, t), t) ; \tag{B.4}$$

$q(x, t)$ and $\tilde{q}(X, t)$ are the spatial and material descriptions of a given quantity.

Let us consider the balance of linear momentum: the temporal rate of change of momentum of a portion of a continuum is equal to the sum of forces acting on and within this part of the continuum. Mathematically, we can express this statement as

$$\frac{\mathrm{d}}{\mathrm{d}t} \iiint\limits_{V(t)} \rho(x, t)\, v(x, t)\, \mathrm{d}V = \iint\limits_{S(t)} T(x, t)\, \mathrm{d}S + \iiint\limits_{V(t)} F(x, t)\, \mathrm{d}V , \tag{B.5}$$

where $\mathrm{d}V$ and $\mathrm{d}S$ denote the volume and surface elements, respectively, expressed in terms of $\mathrm{d}x_i$, which are the spatial coordinates. $V(t)$ and $S(t)$ are the domains of integration, which—being expressed in terms of spatial coordinates—are time-dependent; in other words, the spatial position, x, of a material point, X, might change in time.

The volume integral on the left-hand side is the momentum of the portion of the continuum contained in volume $V(t)$. The two integrals on the right-hand side describe all the forces that we consider as acting on, and within, the portion of the continuum. T is the traction accounting for the surface forces, with S being the surface enclosing volume V, and F is a body force per unit volume accounting for forces such as gravity. Thus, the left-hand side of equation (B.5) is the rate of change of momentum, and the right-hand side is the sum of all forces.

We wish to combine the three integrals into a single integral by invoking the Divergence Theorem, which is Theorem A.1, on page 215. However, to combine these integrals, we need first to incorporate the time derivative within the integral on the left-hand side of equation (B.5). Since the domain of integration is time-dependent, we cannot interchange the order of integration and differentiation with respect to time.

To obtain the desired result, let us choose $V(t)$ to be such that, as it moves through space and changes size and shape, it contains the same portion of continuum. Expressing this volume in material coordinates, $\tilde{V}$, we have the domain of integration that is time-independent, as a consequence of the definition of material coordinates, X, which are attached to material points, and the volume containing the same portion of the continuum. In other words, $\tilde{V}$ contains the same material points at all times.

Following relation (B.4), where $X = X(x,t)$, we write the left-hand side of equation (B.5) as

$$\frac{\mathrm{d}}{\mathrm{d}t}\iiint\limits_{\tilde{V}} \tilde{\rho}(X,t)\,\tilde{v}(X,t)\,\mathrm{d}\tilde{V}.$$

Denoting $\tilde{\rho}\,\mathrm{d}\tilde{V}$ by $\mathrm{d}\tilde{m}$, which is the element of mass expressed in material coordinates, we write

$$\frac{\mathrm{d}}{\mathrm{d}t}\iiint\limits_{\tilde{V}} \tilde{v}(X,t)\,\mathrm{d}\tilde{m}. \tag{B.6}$$

Since the domain of integration does not depend on time, we can interchange the operations of differentiation and integration to write

$$\iiint\limits_{\tilde{V}} \frac{\mathrm{d}}{\mathrm{d}t}\tilde{v}(X,t)\,\mathrm{d}\tilde{m}. \tag{B.7}$$

Replacing $\mathrm{d}\tilde{m}$ by $\tilde{\rho}\,\mathrm{d}\tilde{V}$, we rewrite expression (B.7) as

$$\iiint_{\tilde{V}} \frac{\mathrm{d}}{\mathrm{d}t} \left\{\tilde{v}\left(X,t\right)\right\} \tilde{\rho}\left(X,t\right) \mathrm{d}\tilde{V}\,, \tag{B.8}$$

where—in view of the integrand in expression (B.7)—the braces emphasize the fact that the differentiation refers only to the term within them.

Having incorporated the derivative within the integral, we proceed—invoking expressions (B.1)–(B.4)—to express result (B.8) in spatial coordinates. First, we write

$$\iiint_{\tilde{V}} \frac{\mathrm{d}}{\mathrm{d}t} \left\{v\left(x\left(X,t\right),t\right)\right\} \rho\left(x\left(X,t\right),t\right) \mathrm{d}\tilde{V}\,.$$

Using the chain-rule, we write

$$\iiint_{\tilde{V}} \left\{\frac{\partial}{\partial x} v\left(x\left(X,t\right),t\right) \frac{\partial x}{\partial t} + \frac{\partial}{\partial t} v\left(x\left(X,t\right),t\right) \frac{\partial t}{\partial t}\right\} \rho\left(x\left(X,t\right),t\right) \mathrm{d}\tilde{V}\,.$$

Since $\partial t/\partial t \equiv 1$ and $\partial x/\partial t$ is the velocity expressed in spatial coordinates, v, we write

$$\iiint_{\tilde{V}} \left\{\frac{\partial}{\partial t} v\left(x\left(X,t\right),t\right) + v\left(x\left(X,t\right),t\right) \frac{\partial}{\partial x} v\left(x\left(X,t\right),t\right)\right\} \rho\left(x\left(X,t\right),t\right) \mathrm{d}\tilde{V}\,; \tag{B.9}$$

note that since the first summand in braces is a vector whose components are $\partial v_i/\partial t$, with $i \in \{1,2,3\}$, so is the second summand; it can be written explicitly as a vector whose three components are

$$\left(v_1 \frac{\partial}{\partial x_1} + v_2 \frac{\partial}{\partial x_2} + v_3 \frac{\partial}{\partial x_3}\right) v_i\,, \qquad i \in \{1,2,3\}\,.$$

Since both $\tilde{V}$ and $V(t)$ contain the same portion of the continuum, their volume integrals are equal to one another. Hence, we can state expression (B.9) explicitly in the spatial coordinates,

$$\iiint_{V(t)} \left\{\frac{\partial}{\partial t} v\left(x,t\right) + v\left(x,t\right) \frac{\partial}{\partial x} v\left(x,t\right)\right\} \rho\left(x,t\right) \mathrm{d}V\,, \tag{B.10}$$

which is equivalent to the left-hand side of equation (B.5), with the differentiation included within the integral, as required for subsequent steps.

The term in braces in expression (B.10) is an example of the material time derivative. With a certain abuse of notation, such an operator can be written formally as $(\partial/\partial t + v \cdot \nabla)\,q$, where q stands for a scalar-, vector- or tensor-valued quantity. To gain an insight into the formulation presented between expressions (B.5)–(B.10) and the concept of the material-time-derivative operator acting therein on a vector-valued quantity expressed in spatial coordinates, $v(x,t)$, let us quote Epstein (2010, Appendix A) and consider an example suggested therein.

> [t]he material time derivative corrects the partial time derivative by means of a term equal to the product of the spatial gradient of the field contracted with the velocity [...] A moment's reflection reveals that this correction is exactly what one should expect.

Example B.1. Consider water flowing through a tube and cooling as it travels. Two observers separated by Δx are recording its temperature, q. We assume that the difference between the temperatures recorded at two points, Δq, is constant; hence, we infer that the temperature is only a function of traveltime between the observation points, $\Delta t = \Delta x/v$. Thus, the temporal rate of change of temperature measured by a hypothetical observer moving with the flow is $\Delta q/(\Delta x/v)$. Rearranging, we write $v\,(\Delta q/\Delta x)$, which is the correction term in one spatial dimension; in general, it is given by $v \cdot \nabla q$.

In this example—to isolate the effect of the correction term—we assume that $\partial q/\partial t = 0$, which means that there is no change in temperature at any given location. The change is due to traveltime only; if there is no flow, $v = 0$, and the temperature remains constant. To include cooling as a function of the passage of time at a given location requires $\partial q/\partial t \neq 0$.

The appearance of the material time derivative in expression (B.10) is a consequence of the fact that we cannot interchange the order of integration and differentiation on the left-hand side of equation (B.5), and results from the form of the integrand, which contains mass, as well as from the properties of $V(t)$. The integrand of integral (B.10) is not a result of a standard differentiation of the integrand of the left-hand-side integral in equation (B.5); in other words,

$$\left(\frac{\partial}{\partial t}v(x,t) + v(x,t)\frac{\partial}{\partial x}v(x,t)\right)\rho(x,t) \neq \frac{\mathrm{d}}{\mathrm{d}t}\Big(v(x,t)\,\rho(x,t)\Big)\,, \qquad \text{(B.11)}$$

as we can verify by considering the product and chain rules on the right-hand side of expression (B.11), which result in

$$\left(\frac{\partial v}{\partial t} + v\frac{\partial v}{\partial x}\right)\rho + \left(\frac{\partial \rho}{\partial t} + v\frac{\partial \rho}{\partial x}\right)v\,;$$

only the first summand, which contains no derivatives of ρ, is the left-hand side of expression (B.11). As a consequence of our choice of volume, whose total mass remains constant even though its size and shape change as it moves through space, there are no derivatives of ρ on the left-hand side of expression (B.11). Thus, the temporal rate of change of momentum of the selected portion of a continuum, $V(t)$, is a function of the change of its velocity only; this property is stated explicitly in expressions (B.6) and (B.7).

Regardless of the issue of interchanging the order of integration and differentiation or the consideration of a particular form of a function to be differentiated or the volume over which the integration is to be performed, we can write

$$\begin{aligned}\frac{\mathrm{d}}{\mathrm{d}t}q(x,t) &= \frac{\partial q(x,t)}{\partial t} + \sum_{i=1}^{3}\frac{\partial q(x,t)}{\partial x_i}\frac{\mathrm{d}x_i}{\mathrm{d}t} \\ &\equiv \frac{\partial q(x,t)}{\partial t} + v\cdot\nabla q(x,t)\,, \end{aligned} \tag{B.12}$$

which, if q is in a single spatial dimension, is the material time derivative discussed in Example B.1. In general, this expression is a consequence of the chain rule of differentiation together with endowing x and t with the physical meaning of space and time.

Let us return to our derivation of Cauchy's equations of motion. Using result (B.10), we write equation (B.5) as

$$\begin{aligned}&\iiint\limits_{V(t)}\left(\frac{\partial v\,(x,t)}{\partial t} + v\,(x,t)\,\frac{\partial}{\partial x}v\,(x,t)\right)\rho\,(x,t)\;\mathrm{d}V \\ &\quad = \iint\limits_{S(t)} T\,(x,t)\;\mathrm{d}S + \iiint\limits_{V(t)} F\,(x,t)\;\mathrm{d}V\,. \end{aligned} \tag{B.13}$$

To combine the three integrals of equation (B.13) into a single integral, we express the surface integral as the volume integral using the Divergence Theorem, which is Theorem A.1, stated on page 215. To do so, we introduce an important concept of elasticity theory. We write the components of traction as

$$T_i = \sum_{j=1}^{3}\sigma_{ij}N_j\,, \qquad i\in\{1,2,3\}\,. \tag{B.14}$$

Herein, σ_{ij} are the components of the stress tensor, which is an intrinsic entity of continuum mechanics. N_j are the components of a unit vector normal to the element of surface within the continuum to which traction is applied. This expression can be viewed as the definition of the stress tensor, which allows us to study forces acting on an arbitrarily oriented plane in the continuum. The justification for writing traction in this way is rooted also in the balance of linear momentum, as discussed by Slawinski (2015, Section 2.5.2).

Inserting expression (B.14) into equation (B.13) and invoking Theorem A.1, we write equation (B.13)—using components—as

$$\iiint\limits_{V(t)} \left(\frac{\partial v_i(x,t)}{\partial t} + \sum_{j=1}^{3} v_j(x,t) \frac{\partial}{\partial x_j} v_i(x,t) \right) \rho(x,t)\, \mathrm{d}V$$
$$= \iiint\limits_{V(t)} \sum_{j=1}^{3} \frac{\partial \sigma_{ij}}{\partial x_j} \mathrm{d}V + \iiint\limits_{V(t)} F_i(x)\, \mathrm{d}V\,, \qquad i \in \{1,2,3\}\,.$$

Combining the three integrals, we obtain

$$\iiint\limits_{V(t)} \left(\sum_{j=1}^{3} \frac{\partial \sigma_{ij}}{\partial x_j} + \left(\frac{F_i}{\rho} - \frac{\partial v_i}{\partial t} - \sum_{j=1}^{3} v_j \frac{\partial v_i}{\partial x_j} \right) \rho \right) \mathrm{d}V = 0\,, \qquad \text{(B.15)}$$

where $i \in \{1,2,3\}$.

We wish to write the balance of linear momentum as differential equations. To do so, we use the fact that—for equations (B.15) to be a general statement of the balance of linear momentum—equations (B.15) must be satisfied for an arbitrary volume. Mathematically, this requirement implies that the integrands must be identically zero. Thus, we require

$$\sum_{j=1}^{3} \frac{\partial \sigma_{ij}}{\partial x_j} + \left(\frac{F_i}{\rho} - \frac{\partial v_i}{\partial t} - \sum_{j=1}^{3} v_j \frac{\partial v_i}{\partial x_j} \right) \rho = 0\,, \qquad i \in \{1,2,3\}\,. \qquad \text{(B.16)}$$

Cauchy's equations of motion used in studies of wave propagation result from these differential equations with several approximations. We ignore the effect of body forces by assuming that the effect of elasticity is much greater than the effect of gravity, which we assume to be the only body force under consideration.[5] Moreover, we neglect the product of the velocity of displacement with the gradient of that velocity, assuming that it is a product of two small quantities. Hence, we write

$$\sum_{j=1}^{3} \frac{\partial \sigma_{ij}}{\partial x_j} - \rho \frac{\partial v_i}{\partial t} = 0, \qquad i \in \{1,2,3\}\,, \qquad \text{(B.17)}$$

[5] Readers interested in the estimate of magnitudes of the surface as opposed to the body forces in seismology might refer to Udías (1999, Section 3.5).

which are Cauchy's equations of motion whose validity is restricted by several approximations. Notably, neglecting the product of velocity and its gradient, is neglecting $v \cdot \nabla q$ in expression (B.12), which, as stated in the quote on page 238, is the correction term within the material time derivative. Thus, revisiting the formulation presented between expressions (B.5)–(B.10) and the left-hand side of equation (B.13), we see that neglecting that term is tantamount to letting

$$\frac{\mathrm{d}}{\mathrm{d}t} \iiint\limits_{V(t)} \rho\,(x,t)\, v\,(x,t)\ \mathrm{d}V = \iiint\limits_{V(t)} \frac{\partial v\,(x,t)}{\partial t} \rho\,(x,t)\ \mathrm{d}V$$

in the derivation of equation (B.17).

We wish to express equation (B.17) in terms of displacements not their velocities. Such a formulation allows us to study Cauchy's equations of motion together with constitutive equations, which are discussed in Appendix B.2, below. Examining equation (B.17), in the context of material and spatial descriptions discussed in expressions (B.1)–(B.4), let us express velocity as the derivative of displacement of a portion of a continuum, which is the difference between its positions at time t and $t = 0$. In material coordinates, we write the displacement as

$$\tilde{u}\,(X,t) = x\,(X,t) - x\,(X,0)\ . \tag{B.18}$$

Taking the time derivative, using the fact that—according to expression (B.1)—$x\,(X,0)$ is the position of X at $t = 0$, which is fixed, and recalling expression (B.2), we obtain

$$\frac{\mathrm{d}}{\mathrm{d}t}\tilde{u}\,(X,t) = \frac{\mathrm{d}}{\mathrm{d}t}x\,(X,t) =: \tilde{v}\,(X,t)\ .$$

To use this result in equation (B.17), we need to express $\tilde{v}$ in terms of spatial coordinates. Following expression (B.4), we write $u\,(x,t)$ as $\tilde{u}\,(X\,(x,t)\,,t)$, and we use the chain rule to find its second derivative

$$\frac{\mathrm{d}^2}{\mathrm{d}t^2}\tilde{u}\,(X\,(x,t)\,,t) = \frac{\mathrm{d}}{\mathrm{d}t}\left(\tilde{v}(X(x,t),t) + \frac{\partial \tilde{u}}{\partial X}\frac{\partial X}{\partial t}\right)\,, \tag{B.19}$$

where we let $\partial\tilde{u}/\partial t = \tilde{v}$. Proceeding with the chain and product rules, we obtain

$$\frac{\partial \tilde{v}}{\partial t} + \frac{\partial \tilde{v}}{\partial X}\frac{\partial X}{\partial t} + \frac{\partial^2 \tilde{u}}{\partial X \partial t}\frac{\partial X}{\partial t} + \frac{\partial^2 \tilde{u}}{\partial X^2}\left(\frac{\partial X}{\partial t}\right)^2 + \frac{\partial \tilde{u}}{\partial X}\frac{\partial^2 X}{\partial t^2}\,. \tag{B.20}$$

Examining this expression together with expression (B.3) in the context of the chain rule, we see that the sum of the first two terms in expression (B.20) is $\partial v/\partial t$. Also, we neglect the last three terms in expression (B.20) by assuming them to be products of small quantities. Thus, we infer that—in a linearized theory— $\partial v/\partial t = \partial^2 u/\partial t^2$, which allows us to rewrite Cauchy's equations (B.17) as

$$\sum_{j=1}^{3} \frac{\partial \sigma_{ij}}{\partial x_j} = \rho(x,t) \frac{\partial^2 u_i(x,t)}{\partial t^2}, \qquad i \in \{1,2,3\}. \tag{B.21}$$

In view of justifications for linearization presented between expressions (B.16) and (B.17) and between expressions (B.18) and (B.21), we note that the smallness of displacements does not imply smallnesses of velocities, accelerations or their derivatives represented by ∇ in expression (B.12); nor the smallness of these derivatives of displacement implies smallness of the displacement itself. These are independent assumptions.[6]

Since expression (B.21) is a statement of the balance of linear momentum, which is a general principle, these equations are valid—within a linearized theory—for all continua. They do not refer to particular properties of a continuum. To apply these equations to a Hookean solid, we introduce its properties in the next section.

B.2 Stress-strain equations: Hookean solids

In this section, we formulate a mathematical description of a particular continuum, namely, the constitutive equation of a linearly elastic solid. These equations relate forces and deformations. Herein, we relate the stress tensor, σ_{ij}, to the displacement vector, u.

In our study of the elasticity theory, which focuses on wave propagation, we limit our interests to infinitesimal displacements, since it is reasonable to assume that deformations—due to wave propagation within the continuum—are small. In view of small displacements, we choose to describe the deformation by a tensor that is defined by

$$\varepsilon_{kl} = \frac{1}{2}\left(\frac{\partial u_k}{\partial x_l} + \frac{\partial u_l}{\partial x_k}\right), \qquad k,l \in \{1,2,3\}, \tag{B.22}$$

as discussed by Slawinski (2015, Section 1.4); we refer to it as the strain tensor. By definition, the values of ε_{kl} depend explicitly on position, not on time.

[6] Readers interested in details of the linearization process might refer to Achenbach (1984, Section 1.2.7) and Thurston (1969).

To incorporate the study of deformations into Cauchy's equations of motion, we relate σ_{ij} and ε_{kl} in a way that is physically justified and mathematically convenient. For elastic materials, as observed by Robert Hooke in the seventeenth century, the relation between the forces and deformations can be described accurately by a linear relation. This description is accurate as long as the deformations are small. In mathematical language, we write

$$\sigma_{ij} = \sum_{k=1}^{3} \sum_{l=1}^{3} c_{ijkl}(x)\, \varepsilon_{kl}, \qquad i,j \in \{1,2,3\}\,, \tag{B.23}$$

where c_{ijkl} are the components of a fourth-rank tensor: the elasticity tensor. This tensor is stated explicitly as a function of position, x, which allows us to consider inhomogeneous continua. Time dependence is not considered; material properties are assumed to be unaffected by the temporal aspect of deformations.

In the context of elasticity theory, we refer to equations (B.23) as stress-strain equations, and to the components of c_{ijkl} as the elasticity parameters. Since values of c_{ijkl} change with direction, by a virtue of tensorial properties, and components c_{ijkl} depend on position, x, we have constitutive equations of an anisotropic inhomogeneous continuum, to which—with inclusion of mass density—we refer as a Hookean solid. Every point of such a solid is described fully by the values of c_{ijkl} and ρ, which are functions of position.

To consider motion within Hookean solids, we combine equations (B.23) and (B.21).

B.3 Elastodynamic equations: anisotropy, inhomogeneity

In this section, we combine Cauchy's equations of motion (B.21) with stress-strain equations (B.23). Thus, we obtain the equations that describe propagation of deformation in an elastic continuum.

Inserting the expression for the stress-tensor given in equation (B.23) into Cauchy's equations of motion (B.21), we write

$$\rho(x)\,\frac{\partial^2 u_i(x,t)}{\partial t^2} = \sum_{j=1}^{3} \frac{\partial}{\partial x_j} \sum_{k=1}^{3} \sum_{l=1}^{3} c_{ijkl}(x)\, \varepsilon_{kl}\,, \qquad i \in \{1,2,3\}\,.$$

Note that in invoking equations (B.21), we ignore the temporal dependence of $\rho(x,t)$; mass density is assumed to be unaffected by propagation of small deformations.

Expressing the strain tensor in terms of the displacement vector as shown in expression (B.22), we get

$$\rho(x)\frac{\partial^2 u_i(x,t)}{\partial t^2}=\frac{1}{2}\sum_{j=1}^{3}\frac{\partial}{\partial x_j}\sum_{k=1}^{3}\sum_{l=1}^{3}c_{ijkl}(x)\left(\frac{\partial u_k}{\partial x_l}+\frac{\partial u_l}{\partial x_k}\right),\quad i\in\{1,2,3\}.$$

Using the linearity of the differential operator and following the product rule, we obtain the desired elastodynamic equations. They are

$$\rho(x)\frac{\partial^2 u_i(x,t)}{\partial t^2}=\frac{1}{2}\sum_{j=1}^{3}\sum_{k=1}^{3}\sum_{l=1}^{3}\frac{\partial c_{ijkl}(x)}{\partial x_j}\left(\frac{\partial u_k(x,t)}{\partial x_l}+\frac{\partial u_l(x,t)}{\partial x_k}\right)$$
$$+\frac{1}{2}\sum_{j=1}^{3}\sum_{k=1}^{3}\sum_{l=1}^{3}c_{ijkl}(x)\left(\frac{\partial^2 u_k(x,t)}{\partial x_j\partial x_l}+\frac{\partial^2 u_l(x,t)}{\partial x_j\partial x_k}\right),$$

where $i\in\{1,2,3\}$. We can write these equations more concisely. Renaming the summation indices, we write

$$\rho(x)\frac{\partial^2 u_i(x,t)}{\partial t^2}=\sum_{j=1}^{3}\sum_{k=1}^{3}\sum_{l=1}^{3}\frac{1}{2}\left(\frac{\partial c_{ijkl}(x)}{\partial x_j}+\frac{\partial c_{ijlk}(x)}{\partial x_j}\right)\frac{\partial u_k(x,t)}{\partial x_l}$$
$$+\sum_{j=1}^{3}\sum_{k=1}^{3}\sum_{l=1}^{3}\frac{1}{2}\left(c_{ijkl}(x)+c_{ijlk}(x)\right)\frac{\partial^2 u_k(x,t)}{\partial x_j\partial x_l}.$$

Using the index symmetries of the elasticity tensor, which stem from intrinsic properties of Hookean solids and are given by $c_{ijkl}=c_{jikl}=c_{klij}$,[7] we write

$$\rho(x)\frac{\partial^2 u_i(x,t)}{\partial t^2}=\sum_{j=1}^{3}\sum_{k=1}^{3}\sum_{l=1}^{3}\frac{\partial c_{ijkl}(x)}{\partial x_j}\frac{\partial u_k(x,t)}{\partial x_l}$$
$$+\sum_{j=1}^{3}\sum_{k=1}^{3}\sum_{l=1}^{3}c_{ijkl}(x)\frac{\partial^2 u_k(x,t)}{\partial x_j\partial x_l},\qquad\text{(B.24)}$$

where $i\in\{1,2,3\}$. These are linear second-order partial differential equations that describe propagation of displacement in an anisotropic inhomogeneous linearly elastic continuum, which we use in Section 2.3.5.1 to study characteristic hypersurfaces. In contrast to Cauchy's equations of motion (B.21), equations (B.24) refer to properties of a given continuum stated by c_{ijkl}.

[7] *See also*: Slawinski (2015, Chapters 3 and 4).

B.4 Elastodynamic equations: isotropy, homogeneity

B.4.1 *Equations of motion*

An important aspect of the study of wave propagation consists of examining the elastodynamic equations in the case of isotropy and homogeneity. In such a case, we can formulate the concept of wave propagation in terms of scalar and vector potentials.

To discuss the case of equations (B.24) that correspond to isotropic and homogeneous solids, we begin with equations (B.21), whose formulation is prior to introducing the definition of Hookean solids. The derivation presented herein possesses several similarities with discussions of the Maxwell equations in Appendix C.

To formulate the elastodynamic equations for an isotropic homogeneous continuum, consider Cauchy's equations (B.21),

$$\sum_{j=1}^{3}\frac{\partial\sigma_{ij}}{\partial x_j}=\rho\frac{\partial^2 u_i}{\partial t^2},\qquad i\in\{1,2,3\}\ , \tag{B.25}$$

where $u(x,t)$ is the displacement vector and ρ is the mass density, which herein is a constant; it does not depend on position. The stress tensor, σ_{ij}, can be expressed in terms of the strain tensor, ε_{ij}, by invoking equations (B.23), which allows us to study anisotropic inhomogeneous continua. To consider homogeneity, we use the elasticity tensor, c_{ijkl}, whose components do not depend on position, x; we write

$$\sigma_{ij}=\sum_{k=1}^{3}\sum_{l=1}^{3}c_{ijkl}\varepsilon_{kl},\qquad i,j\in\{1,2,3\}\ . \tag{B.26}$$

To consider isotropy, we need a form of the tensor whose components have the same value for all orientations of a coordinate system. In other words,

$$c_{ijkl}=\sum_{r=1}^{3}\sum_{s=1}^{3}\sum_{t=1}^{3}\sum_{w=1}^{3}A_{ir}A_{js}A_{kt}A_{lw}c_{rstw}=\tilde{c}_{ijkl}\ ,$$

where c and $\tilde{c}$ denote the same elasticity tensor in two orientations, and A is the orthogonal transformation relating these orientations. It can be shown that the general form of an isotropic fourth-rank tensor in $\mathbb{R}^3$ is[8]

$$a_{ijkl}=\lambda\delta_{ij}\delta_{kl}+\xi\delta_{ik}\delta_{jl}+\eta\delta_{il}\delta_{jk}\ ,\qquad i,j,k,l\in\{1,2,3\}\ , \tag{B.27}$$

[8]Readers interested in the formulation of this form might refer to Synge and Schild (1978, pp. 210–211).

where δ_{ij} is a second-rank tensor that transforms as an identity and is referred to as the Kronecker delta; λ, ξ and η are constants, which do not depend on the orientation of the coordinate system.

In elasticity theory—due to the index symmetries,[9] $c_{ijkl} = c_{jikl} = c_{klij}$—the isotropic elasticity tensor is

$$c_{ijkl} = \lambda\delta_{ij}\delta_{kl} + \mu(\delta_{il}\delta_{jk} + \delta_{ik}\delta_{jl}), \qquad i,j,k,l \in \{1,2,3\}, \tag{B.28}$$

where $\mu := (\xi + \eta)/2$. In the context of elasticity, λ and μ are called the Lamé parameters.

Inserting expressions (B.28) into equations (B.26) and using the properties of the Kronecker delta, $\delta_{ij} = 1$ if $i = j$ and $\delta_{ij} = 0$ if $i \neq j$, we obtain the stress-strain equations for an isotropic homogeneous continuum; they are

$$\sigma_{ij} = \lambda\delta_{ij}\sum_{k=1}^{3}\varepsilon_{kk} + 2\mu\varepsilon_{ij}, \qquad i,j \in \{1,2,3\}. \tag{B.29}$$

Expressing the strain tensor in equation (B.29) using expression (B.22) and inserting the right-hand side of the resulting equation into equation (B.25), we get—following standard vector-calculus operations[10]—the equations of motion in an elastic isotropic homogeneous medium, namely,

$$\rho\frac{\partial^2 u}{\partial t^2} = (\lambda + \mu)\nabla(\nabla\cdot u) + \mu\nabla^2 u. \tag{B.30}$$

In view of such vector-calculus operations, we might mention that, as discussed by Feynman *et al.* (1964) in Volume II on page 2-12, in general, for a vector field, the ith component of $\nabla^2 u$ is not equal to the Laplace operator acting on the ith component of u; this equality holds only for the Cartesian coordinates, where $(\nabla^2 u)_i = \nabla^2 u_i$. Even though such an equality does not hold for the Laplace operator applied to vectors expressed in curvilinear coordinates, the derived equations remain valid, provided that a proper form of the Laplace operator is used.

To express equation (B.30) in a form that allows the use of curvilinear coordinates, one could invoke the vector-calculus identity used on page 248, below, as discussed by Slawinski (2015, Section 6.1.1).

Note that—in contrast to equations (B.25), which do not specify properties of a continuum—equations (B.30) refer to properties of an isotropic homogenous Hookean solid defined by λ and μ.

[9] *See also*: Slawinski (2015, Chapters 3 and 4).
[10] *See also*: Slawinski (2015, Section 6.1.1).

Also, note that equation (B.30) can be obtained also by substituting expression (B.28) into the elastodynamic equations (B.24) and following a series of mathematical steps. The presented derivation, however, introduces explicitly the expression for the stress-strain equations of an isotropic homogeneous continuum.

B.4.2 *Scalar and vector potentials*

To obtain the wave equations for the P and S waves, we follow the Helmholtz Theorem,[11] which allows us to separate a vector function, u, into its scalar potential, ϕ, and vector potential, A.[12] We write the displacement as

$$u\left(x,t\right)=\nabla\phi+\nabla\times A\,. \tag{B.31}$$

In terms of u_1, u_2 and u_3, expression (B.31) constitutes three equations for four unknowns, namely, A_1, A_2, A_3 and ϕ.

We formulate another equation by using the fact that the divergence of vector A is arbitrary up to the gradient of function f. This means that we can change A in equation (B.31) by adding ∇f to it. If we choose f in such a way that $\nabla^2 f=-\nabla\cdot A$, we obtain a new vector potential, $\tilde{A}$, whose divergence vanishes, namely,

$$\nabla\cdot\tilde{A}=\nabla\cdot(A+\nabla f)=\nabla\cdot A+\nabla^2 f=0.$$

Since A is arbitrary, we conclude that the equation we seek is

$$\nabla\cdot A=0. \tag{B.32}$$

Equation (B.32) is analogous to equation (C.22) with the assumption that $\partial\phi/\partial t=0$. This choice of the potential is called the gauge.

Inserting expression (B.31) into equation (B.30) and rearranging, we get

$$\nabla\left[(\lambda+2\mu)\,\nabla^2\phi-\rho\frac{\partial^2\phi}{\partial t^2}\right]+\nabla\times\left[\mu\nabla^2 A-\rho\frac{\partial^2 A}{\partial t^2}\right]=0\,. \tag{B.33}$$

Examining equation (B.33), we see that if we take the divergence of this equation, the second term disappears; if we take the curl, the first term disappears. We proceed in this way to obtain the two wave equations that correspond to P waves and S waves.

[11] Readers interested in the Helmholtz Theorem might refer to Arfken *et al.* (2013, pp. 96–101).

[12] Readers interested in an elegant formulation of the P and S waves without invoking the scalar and vector potentials might refer to Rochester (2010).

B.4.3 *Wave equations*

***P* waves** Taking the divergence of equation (B.33) and using the fact that the divergence of a gradient is the Laplace operator, ∇^2, and the divergence of a curl is zero, we get

$$\nabla^2\left[(\lambda+2\mu)\,\nabla^2\phi-\rho\frac{\partial^2\phi}{\partial t^2}\right]=0\,.$$

Using the linearity of the differential operators and the fact that λ and μ are constants, we write

$$\left[(\lambda+2\mu)\,\nabla^2\left(\nabla^2\phi\right)-\rho\frac{\partial^2}{\partial t^2}\left(\nabla^2\phi\right)\right]=0\,. \tag{B.34}$$

To examine the meaning of $\nabla^2\phi$ in terms of the displacement vector, let us take the divergence of equation (B.31). Again, using the fact that the divergence of a gradient is the Laplace operator, ∇^2, and the divergence of a curl is zero, we obtain

$$\nabla^2\phi=\nabla\cdot u\,. \tag{B.35}$$

It is common to denote $\nabla\cdot u$ by φ, and to refer to it as dilatation. This name results from the fact that φ is a scalar quantity that we can write as

$$\varphi:=\nabla\cdot u=\frac{\partial u_1}{\partial x_1}+\frac{\partial u_2}{\partial x_2}+\frac{\partial u_3}{\partial x_3}\,,$$

and which allows us to view φ as the volume change of an infinitesimal cube.[13] Using expression (B.35), we rewrite equation (B.34) as

$$\nabla^2\varphi-\frac{1}{\frac{\lambda+2\mu}{\rho}}\frac{\partial^2\varphi}{\partial t^2}=0\,. \tag{B.36}$$

This equation is a linear second-order partial differential equation for φ; it is a wave equation, where the square root of the denominator is the speed of wave propagation. Herein, it is the wave equation for dilatational waves, which are commonly referred to as *P* waves.

***S* waves** Taking the curl of equation (B.33) and using the fact that the curl of the gradient is zero, we get

$$\nabla\times\nabla\times\left[\mu\nabla^2 A-\rho\frac{\partial^2 A}{\partial t^2}\right]=0\,. \tag{B.37}$$

[13] *See also*: Slawinski (2015, Section 1.4.3).

Let us invoke a vector-calculus identity—$\nabla \times \nabla \times V = \nabla\left(\nabla \cdot V\right) - \nabla^2 V$, where V is a vector—to write equation (B.37) as

$$\nabla\left(\nabla \cdot \left[\mu\nabla^2 A - \rho\frac{\partial^2 A}{\partial t^2}\right]\right) - \nabla^2\left[\mu\nabla^2 A - \rho\frac{\partial^2 A}{\partial t^2}\right] = 0\,.$$

This equation is satisfied if the term in brackets is zero. Considering the second term, we write

$$\mu\nabla^2\left(\nabla^2 A\right) - \rho\frac{\partial^2}{\partial t^2}\left(\nabla^2 A\right) = 0\,, \tag{B.38}$$

where we use the linearity of the differential operators and the fact that μ is constant. To examine the meaning of $\nabla^2 A$ in terms of the displacement vector, let us take the curl of equation (B.31). Again, using the fact that the curl of a gradient is zero, we obtain

$$\nabla \times u = \nabla \times \nabla \times A\,.$$

Using the above vector-calculus identity, we write

$$\nabla \times u = \nabla\left(\nabla \cdot A\right) - \nabla^2 A\,.$$

In view of equation (B.32), we simplify the right-hand side to get

$$\nabla \times u = -\nabla^2 A\,.$$

Using this expression in equation (B.38), we write

$$-\,\mu\nabla^2\left(\nabla \times u\right) + \rho\frac{\partial^2}{\partial t^2}\left(\nabla \times u\right) = 0\,. \tag{B.39}$$

Denoting $\nabla \times u$ by $\mathcal{A}$ and rearranging expressions on the left-hand side, we rewrite equation (B.39) as

$$\nabla^2\mathcal{A} - \frac{1}{\frac{\mu}{\rho}}\frac{\partial^2\mathcal{A}}{\partial t^2} = 0\,. \tag{B.40}$$

This equation is a linear second-order partial differential equation for $\mathcal{A}$; it is a wave equation, where the square root of the denominator is the speed of wave propagation. It is the wave equation for rotational waves, which are commonly referred to as the S waves. Since $\mathcal{A}$ is a vectorial quantity, equation (B.40) contains three equations to be solved for $\mathcal{A}_1$, $\mathcal{A}_2$ and $\mathcal{A}_3$.

B.5 Equations of motion versus wave equations

The three equations of motion given in expression (B.30) are equivalent to the four wave equations given by expressions (B.36) and (B.40) under conditions of differentiability of u, whose second derivatives appear in expression (B.30), and, in view of $\varphi = \nabla \cdot u$ and $\mathcal{A} = \nabla \times u$, whose third derivatives appear in expressions (B.36) and (B.40). Within the wave equations, we notice the separation of physical quantities. Equation (B.36) deals with the scalar potential, which is associated with P waves, and equations (B.40) deal with the vector potential, which is associated with S waves.

If we solve the system given by expressions (B.36) and (B.40) for $\mathcal{A} = \nabla \times u$ and $\varphi = \nabla \cdot u$, we find u. Thus, at the end, we obtain $u(x,t)$, which is tantamount to solving equation (B.30).

It is possible to use equation (B.30) to derive the wave equation for P waves and the wave equation for S waves directly in terms of the displacement, u, without invoking the potentials, A and ϕ, and their derivatives, $\mathcal{A}$ and φ.[14]

Expressions (B.36) and (B.40) constitute a system of four linear partial differential equations for four unknowns, $\mathcal{A}_1$, $\mathcal{A}_2$, $\mathcal{A}_3$ and φ. We study the characteristic hypersurfaces of this system in Section 2.3.5.1.

Closing remarks

The elastodynamic equations are a particular form of Cauchy's equations of motion. They are Cauchy's equations wherein the stress tensor is defined by the stress-strain equations of the Hookean solids. In turn, the elastodynamic equations contain the wave equations for the waves that propagate in such solids. In anisotropic Hookean solids, there are three types of waves, as discussed in Sections 2.3.5.1 and 3.4.1. In the case of isotropy, there are only two waves; their expressions are formulated in Appendix B.4.3.

In obtaining Cauchy's equations from which we formulate the elastodynamic equations, we use both the intrinsic equivalence and distinct properties of the material and spatial descriptions in continuum mechanics; the former stems from their being—at any instant—inverses of one another, and the latter from the interchange of their dependent and independent variables. Also, this derivation illustrates justifications used to neglect cer-

[14]Readers interested in a derivation of the wave equations in terms of displacements might refer to Rochester (2010).

tain quantities in arriving at a linearized theory, and—hence—gives us an insight into limitations of such a theory.

Elastodynamic equations, which are hyperbolic partial differential equations, underlie the study of motion in the context of the theory of elasticity. We examine the characteristics of these equations in Chapters 2 and 3.

Appendix C

Maxwell equations *in vacuo*

Three of the four Maxwell equations are direct restatements of expressions known before the time of Maxwell: Coulomb's law, Faraday's law, and the absence of free magnetic poles. When he combined these laws with Ampère's law from magnetostatics, Maxwell realized that this equation was inconsistent with the rest since it violated the continuity of charge and current,[1] *and, therefore, it had to be modified. This was the crucial insight that gave birth to the concept of electromagnetic waves.*

Martin Moskovits (1977)

Preliminary remarks

In this appendix, we formulate the Maxwell equations, which are rooted in the fundamental laws of electricity and magnetism. They are the quantitative basis of the theory of electromagnetism. For our purposes, we express them as wave equations, whose characteristics we study in this book. In this formulation we assume that the sources, which are charges and currents, are located in an empty space, *in vacuo*; in other words, we assume the absence of a magnetic or polarizable medium; herein, *in vacuo* does not imply the source-free Maxwell equations.

We begin this appendix by discussing Coulomb's law, no-monopole law, Faraday's law and Ampère's law. We combine these laws into the Maxwell equations. Subsequently, invoking the scalar and vector potentials, we express these equations as wave equations.

Readers might find it useful to study this appendix together with Appendix A, where we discuss the Divergence Theorem and the Curl Theorem.

[1] In other words, *the conservation of electric charge.*

C.1 Formulation

C.1.1 *Fundamental equations*

The electromagnetic field manifests its presence by the electric-field intensity, E, and by the magnetic induction, B. We measure the effect of this field by force F that is exerted on a particle bearing charge q and moving with velocity v. This force is

$$F = q\,(E + v \times B)\;;$$

it is commonly referred to as the Lorentz force. The equation of continuity for electric charges states that

$$\nabla \cdot J = -\frac{\partial \rho}{\partial t}\,, \tag{C.1}$$

with ρ denoting the electric-charge density, which is the amount of charge per unit volume. This equation states the conservation of charge and can be viewed as a definition of J, which is the electric current density: the rate at which charge flows through a unit area per second.

In Appendices C.1.2–C.1.5, we consider equations involving E and B, as well as ρ and J.

C.1.2 *Coulomb's law*

According to Coulomb's law, the flux of E through any closed surface, S, is proportional to the charge that is enclosed within this surface. Mathematically, we can write it as the integral equation given by

$$\iint\limits_S E \cdot N\,\mathrm{d}S = \iiint\limits_V \frac{\rho}{\epsilon_0}\,\mathrm{d}V\,, \tag{C.2}$$

where N is the unit vector normal to S and pointing away from volume V. Parameter ϵ_0 is a proportionality constant whose value can be determined experimentally by measuring the force between two unit charges. This value depends on our definition of unit charge.

Returning to the integral equation and invoking the Divergence Theorem, stated on page 215, as well as using the linearity of the integral operator, we get

$$\iiint\limits_V \left(\nabla \cdot E - \frac{\rho}{\epsilon_0}\right)\mathrm{d}V = 0\,.$$

For the integral to vanish for an arbitrary volume, we require that the integrand vanishes. Thus, we obtain a differential equation given by

$$\nabla \cdot E = \frac{\rho}{\epsilon_0}, \tag{C.3}$$

wherein the divergence of E is equal to the charge density scaled by parameter ϵ_0. Equation (C.3) expresses locally—for an infinitesimal region—the properties of E, stated in equation (C.2), for a finite volume, V, enclosed within a surface, S.

C.1.3 *No-monopole law*

Having obtained equation (C.3), we search for an analogous equation to quantify intrinsic properties of B. Since there is no experimental indication of magnetic monopoles, which for B would be analogous to the isolated electric charge for E, the net flux of magnetic induction through S is zero. We can refer to this statement as the no-monopole law. Mathematically, we can write it as the integral equation given by

$$\iint_S B \cdot N \, \mathrm{d}S = 0; \tag{C.4}$$

in other words, the flux of B through any closed surface is zero.

To write equation (C.4) as a differential equation, we invoke the Divergence Theorem, stated on page 215, to get

$$\iiint_V \nabla \cdot B \, \mathrm{d}V = 0.$$

For the integral to vanish for an arbitrary volume, we require that

$$\nabla \cdot B = 0, \tag{C.5}$$

which is a differential equation that is tantamount to equation (C.4); in other words, the zero flux—in the integral equation—is equivalent to the zero divergence, in the differential equation. Herein, both imply that there are no magnetic monopoles.

Note that we cannot use the argument of vanishing integrand in equation (C.4), since the surface is not arbitrary; it is closed to contain volume V.

C.1.4 *Faraday's law*

Herein, we formulate the law of electromagnetic induction, which is called Faraday's law. This law states that the circulation of E around a loop, C, is equal to the negative of the time rate of change of the magnetic flux, B, through this loop. Mathematically, we can write this law as the integral equation given by

$$\int_C E \cdot \mathrm{d}\ell = -\frac{\partial}{\partial t} \iint_S B \cdot N \,\mathrm{d}S\,, \tag{C.6}$$

where S is the surface bounded by curve C, $\mathrm{d}\ell$ is a vector element along this curve and N is the unit vector normal to S. Note that the curve is a loop but the surface, in contrast to Coulomb's and no-monopole laws, is not closed. In all cases, however, there is a closure of a boundary, as discussed on page 224.

To combine the two integrals in equation (C.6), we write this equation as

$$\int_C E \cdot \underbrace{n\,\mathrm{d}s}_{\mathrm{d}\ell} = -\frac{\partial}{\partial t} \iint_S B \cdot N \,\mathrm{d}S\,,$$

where n is the unit vector tangent to C and $\mathrm{d}s$ is the element of length along C. Thus, following the Curl Theorem, stated on page 223, we write

$$\iint_S \left[(\nabla \times E) \cdot N\right] \mathrm{d}S = -\frac{\partial}{\partial t} \iint_S B \cdot N \,\mathrm{d}S\,.$$

Using the linearity of the integral operator and of the scalar product, we get

$$\iint_S \left[\nabla \times E + \frac{\partial B}{\partial t}\right] \cdot N \,\mathrm{d}S = 0\,.$$

For the integral to vanish for an arbitrary surface, we require the integrand to vanish. Since $N \neq 0$, we conclude that

$$\nabla \times E = -\frac{\partial B}{\partial t}\,, \tag{C.7}$$

which is tantamount to equation (C.6) stated herein as a differential equation. Equation (C.7) expresses locally—for an infinitesimal region—the relation between E and B, stated in equation (C.6), for a finite surface, S, bounded by a curve, C.

C.1.5 *Ampère's law*

We search for an equation for B that is analogous to equation (C.6). Such an equation results from Ampère's law whose form, as an integral equation, is

$$\int_C B \cdot \mathrm{d}\ell = \frac{1}{c^2 \epsilon_0} \iint_S J \cdot N \, \mathrm{d}S \,, \tag{C.8}$$

where $c^2\epsilon_0$ is a proportionality constant whose value can be determined experimentally by measuring the force between two unit currents, which are defined as a unit charge per second.

Invoking the Curl Theorem, stated on page 223, and requiring the vanishing of the integral for an arbitrary surface, we write equation (C.8) as a differential equation,

$$\nabla \times B = \frac{J}{c^2 \epsilon_0} \,. \tag{C.9}$$

This equation, however, is not consistent with equation (C.1), which is a fundamental equation expressing the conservation of charge. We can see the contradiction between equations (C.1) and (C.9) by taking the divergence of the latter equation, which results in $\nabla \cdot J = 0$, and which means that the total flux of current out of any closed surface is equal to zero, not to the rate of change of the charge inside this surface, as required by equation (C.1). Maxwell resolved this contradiction by adding, to equation (C.9), a term that contains E,

$$c^2 \nabla \times B = \frac{J}{\epsilon_0} + \frac{\partial E}{\partial t} \,. \tag{C.10}$$

To verify the consistency, we take the divergence of equation (C.10) and use equation (C.3), which is a statement of Coulomb's law, to get

$$\nabla \cdot J = -\epsilon_0 \frac{\partial}{\partial t} \nabla \cdot E = -\frac{\partial \rho}{\partial t} \,,$$

which is equation (C.1), as required.

Examining equation (C.10) in the context of its derivation from equation (C.8), we see that the circulation of B around loop C is proportional to the sum of the flux of the electric current, J, through surface S, and the temporal rate of change of flux of the electric intensity, E, through this surface. For consistency with equation (C.10), the integral equation can be modified to

$$\int_C B \cdot \mathrm{d}\ell = \frac{1}{c^2 \epsilon_0} \iint_S J \cdot N \, \mathrm{d}S + \frac{1}{c^2} \frac{\partial}{\partial t} \iint_S E \cdot N \, \mathrm{d}S \,.$$

C.1.6 *Speed of light*

Examining constant ϵ_0 used in Coulomb's law and constant $c^2\epsilon_0$ used in Ampère's law, we realize that—although their values depend on our definition of the unit charge—their ratio is the same, namely, c^2. As originally remarked by Maxwell and confirmed by subsequent studies, c is the speed of light. Furthermore, the relation between the magnitudes of E and B is $\|E\| = c\|B\|$.

C.1.7 *Maxwell equations*

Equations (C.3), (C.5), (C.7) and (C.10) are the Maxwell equations. Formally, they constitute the entire theory of electrodynamics, where concepts of electricity and magnetism are intimately linked. They also show that the propagation of light belongs to this theory. To appreciate the conciseness of Maxwell's formulations, let us write all four equations.

$$\nabla \cdot E = \frac{\rho}{\epsilon_0}, \tag{C.11}$$

$$\nabla \cdot B = 0, \tag{C.12}$$

$$\nabla \times E = -\frac{\partial B}{\partial t}, \tag{C.13}$$

and

$$c^2 \nabla \times B = \frac{J}{\epsilon_0} + \frac{\partial E}{\partial t}. \tag{C.14}$$

We note that if we consider a static case—which is the case where E and B do not depend on time and, hence, $\partial B/\partial t = \partial E/\partial t = 0$—equations (C.11) and (C.13) deal only with electricity, and equations (C.12) and (C.14) deal only with magnetism. In other words, the Maxwell equations can be separated into the electrostatic and magnetostatic equations. In particular, from this argument we see that equation (C.9) is the magnetostatic analogue of equation (C.14), which incorporates time dependence. Thus, in the static case, there is no contradiction between equations (C.1) and (C.9).

It is time dependence that allows for the descriptions of electricity and magnetism to be combined into a single coherent theory of electromagnetism. We can also refer to this theory as electrodynamics.

Equations (C.11)–(C.14) can be written explicitly in coordinates as

$$\left\{\begin{array}{l}\dfrac{\partial E_1}{\partial x_1}+\dfrac{\partial E_2}{\partial x_2}+\dfrac{\partial E_3}{\partial x_3}=\dfrac{\rho}{\epsilon_0}\\ \dfrac{\partial B_1}{\partial x_1}+\dfrac{\partial B_2}{\partial x_2}+\dfrac{\partial B_3}{\partial x_3}=0\\ \dfrac{\partial E_3}{\partial x_2}-\dfrac{\partial E_2}{\partial x_3}=-\dfrac{\partial B_1}{\partial t}\\ \dfrac{\partial E_1}{\partial x_3}-\dfrac{\partial E_3}{\partial x_1}=-\dfrac{\partial B_2}{\partial t}\\ \dfrac{\partial E_2}{\partial x_1}-\dfrac{\partial E_1}{\partial x_2}=-\dfrac{\partial B_3}{\partial t}\\ c^2\dfrac{\partial B_3}{\partial x_2}-c^2\dfrac{\partial B_2}{\partial x_3}=\dfrac{J_1}{\epsilon_0}+\dfrac{\partial E_1}{\partial t}\\ c^2\dfrac{\partial B_1}{\partial x_3}-c^2\dfrac{\partial B_3}{\partial x_1}=\dfrac{J_2}{\epsilon_0}+\dfrac{\partial E_2}{\partial t}\\ c^2\dfrac{\partial B_2}{\partial x_1}-c^2\dfrac{\partial B_1}{\partial x_2}=\dfrac{J_3}{\epsilon_0}+\dfrac{\partial E_3}{\partial t}\end{array}\right. . \tag{C.15}$$

This system of eight equations for six unknowns, B_i , E_i , where $i = 1, 2, 3$, can be written in a matrix form, as shown on page 27 , in expression (1.57).

C.2 Scalar and vector potentials

We wish to write the four Maxwell equations in a way that allows us to study them in the context of the characteristic hypersurfaces. Herein, we do it by formulating the scalar and vector potentials to write the Maxwell equations as the wave equations in terms of these potentials. Also, we can proceed with system (C.15) to obtain wave equations in terms of the electric and magnetic fields, as discussed in Example 1.7. In both cases, the resulting characteristics are the same, since the differential equations in question exhibit the same form.

Consider equation (C.12). In the context of the properties of vector operators, there is a theorem stating that if the divergence of a vector field is zero, it follows that this field is a curl of a vector field. Thus, equation (C.12) implies that

$$B = \nabla \times A, \tag{C.16}$$

where A is a vector field. We can view this expression as a solution of equation (C.12), and we refer to A as a vector potential. A is not unique;

in view of identity $\nabla \times \nabla\xi = 0$, we can rewrite it as $B = \nabla \times \breve{A}$, where $\breve{A} = A + \nabla\xi$ with ξ being an arbitrary scalar function.[2]

Using expression (C.16), we can write equation (C.13) as

$$\nabla \times E = -\frac{\partial}{\partial t}\nabla \times A\,.$$

Exchanging the order of spatial and temporal derivatives, we rewrite it as

$$\nabla \times E = -\nabla \times \frac{\partial A}{\partial t}\,,$$

to get

$$\nabla \times \left(E + \frac{\partial A}{\partial t}\right) = 0\,.$$

In the context of the properties of vector operators, there is a theorem stating that if the curl of a vector field is zero, it follows that this field is a gradient of a scalar function. Thus, we can write the term in parentheses as

$$E + \frac{\partial A}{\partial t} = -\nabla\phi. \tag{C.17}$$

We refer to ϕ as a scalar potential. Rearranging expression (C.17), we write

$$E = -\nabla\phi - \frac{\partial A}{\partial t}, \tag{C.18}$$

which is a solution of equation (C.13). Examining equations (C.16) and (C.18), we see that vector potential A appears in the expressions for both B and E. Since, as shown following equation (C.16), A is not unique, we must constrain A in such a way that equations (C.16) and (C.18) are compatible with one another. To obtain the required compatibility, we set $\breve{A} = A + \nabla\xi$ in equation (C.18) to get

$$E = -\nabla\phi - \frac{\partial}{\partial t}\left(\breve{A} - \nabla\xi\right) = -\nabla\left(\phi - \frac{\partial\xi}{\partial t}\right) - \frac{\partial\breve{A}}{\partial t}\,, \tag{C.19}$$

where we use the linearity of differential operators and exchange the order of spatial and temporal derivatives. Examining expression (C.18) and the second equality of expression (C.19), we conclude that if we use $\breve{A} = A + \nabla\xi$ in equations (C.16), we must set $\breve{\phi} = \phi - \partial\xi/\partial t$ in equation (C.18). Also, this result means that ξ relates the vector and scalar potentials to one another.

[2]Readers interested in philosophical implications of nonuniqueness of the vector potential might refer to Brown (1994, pp. 142–159).

Let us consider the two remaining Maxwell equations. Using expression (C.18), we can write equation (C.11) as

$$\nabla \cdot \left(\nabla \phi + \frac{\partial A}{\partial t} \right) = -\frac{\rho}{\epsilon_0} ,$$

and rewrite it as

$$\nabla^2 \phi + \frac{\partial}{\partial t} \nabla \cdot A = -\frac{\rho}{\epsilon_0} . \tag{C.20}$$

This equation relates the two potentials to the source.

Consider the last Maxwell equation, namely equation (C.14). We can rewrite it as

$$c^2 \nabla \times B - \frac{\partial E}{\partial t} = \frac{J}{\epsilon_0} . \tag{C.21}$$

Using expressions (C.16) and (C.18), we can rewrite equation (C.21) in terms of the potentials as

$$c^2 \nabla \times (\nabla \times A) + \frac{\partial}{\partial t} \left(\nabla \phi + \frac{\partial A}{\partial t} \right) = \frac{J}{\epsilon_0} .$$

Invoking the identity given by

$$\nabla \times (\nabla \times A) = \nabla (\nabla \cdot A) - \nabla^2 A ,$$

and using the linearity of the differential operator, we get

$$c^2 \nabla (\nabla \cdot A) - c^2 \nabla^2 A + \frac{\partial}{\partial t} \nabla \phi + \frac{\partial^2 A}{\partial t^2} = \frac{J}{\epsilon_0} .$$

Exchanging the order of spatial and temporal derivatives, we can combine the first and the third terms on the left-hand side to write

$$-c^2 \nabla^2 A + \nabla \left(c^2 \nabla \cdot A + \frac{\partial \phi}{\partial t} \right) + \frac{\partial^2 A}{\partial t^2} = \frac{J}{\epsilon_0} .$$

Since the divergence of A is arbitrary due to the freedom of adding the gradient of any function, as discussed on page 260, let us set

$$\nabla \cdot A = -\frac{1}{c^2} \frac{\partial \phi}{\partial t} . \tag{C.22}$$

This way, the term in parentheses disappears and we get

$$-c^2 \nabla^2 A + \frac{\partial^2 A}{\partial t^2} = \frac{J}{\epsilon_0} ,$$

which we can rewrite as

$$\nabla^2 A - \frac{1}{c^2} \frac{\partial^2 A}{\partial t^2} = -\frac{J}{c^2 \epsilon_0} ; \tag{C.23}$$

this is a vector equation that contains three equations that relate A_1, A_2 and A_3 to J_1, J_2 and J_3.

To complete our formulations of the Maxwell equations in terms of potentials, let us substitute expression (C.22) into equation (C.20) to obtain

$$\nabla^2\phi - \frac{1}{c^2}\frac{\partial^2\phi}{\partial t^2} = -\frac{\rho}{\epsilon_0}. \tag{C.24}$$

Thus, the classical form of the Maxwell equations given by expressions (C.11), (C.12), (C.13) and (C.14), which comprise eight equations, is equivalent to equations (C.23) and (C.24) together with gauge equation (C.22). We notice that expressions (C.23) and (C.24) possess a similarity of form; they are wave equations. Also, we notice the separation of physical quantities. In equations (C.23), the vector potential, A, is related to the current, J, while in equation (C.24), the scalar potential, ϕ, is related to the charge density, ρ.

Expressions (C.23) and (C.24) are four linear second-order partial differential equations for four unknowns: A_1, A_2, A_3 and ϕ. Together with equation (C.22), which is a linear first-order partial differential equation that relates equations (C.23) and (C.24), we have a system of five equations to study its characteristic hypersurfaces in Sections 3.4.2 and 4.7.2.

If we solve the system given by expressions (C.23) and (C.24) for A and ϕ, we could get B using equation (C.16) and E using equation (C.18).

Closing remarks

Having discussed the Maxwell equations *in vacuo*, let us comment on the Maxwell equations in matter. Instead of the vacuum, one could consider a polarizable and magnetizable continuum. We note that neither set of the Maxwell equations is a particular case of the other set. Fundamentally, they both stand at the same level of generality. The Maxwell equations in matter can be derived from the equations in vacuum by averaging the electric and magnetic dipoles over the medium, and the equations in vacuum can be obtained from the equations in matter by setting the polarization and magnetization vectors to zero.

Since ρ and J are the sources of the electromagnetic fields that extend far beyond these sources, we can—at large distances—consider a simpler form of the Maxwell equations by setting both ρ and J to zero. Certain authors refer to this case as the Maxwell equations *in vacuo*; we refer to it as the source-free Maxwell equations.

We conclude this appendix by commenting on analogies between equations (C.24) and (B.36), and between equations (C.23) and (B.40), in other words, between our formulation of wave equations in elastodynamics and electrodynamics. In both cases we invoke the scalar and vector wave functions, as shown by equations (C.24) and (B.36) and by equations (C.23) and (B.40), respectively. However, the invoked functions for elastodynamics and electrodynamics have distinct origins. In the former case, the potentials, A and ϕ, are related to the wave functions by derivatives of the displacement, $\varphi = \nabla \cdot u$ and $\mathcal{A} = \nabla \times u$; in the latter case, the fields are derivatives of the potentials, $B = \nabla \times A$ and $E = -\nabla\phi - \partial A/\partial t$. In the former case, the potentials are a consequence of the Helmholtz Theorem; in the latter case, they are a consequence of the Maxwell equations.

The left-hand sides of all four equations possess the identical form in which c, $\sqrt{(\lambda + 2\mu)/\rho}$ and $\sqrt{\mu/\rho}$ are the velocities with which the electromagnetic waves, P waves and S waves, propagate respectively. The right-hand sides of equations (B.36) and (B.40) are zero, which means that we consider the case where there is no source generating waves in an elastic medium. The right-hand sides of equations (C.23) and (C.24) indicate that the magnetic field is generated by the electric current while the electric field is generated by the electric charge. We could make the formal analogy even closer by including sources in equations (B.36) and (B.40). However, since the characteristic hypersurfaces, which are the focus of our study, are associated with the highest-order derivatives, adding source functions on the right-hand sides of equations (B.36) and (B.40) does not affect the characteristics. In contrast to equations (B.36) and (B.40), equations (C.23) and (C.24) belong to a single system of equations. In the former case, the gauge equation is $\nabla \cdot A = 0$, while in the latter case, the analogous equation is $c^2\nabla \cdot A = -\partial\phi/\partial t$, which relates the equations for A to the equation for ϕ.

The characteristics of the Maxwell equations are examined in Chapters 1, 2 and 3.

Appendix D

Fourier series and transforms

Fourier analysis [...] may be described as a collection of related techniques for resolving general functions into sums or integrals of simple functions or functions with certain special properties. [...] Fourier series give expansions of periodic functions on the line in terms of trigonometric functions. [...] Fourier transform [...] provides a way of expanding functions on the whole real line $\mathbb{R} = (-\infty, \infty)$ as a (continuous) superposition of the basic oscillatory functions $e^{\iota\xi x}$, ($\xi \in \mathbb{R}$) in much the same way that Fourier series are used to expand functions on a finite interval.[1]

Gerald B. Folland (1992)

Preliminary remarks

Solutions of differential equations are functions. It might be necessary or convenient to restrict a space of functions to a subspace that exhibits particular properties. For instance, we might consider differentiable functions—the space of all differentiable functions is an example of a functional space. It is possible to approximate an element of the space by an element of the subspace, which could allow us to obtain convenient expressions to study a given function in terms of its simpler counterparts. For instance, as suggested in footnote 1, we might approximate "generic" functions by "symmetric" ones.

[1]An insightful description of a prototype of the Fourier series and transforms is presented by Tao (2008), where he writes that

> broadly speaking, a Fourier transform is a systematic way to decompose "generic" functions into a superposition of "symmetric" functions.

To consider such approximations, we need a notion of similarity between functions, which leads us to the notion of a norm, and, hence, to the concept of distance and perpendicularity.

We begin this appendix by examining the concept of similarity of functions. Subsequently, we introduce the Fourier series, which results in the Fourier transform.

D.1 Similarity of functions

To study similarity of functions, we need to examine if pertinent differences between functions are small enough to be acceptable for our purposes. To do so, we invoke the concept of a norm. If we consider real-valued functions, a norm can be defined by

$$\|f\|_2 = \sqrt{\int_I (f(x))^2 \, \mathrm{d}x}\,,$$

where the subscript indicates a 2-norm, also known as the Euclidean norm, and I stands for an interval. To include complex-valued functions in our considerations, we extend the above formula to

$$\|f\|_2 = \sqrt{\int_I f(x)\,\overline{f(x)}\,\mathrm{d}x} = \sqrt{\int_I |f(x)|^2 \, \mathrm{d}x}\,, \tag{D.1}$$

where $\bar{f}$ represents the complex conjugate of f and $|f|$ represents its modulus. We note that, for real-valued functions, $f(x) = \overline{f(x)}$. As required, this definition of a norm satisfies the following properties.

nondegeneracy: $\|f\|_2 = 0$ only if $f = 0$,
triangular inequality: $\|f + g\|_2 \leq \|f\|_2 + \|g\|_2$,
absolute homogeneity: $\|af\|_2 = |a|\,\|f\|_2$, for any number a.

We restrict our attention to functions whose norm is a real number; in other words, functions belonging to the space defined by

$$L^2(I) = \{f(x) \mid \|f\|_2 < \infty\}\,, \tag{D.2}$$

which is the space of square-integrable functions. Notably, the continuous functions on a closed interval, $C(I)$, are square-integrable, namely, $C(I) \subset L^2(I)$.

We can view function $f(x)$ as a vector with an infinite number of components. Since an integral is a summation, we define the scalar product of functions f and g in $L^2(I)$ as

$$\langle f, g\rangle = \int_I f(x)\,\overline{g(x)}\,\mathrm{d}x\,, \tag{D.3}$$

which has the required properties of a scalar product, namely,

commutativity: $\langle f, g\rangle = \langle g, f\rangle$,
linearity: $\langle af + g, h\rangle = a\langle f, h\rangle + \langle g, h\rangle$, for any real number a ,
positive definiteness: $\langle f, f\rangle \geqslant 0$, for all f , where $\langle f, f\rangle = 0$ if and only if $f = 0$.

The scalar product allows us to define the distance between functions, and the perpendicularity of functions.

The norm is induced by the scalar product. They are related uniquely by

$$\|f\|_2^2 = \langle f, f\rangle\,,$$

provided the norm itself satisfies the parallelogram law, which we can write as

$$\|f + g\|_2^2 + \|f - g\|_2^2 = 2\,\|f\|_2^2 + 2\,\|g\|_2^2\,;$$

the uniqueness results from the polarization identity.[2]

We note that, by analogy to expression (D.1), we can define other norms,

$$\|f\|_p = \sqrt[p]{\int_I |f(x)|^p\,\mathrm{d}x}\,; \tag{D.4}$$

functions that satisfy $\|f\|_p < \infty$ form a space denoted by $L^p(I)$. Herein, we limit our attention to $\|f\|_2$ and $L^2(I)$. Consequently, for notational convenience, we omit the subscript; in other words, for the remainder of this appendix, $\|f\| \equiv \|f\|_2$.

For our formulations, we consider the space of compactly supported functions in $\mathbb{R}^n$, which we denote by $C_0(\mathbb{R}^n)$. A compactly supported function in $\mathbb{R}^n$ is a function that is nonzero only on a closed bounded set. An example of such a set is a closed interval, such that the function is zero at positive and negative infinities; it is a stronger condition than a function tending to zero at infinities. If we require that a function on $\mathbb{R}^n$ be compactly supported with k continuous derivatives, we write $C_0^k(\mathbb{R}^n)$, and for infinitely differentiable functions with compact support, $C_0^\infty(\mathbb{R}^n)$.

[2]Readers interested in this identity might refer to textbooks on linear algebra, including Lang (1968, Chapter VIII, §4).

D.2 Fourier series

Since we can view $f(x)$ as a vector with an infinite number of components, we wish to express it in terms of such components. To do so, we invoke the concept of linearly independent and mutually orthogonal functions. An important example of such functions on interval $[0, L]$ is given by

$$\phi_0 = 1\,,$$

$$\phi_k = \cos\left(\frac{2\pi}{L}kx\right),$$

$$\psi_k = \sin\left(\frac{2\pi}{L}kx\right),$$

where $k = 1, 2, 3, \ldots$. Orthogonality means that, in view of definition (D.3) and its commutativity, $\langle\phi_0, \phi_k\rangle = \langle\phi_k, \psi_k\rangle = \langle\psi_k, \phi_0\rangle = 0$. Linear independence means that none of the three functions can be written as a linear combination of the other two.

To examine a function, we often wish to expand it in term of orthonormal functions. In the following example, let us consider a set of such functions to be used in the Fourier series.

Example D.1. Let us show that

$$\{1, \cos x, \cos(2x), \ldots, \cos(kx), \sin x, \sin(2x), \ldots, \sin(kx)\}\,, \tag{D.5}$$

whose period is 2π, form an orthogonal family. Considering all pairs of set (D.5), we obtain

$$\int_{-\pi}^{\pi} 1 \sin(mx)\,\mathrm{d}x = \int_{-\pi}^{\pi} 1 \cos(mx)\,\mathrm{d}x = 0,$$

where $m \in \mathbb{N}\setminus\{0\}$, which is the set of the natural numbers excluding zero.

$$\int_{-\pi}^{\pi} \sin(mx)\sin(\ell x)\,\mathrm{d}x = \int_{-\pi}^{\pi} \cos(mx)\cos(\ell x)\,\mathrm{d}x = 0\,,$$

where $m \in \mathbb{N}\setminus\{0\}$, $\ell \in \mathbb{N}\setminus\{0\}$.

$$\int_{-\pi}^{\pi} \cos(mx)\sin(\ell x)\,\mathrm{d}x = 0\,,$$

where $m \in \mathbb{N}\setminus\{0\}$, $\ell \in \mathbb{N}\setminus\{0\}$. Since every two distinct members of set (D.5) are orthogonal to one another, this set forms a family of orthogonal functions.

It is possible to express any continuous function on interval $[0, L]$ as a sum of coefficients times these orthogonal functions. We can find coefficients $a_0, a_1, \ldots$ and $b_1, b_2, \ldots$ for function $f \in L^2(I)$, such that f can be expressed as

$$f(x) = a_0 + \sum_{k=1}^{\infty} a_k \cos\left(\frac{2\pi}{L}kx\right) + \sum_{k=1}^{\infty} b_k \sin\left(\frac{2\pi}{L}kx\right). \tag{D.6}$$

We proceed in a manner analogous to finding components of a vector expressed in an orthogonal basis in $\mathbb{R}^n$. If the basis functions are orthonormal, the coefficients are the orthogonal projections of function f onto the directions of functions ϕ_0, ϕ_k and ψ_k. These projections are given by the scalar products of f with a basis function. Since the norms of ϕ_0, ϕ_k and ψ_k are not equal to unity, we scale these functions by an appropriate factor to obtain unit norms. Hence, in view of linearity of the scalar product and taking into account the factors, we write

$$\begin{aligned} a_0 &= \frac{\langle f, \phi_0 \rangle}{\|\phi_0\|^2}, \\ a_k &= \frac{\langle f, \phi_k \rangle}{\|\phi_k\|^2}, \\ b_k &= \frac{\langle f, \psi_k \rangle}{\|\psi_k\|^2}, \end{aligned} \tag{D.7}$$

where, following expression (D.1), the corresponding norms are

$$\|\phi_0\| = \sqrt{\int_0^L 1 \, \mathrm{d}x} = \sqrt{L},$$

$$\|\phi_k\| = \sqrt{\int_0^L \left|\cos\left(\frac{2\pi}{L}kx\right)\right|^2 \mathrm{d}x} = \sqrt{\frac{L}{2}}$$

and

$$\|\psi_k\| = \sqrt{\int_0^L \left|\sin\left(\frac{2\pi}{L}kx\right)\right|^2 \mathrm{d}x} = \sqrt{\frac{L}{2}}.$$

Example D.2. Let us obtain an orthonormal family corresponding to set (D.5). Since $L = 2\pi$, we write

$$\|1\| = \sqrt{2\pi},$$

$$\|\cos(kx)\| = \sqrt{\pi}$$

and

$$\|\sin(kx)\| = \sqrt{\pi}\,,$$

where $k \in \mathbb{N}\setminus\{0\}$. Hence, normalizing, we rewrite set (D.5) as

$$\left\{\frac{1}{\sqrt{2\pi}}, \frac{\cos x}{\sqrt{\pi}}, \frac{\cos(2x)}{\sqrt{\pi}}, \ldots, \frac{\cos(kx)}{\sqrt{\pi}}, \frac{\sin x}{\sqrt{\pi}}, \frac{\sin(2x)}{\sqrt{\pi}}, \ldots, \frac{\sin(kx)}{\sqrt{\pi}}\right\}. \tag{D.8}$$

Set (D.8) consists of $2k+1$ orthonormal functions.

In general, we can write coefficients (D.7) as

$$\begin{aligned} a_0 &= \frac{\langle f, \phi_0\rangle}{L} = \frac{1}{L}\int_0^L f(x)\,\mathrm{d}x, \\ a_k &= \frac{2\langle f, \phi_k\rangle}{L} = \frac{2}{L}\int_0^L f(x)\cos\left(\frac{2\pi}{L}kx\right)\mathrm{d}x, \\ b_k &= \frac{2\langle f, \psi_k\rangle}{L} = \frac{2}{L}\int_0^L f(x)\sin\left(\frac{2\pi}{L}kx\right)\mathrm{d}x\,. \end{aligned} \tag{D.9}$$

Decomposition (D.6) of a function into the directions of ϕ_0 ϕ_k and ψ_k is the Fourier series of f, and coefficients (D.9) are the Fourier coefficients.

Example D.3. Let us develop a 2π-periodic function,

$$f(x) = |x|\,, \qquad x \in [-\pi, \pi]\,, \tag{D.10}$$

into its Fourier series. Since $L = 2\pi$ and f is an even function, using expressions (D.9), we obtain

$$a_0 = 2\left(\frac{1}{2\pi}\int_0^\pi x\,\mathrm{d}x\right) = \frac{\pi}{2}\,;$$

similarly,

$$a_k = \frac{2}{\pi}\int_0^\pi x\cos(kx)\,\mathrm{d}x = \frac{2}{\pi}\left(\frac{x\sin(kx)}{k} + \frac{\cos(kx)}{k^2}\right)\Bigg|_0^\pi = \begin{cases} 0\,, & \text{even } k \\ -\dfrac{4}{\pi}\dfrac{1}{k^2}\,, & \text{odd } k \end{cases}$$

and

$$b_k = 0\,.$$

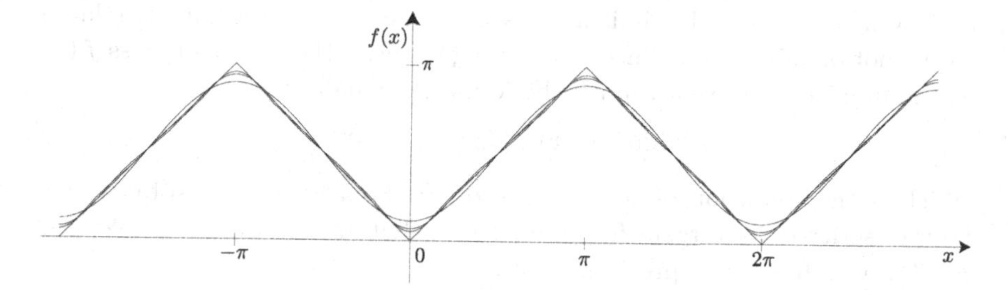

Fig. D.1 Function $|x|$ and three partial sums of its Fourier series, as discussed in Example D.3. The accuracy of approximation, which is already high for $m = 1$, increases for $m = 2$ and $m = 3$, as expected.

Thus,

$$|x| = \frac{\pi}{2} - \frac{4}{\pi}\left(\cos x + \frac{\cos 3x}{3^2} + \frac{\cos 5x}{5^2} + \dots\right) = \frac{\pi}{2} - \frac{4}{\pi}\sum_{m=1}^{\infty}\frac{\cos\left((2m-1)\,x\right)}{(2m-1)^2}\,,$$

as shown in Figure D.1.

We note that, in general, if f is even, $b_k = 0$, and if f is odd, $a_k = 0$, for all k.

To gain further insight into the Fourier series, we can invoke a trigonometric identity,

$$A\cos\theta + B\sin\theta = \sqrt{A^2 + B^2}\cos\left(\theta - \arctan\frac{B}{A}\right)\,,$$

to rewrite series (D.6) as

$$f(x) = a_0 + \sum_{k=1}^{\infty} d_k \cos\left(\frac{2\pi}{L}kx - \Phi_k\right)\,,$$

where

$$d_k = \sqrt{a_k^2 + b_k^2}$$

and

$$\Phi_k = \arctan\frac{b_k}{a_k}\,.$$

Thus a function on interval I can be expanded as a sum of cosines. The kth cosine has amplitude d_k, angular frequency $2k\pi/L$, and phase Φ_k.

Also, let us examine the meaning of infinite sums in expression (D.6) by considering partial sums, namely,

$$f_n(x) = a_0 + \sum_{k=1}^{n} a_k \cos\left(\frac{2\pi}{L}kx\right) + \sum_{k=1}^{n} b_k \sin\left(\frac{2\pi}{L}kx\right)\,, \tag{D.11}$$

and by investigating the limit as $n \to \infty$. The notion of a limit is related to the notion of a norm. Since—according to series (D.6)—we express $f(x)$ as the infinite sum, we establish the following equality:

$$\lim_{n\to\infty} \|f(x) - f_n(x)\|_2 = 0\,,$$

which is the condition of convergence of the Fourier series. Notably, the Fourier series is convergent for all functions that are square-integrable, $f \in L^2(I)$, as defined in expression (D.2).

D.3 Fourier transform

D.3.1 *Formulation*

Let us introduce a concise formula for the Fourier series and its coefficients; this formula allows for an important generalization.

The relation between the trigonometric functions and the complex exponential function is given by the Euler formula,

$$e^{\iota x} = \cos(x) + \iota \sin(x)\,, \tag{D.12}$$

which can be written also as

$$\cos(x) = \frac{e^{\iota x} + e^{-\iota x}}{2}\,,$$

$$\sin(x) = \frac{e^{\iota x} - e^{-\iota x}}{2\iota}\,.$$

Using these expressions, we write series (D.6) as

$$f(x) = \sum_{k=-\infty}^{\infty} c_k e^{\iota k x}\,,$$

where

$$c_0 = a_0, \quad c_k = \frac{1}{2}(a_k - \iota b_k)\,, \quad c_{-k} = \frac{1}{2}(a_k + \iota b_k)\,,$$

for $k = 1, 2, 3, \ldots$. In a manner analogous to expression (D.8),

$$\left\{\frac{\exp\{-\iota k x\}}{\sqrt{2\pi}}, \ldots, \frac{\exp\{-\iota x\}}{\sqrt{2\pi}}, \frac{1}{\sqrt{2\pi}}, \frac{\exp\{\iota x\}}{\sqrt{2\pi}}, \ldots, \frac{\exp\{\iota k x\}}{\sqrt{2\pi}}\right\}$$

is the family of $2k+1$ orthonormal functions in terms of which a given function can be expressed. In such a compact notation, coefficients (D.7) are

$$c_k = \frac{1}{L}\int_0^L f(x)\, e^{-\iota k x}\, \mathrm{d}x\,,$$

where $k = \ldots, -3, -2, -1, 0, 1, 2, 3, \ldots$.

Using a compact notation, let us consider the Fourier series for a function defined on a closed interval, $[-a, a]$; we write

$$f(x) = \sum_{k=-\infty}^{\infty} C_k e^{\iota k x \frac{\pi}{a}},$$

where

$$C_k := \frac{1}{2\pi}\frac{\pi}{a}\int_{-a}^{a} f(\xi)\, e^{-\iota k \xi \frac{\pi}{a}}\, \mathrm{d}\xi .$$

We combine these two expressions to write

$$f(x) = \frac{1}{2\pi}\frac{\pi}{a}\sum_{k=-\infty}^{\infty} e^{\iota x k \frac{\pi}{a}} \int_{-a}^{a} f(\xi)\, e^{-\iota k \xi \frac{\pi}{a}}\, \mathrm{d}\xi .$$

Substituting $\Delta p := \pi/a$ and $p_k = k\Delta p = k\pi/a$, we write

$$f(x) = \frac{1}{2\pi}\sum_{k=-\infty}^{\infty} e^{\iota x p_k}\left(\int_{-a}^{a} f(\xi)\, e^{-\iota p_k \xi}\, \mathrm{d}\xi\right)\Delta p .$$

This sum resembles a Riemann sum defining the integral in p. Indeed, if we take a limit of Δp as $a \to \infty$, we obtain $\Delta p \to 0$, and

$$f(x) = \frac{1}{2\pi}\int_{-\infty}^{\infty} e^{\iota x p} \int_{-\infty}^{\infty} f(\xi)\, e^{-\iota p \xi}\, \mathrm{d}\xi\, \mathrm{d}p, \tag{D.13}$$

which is called the Fourier Integral.

Considering the integral with respect to ξ, we write

$$\int_{-\infty}^{\infty} f(\xi)\, e^{-\iota p \xi}\, \mathrm{d}\xi =: \hat{f}(p), \tag{D.14}$$

which is a function of p. Thus, we rewrite expression (D.13) as

$$f(x) = \frac{1}{2\pi}\int_{-\infty}^{\infty} \hat{f}(p)\, e^{\iota x p}\, \mathrm{d}p, \tag{D.15}$$

Denoting, in expression (D.14), the integration variable, ξ, by x, we define a transform,

$$f(x) \mapsto \hat{f}(p) := \frac{1}{\sqrt{2\pi}}\int_{-\infty}^{\infty} f(x)\, e^{-\iota x p}\, \mathrm{d}x, \tag{D.16}$$

and, in view of expression (D.15), its inverse,

$$\hat{f}(p) \mapsto f(x) := \frac{1}{\sqrt{2\pi}} \int_{-\infty}^{\infty} \hat{f}(p)\, e^{\iota x p}\, \mathrm{d}p\,. \tag{D.17}$$

The product of factors $1/\sqrt{2\pi}$ is $1/2\pi$, in agreement with expression (D.13). Expression (D.16) defines the Fourier transform and expression (D.17) defines the inverse Fourier transform.[3]

The Fourier transform transforms absolutely integrable functions to continuous functions, $f \in L^1(\mathbb{R}) \mapsto \hat{f} \in C(\mathbb{R})$. One can extend this definition to the square-integrable functions using the fact that compactly supported functions are dense in $L^2(\mathbb{R})$.

Let us write the Fourier transform for functions of several variables. For $f(x)$, where $x = [x_1, x_2, \ldots, x_n]$, we extend transform (D.16) to

$$\hat{f}(p) = \frac{1}{(2\pi)^{n/2}} \int_{\mathbb{R}^n} f(x) \exp\{-\iota x \cdot p\}\, \mathrm{d}x\,, \tag{D.18}$$

where $p = [p_1, p_2, \ldots, p_n]$ and $\mathrm{d}x := \mathrm{d}x_1\, \mathrm{d}x_2 \ldots \mathrm{d}x_n$. The inverse transform can be written as

$$f(x) = \frac{1}{(2\pi)^{n/2}} \int_{\mathbb{R}^n} \hat{f}(p) \exp\{\iota x \cdot p\}\, \mathrm{d}p\,, \tag{D.19}$$

where $\mathrm{d}p := \mathrm{d}p_1\, \mathrm{d}p_2 \ldots \mathrm{d}p_n$. In these expressions, $\int_{\mathbb{R}^n}$ stands for n integrals from $-\infty$ to ∞.

D.3.2 *Application to differential equations*

To study implications of the Fourier transform on differential equations, we investigate the transform of a derivative of a function, namely,

$$\widehat{\frac{\mathrm{d}f}{\mathrm{d}x}} = \frac{1}{\sqrt{2\pi}} \int_{-\infty}^{\infty} \frac{\mathrm{d}f}{\mathrm{d}x} e^{-\iota x p}\, \mathrm{d}x\,.$$

We integrate by parts to obtain

$$\widehat{\frac{\mathrm{d}f}{\mathrm{d}x}} = \frac{1}{\sqrt{2\pi}} \left(\lim_{a \to \infty} \left[f(x)\, e^{-\iota x p} \right]_{x=-a}^{a} + \iota p \int_{-\infty}^{\infty} f(x)\, e^{-\iota x p}\, \mathrm{d}x \right).$$

[3]Readers interested in the precise meaning of these definitions and a rigorous proof of the inverse might refer to books on functional analysis, such as Reed and Simon (1980) and Rudin (1991).

If $f(x)$ vanishes at the positive and negative infinity, then the limit term vanishes, and we write

$$\widehat{\frac{\mathrm{d}f}{\mathrm{d}x}} = \frac{\iota p}{\sqrt{2\pi}} \int_{-\infty}^{\infty} f(x)\, e^{-\iota x p}\, \mathrm{d}x \,.$$

In view of definition (D.16), we recognize the Fourier transform of function f, and, hence, we write

$$\widehat{\frac{\mathrm{d}f}{\mathrm{d}x}} = \iota p \hat{f} \,.$$

This is an important formula. It states that the Fourier transform transforms the operation of differentiation into the operation of multiplication by the independent variable in the transformed domain. It can be shown that the kth-order derivative transforms as

$$\widehat{\frac{\mathrm{d}^k f}{\mathrm{d}x^k}} = (\iota p)^k \hat{f} \,, \tag{D.20}$$

which allows us to transform derivatives into algebraic expressions.

Example D.4. Let us use the Fourier transform to solve the wave equation given by

$$\frac{\partial^2 u(x,t)}{\partial x^2} = \frac{1}{v^2} \frac{\partial^2 u(x,t)}{\partial t^2} \,. \tag{D.21}$$

Applying transform (D.16), we write

$$\frac{1}{\sqrt{2\pi}} \int_{-\infty}^{\infty} \frac{\partial^2 u(x,t)}{\partial x^2} \exp\{-\iota x p\}\, \mathrm{d}x = \frac{1}{\sqrt{2\pi}} \int_{-\infty}^{\infty} \frac{1}{v^2} \frac{\partial^2 u(x,t)}{\partial t^2} \exp\{-\iota x p\}\, \mathrm{d}x \,, \tag{D.22}$$

which is the transform with respect to x. Exchanging, on the right-hand side, the order of integration and differentiation—which are with respect to different variables—we write equation (D.22) as

$$\frac{1}{\sqrt{2\pi}} \int_{-\infty}^{\infty} \frac{\partial^2 u(x,t)}{\partial x^2} \exp\{-\iota x p\}\, \mathrm{d}x = \frac{1}{v^2} \frac{\partial^2}{\partial t^2} \left(\frac{1}{\sqrt{2\pi}} \int_{-\infty}^{\infty} u(x,t) \exp\{-\iota x p\}\, \mathrm{d}x \right) .$$

In view of definition (D.16), we see that the left-hand side is the Fourier transform of $\partial^2 u / \partial x^2$, and the term in parentheses on right-hand side is the Fourier transform of u. Using relation (D.20) on the left-hand side and definition (D.16) on the right-hand side, we obtain

$$(\iota p)^2 \hat{u}(p,t) = \frac{1}{v^2} \frac{\mathrm{d}^2 \hat{u}(p,t)}{\mathrm{d}t^2} \,,$$

where we use d, not ∂, since this is an ordinary differential equation in t; herein, $p \in \{-\infty, \infty\}$, can be viewed as a parameter. Rearranging, we write

$$\frac{\mathrm{d}^2 \hat{u}(p,t)}{\mathrm{d}t^2} + (vp)^2\, \hat{u}(p,t) = 0\,. \tag{D.23}$$

The solution of equation (D.23) is

$$\hat{u}(p,t) = \hat{u}_1(p) \exp\{\iota v p t\} + \hat{u}_2(p) \exp\{-\iota v p t\}\,,$$

where u_i are constants with respect to x and correspond to given values of $p \in \{-\infty, \infty\}$. To obtain the solution in terms of $u(x,t)$, we apply inverse (D.17) and write

$$\frac{1}{\sqrt{2\pi}} \int\limits_{-\infty}^{\infty} \hat{u}(p,t) \exp\{\iota x p\} \mathrm{d}p \tag{D.24}$$

$$= \frac{1}{\sqrt{2\pi}} \int\limits_{-\infty}^{\infty} \hat{u}_1(p) \exp\{\iota p(x+vt)\} \mathrm{d}p + \frac{1}{\sqrt{2\pi}} \int\limits_{-\infty}^{\infty} \hat{u}_2(p) \exp\{\iota p(x-vt)\} \mathrm{d}p\,.$$

In view of expression (D.17), we recognize the left-hand side to be $u(x,t)$. Denoting $y := x+vt$ and $z := x-vt$, we formally write the right-hand side as

$$\frac{1}{\sqrt{2\pi}} \int\limits_{-\infty}^{\infty} \hat{u}_1(p) \exp\{\iota p y\} \mathrm{d}p + \frac{1}{\sqrt{2\pi}} \int\limits_{-\infty}^{\infty} \hat{u}_2(p) \exp\{\iota p z\} \mathrm{d}p = u_1(y) + u_2(z)\,,$$

and apply inverse (D.17). Recalling definitions of y and z, we write the inverse transform of expression (D.24) as

$$u(x,t) = u_1(x+vt) + u_2(x-vt)\,. \tag{D.25}$$

This is the general solution of equation (D.21): the wave equation in one spatial dimension. Solution (D.25) is equivalent to solution (2.20), which is obtained in the context of the directional derivative. It is also equivalent to the d'Alembert solution of the wave equation (e.g., Slawinski, 2015, Section 6.5.1). However, in contrast to the d'Alembert approach, the Fourier-transform approach is not limited to the single spatial dimension.

Note that, having obtained equation (D.23), we could proceed with the Fourier transform—this time, with respect to t—to obtain

$$r^2 \hat{u}(p,r) = (vp)^2\, \hat{u}(p,r)\,,$$

which is an algebraic equation, whose nontrivial solution requires $r^2 - (vp)^2 = 0$. This is a dispersion relation; it states that all plane waves—the summands in space-time corresponding to each Fourier-transform term, $\hat{u}(p,r)$—travel with the same velocity, regardless of frequency.

Closing remarks

In the context of this book, the Fourier series and the resulting Fourier transform allow us to study differential equations by transforming these equations into algebraic equations. Notably, the transform allows us to examine these equations and their solutions in both the temporal domain and the frequency domain.

The Fourier series and transform are explicitly used in Chapters 4 and 5.

Appendix E

Distributions

Les distributions ont plusieurs propriétés très différentes. D'abord — et essentiellement dirais-je —, elles sont une généralisation de la notion de fonction, qui permet de résoudre le problème de la derivation.[1]

Laurent Schwartz (1997)

Preliminary remarks

As stated in Appendix D, solutions of differential equations are functions. Commonly, it is assumed that these functions are differentiable, which restricts solutions to smooth cases. Since physical phenomena need not exhibit such a smoothness—for instance, waves might have sharp crests—in this appendix we introduce generalized functions, known also as distributions, which allow us to consider nondifferentiable solutions of differential equations.

We begin this appendix with a discussion of the Dirac delta—a well-known example of distributions—which allows us to introduce a concept of distributions. Subsequently, we examine the properties of, and operations

[1] *Distributions have several very different properties. Mainly—and, I would say, essentially—they are a generalization of the notion of function, which allows us to solve the problem of differentiation.*

Hörmander (1983) makes a similar statement.

> In differential calculus one encounters immediately the unpleasant fact that not every function is differentiable. The purpose of distribution theory is to remedy this flaw; indeed, the space of distributions is essentially the smallest extension of the space of continuous functions where differentiation is always well defined.

on, distributions. We conclude this appendix with the definition of symbols of differential operators.

E.1 Definition of distributions

In contrast to a function, a distribution is not defined by itself but by properties that relate it to well-behaved functions. These functions are called test functions and belong to the space of compactly supported and infinitely differentiable functions, $C_0^\infty(\mathbb{R}^n)$, which are discussed on page 267. To define a distribution, we consider its action on the test functions.

We can write the action of the Dirac delta, which is a commonly used distribution, as

$$\langle \delta, f \rangle := f(0) ; \tag{E.1}$$

in other words, δ acts on a test function by giving the value of that function at zero. This action has the property of being a continuous linear function on the space of infinitely differentiable functions. Since an operator that assigns a number to a function is called a functional, the action of the Dirac delta on a function is a linear functional. Let us define a general distribution.

Definition E.1. A distribution on $\mathbb{R}^n$ is a continuous linear functional operating on $C_0^\infty(\mathbb{R}^n)$, which is a test function.

By analogy to product (D.3), the action of distribution h on a test function, $f \in C_0^\infty$, is defined by

$$\langle h, f \rangle := \int_{\mathbb{R}^n} h(x) f(x) \, \mathrm{d}x , \tag{E.2}$$

where $\int_{\mathbb{R}^n} [\,] \, \mathrm{d}x$ stands for the integral over n variables from $-\infty$ to ∞. Following this definition, we can perform operations on distributions using operations on functions, as illustrated by formula (E.11), below.

Example E.1. If we consider $x \in \mathbb{R}$ for expressions (E.1) and (E.2), we write

$$f(0) = \int_{-\infty}^{\infty} \delta(x) f(x) \, \mathrm{d}x , \tag{E.3}$$

which is the defining property of the Dirac delta.

We note that δ has a meaning as part of an integrand only. We can regard δ as a linear operator that acts—under the integral sign—on f to give its value at $x = 0$.

To gain a heuristic insight into δ itself, let us set $f(x) = 1$ in expression (E.3) to obtain

$$1 = \int_{-\infty}^{\infty} \delta(x)\, \mathrm{d}x\,, \tag{E.4}$$

since $f(x) = 1$ for all x, including for $x = 0$. In view of expressions (E.1) and (E.4), we could suggest that value of δ is infinitely large in an infinitesimal neighbourhood of zero, and is zero elsewhere.

Such a behaviour can be modelled by

$$\delta_n(x) = \frac{1}{\pi} \frac{\sin(nx)}{x}\,, \tag{E.5}$$

where δ_n is a delta sequence, which is defined symbolically as

$$\lim_{n\to\infty} \delta_n(x) = \delta(x)\,;$$

there are many sequences satisfying this definition. As illustrated in Figure E.1, as n increases, the central lobe of the graph of δ_n becomes higher and thinner.

To justify the use of δ_n as a representation of δ, let us consider the defining properties of δ. In view of property (E.3), we require that

$$\lim_{n\to\infty} \int_{-n}^{n} \frac{\sin(nx)}{\pi x} f(x)\, \mathrm{d}x = f(0)\,.$$

To examine this requirement, let us consider sequence (E.5) in the context of the Fourier analysis. Let us write

$$\int_{-n}^{n} \cos(\alpha x)\, \mathrm{d}\alpha = \left.\frac{\sin(\alpha x)}{x}\right|_{\alpha=-n}^{\alpha=n} = \frac{2\sin(nx)}{x}\,, \tag{E.6}$$

where we use the fact that the sine function is odd. Recalling formula (D.12) and using the linearity of the integral operator, we write

$$\int_{-n}^{n} e^{\iota\alpha x}\, \mathrm{d}\alpha = \int_{-n}^{n} \cos(\alpha x)\, \mathrm{d}\alpha + \iota \int_{-n}^{n} \sin(\alpha x)\, \mathrm{d}\alpha\,,$$

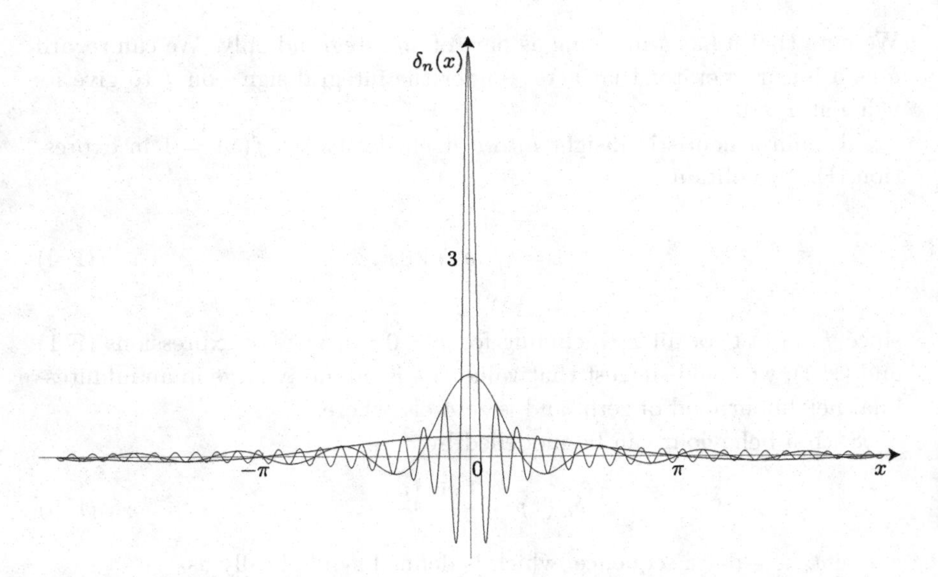

Fig. E.1 Plots of δ_n, given in expression (E.5), for $n = 1$, $n = 4$ and $n = 20$. The width of the central lobe decreases with n.

Again, since the sine function is odd, the second integral on the right-hand side is zero. Thus, in view of expression (E.6), we write

$$\int_{-n}^{n} e^{\iota\alpha x}\,\mathrm{d}\alpha = \frac{2\sin(nx)}{x}.$$

Hence, sequence (E.5) can be written as

$$\frac{\sin(nx)}{\pi x} = \frac{1}{2\pi}\int_{-n}^{n} e^{\iota\alpha x}\,\mathrm{d}\alpha.$$

In view of Figure E.1 and letting $n \to \infty$, we can formally write

$$\delta(x) = \frac{1}{2\pi}\int_{-\infty}^{\infty} e^{\iota\alpha x}\,\mathrm{d}\alpha.$$

Using the translational properties of δ, we can restate this expression as

$$\delta(x-t) = \frac{1}{2\pi}\int_{-\infty}^{\infty} e^{\iota\alpha(x-t)}\,\mathrm{d}\alpha. \tag{E.7}$$

Let us consider the Fourier integral stated in expression (D.13), which—by definition—is the composition of the Fourier transform and its inverse,

$$f(t)=\frac{1}{\sqrt{2\pi}}\int\limits_{-\infty}^{\infty}\left(\frac{1}{\sqrt{2\pi}}\int\limits_{-\infty}^{\infty}e^{\iota\alpha(x-t)}\,\mathrm{d}\alpha\right)f(x)\,\mathrm{d}x\,,\qquad t\in(-\infty,\infty)\,.$$

Using expression (E.7), we write

$$f(t)=\int\limits_{-\infty}^{\infty}\delta\left(x-t\right)f\left(x\right)\,\mathrm{d}x\,,\tag{E.8}$$

which, for $t=0$, is expression (E.3): the defining property of the Dirac delta, as required.

Arnold (2004, p. 79) provides us with interpretations of expression (E.8) that we use in Appendix F.

> An arbitrary function f is a continuous linear combination of translated δ-functions $\delta(x-\cdot)$ concentrated at the points x with coefficients $f(x)$ [...] It is also useful to interpret $\delta(x-\cdot)$ as the density of a unit point mass or charge concentrated at x.

We recognize a heuristic formulation leading to expression (E.8). For instance, considering integral (E.4), we note that—within the realm of functions—the value of the integral of a function that vanishes everywhere except at a single point is zero, regardless of the value of that function at that point. Furthermore, considering sequence (E.5), we note that

$$\lim_{n\to\infty}\frac{\sin\left(nx\right)}{x}$$

does not exist; it diverges for $x\neq k\pi$.

To obtain a rigorous formulation of the Dirac delta, we must consider the theory of distributions, where it is described by its effect on other functions. In general, the effect of h on f is defined by the value of the linear functional

$$\tilde{h}\left(f\right):=\int\limits_{-\infty}^{\infty}h\left(x\right)f\left(x\right)\,\mathrm{d}x\,,\tag{E.9}$$

where we refer to $\tilde{h}$ as a distribution of h. Thus, instead of describing h on $\mathbb{R}$ by its values for $x\in\mathbb{R}$, we describe it by the values of $\tilde{h}\left(f\right)$, for $f\in C_0^\infty$.

E.2 Operations on distributions

E.2.1 *Derivative of distributions*

Let us examine the fact that, since f is a well-defined and compactly supported function, we can perform regular operations on $\int hf\,\mathrm{d}x$, even though h might not be a function. For instance, let us consider the derivative of h. In other words, let us look at the effect of the derivative of h on other functions, namely, $\int h'(x)\,f(x)\,\mathrm{d}x$. Integrating by parts, we obtain

$$\int_{-\infty}^{\infty} h'(x)\,f(x)\,\mathrm{d}x = h(x)\,f(x)\big|_{-\infty}^{\infty} - \int_{-\infty}^{\infty} h(x)\,f'(x)\,\mathrm{d}x\,. \qquad \text{(E.10)}$$

Since f is compactly supported—it tends to zero as x tends to infinity—the first term on the right-hand side vanishes and we state the effect of the derivative of h as

$$\int_{-\infty}^{\infty} h'(x)\,f(x)\,\mathrm{d}x = -\int_{-\infty}^{\infty} h(x)\,f'(x)\,\mathrm{d}x =: \partial\tilde{h}(f)\,, \qquad \text{(E.11)}$$

where $\partial\tilde{h}$ stands for the distributional derivative of h.

Example E.2. Let us use formula (E.11) to find the derivative of the Heaviside step, which is defined by

$$h(x) = \begin{cases} 0\,, & x < 0 \\ 1\,, & x \geqslant 0 \end{cases}\,.$$

Using this definition and expression (E.11), we get

$$\int_{-\infty}^{\infty} h'(x)\,f(x)\,\mathrm{d}x = -\int_{0}^{\infty} f'(x)\,\mathrm{d}x = -\,f(x)\big|_{0}^{\infty} = f(0)\,,$$

where again we use the compact support of f. Thus, we conclude that

$$\int_{-\infty}^{\infty} h'(x)\,f(x)\,\mathrm{d}x = f(0)\,, \qquad \text{(E.12)}$$

which, in contrast to h and its derivative, is a well-defined quantity. To interpret this result, we recall expression (E.3), namely,

$$\int_{-\infty}^{\infty} \delta(x)\,f(x)\,\mathrm{d}x = f(0)\,. \qquad \text{(E.13)}$$

Equating the left-hand sides of equations (E.12) and (E.13), we write

$$\int_{-\infty}^{\infty} h'(x)\, f(x)\, \mathrm{d}x = \int_{-\infty}^{\infty} \delta(x)\, f(x)\, \mathrm{d}x\,,$$

which means that $h'(x)$ behaves like $\delta(x)$. In other words—in a distributional sense—the Dirac delta can be regarded as the derivative of the Heaviside step.

Formula (E.11) can be extended to derivatives of higher order,

$$\int_{-\infty}^{\infty} h^{(n)}(x)\, f(x)\, \mathrm{d}x = (-1)^n \int_{-\infty}^{\infty} h(x)\, f^{(n)}(x)\, \mathrm{d}x =: \partial^n \tilde{h}(f)\,, \qquad \text{(E.14)}$$

where, in a manner analogous to expression (E.11), $\partial^n \tilde{h}$ stands for the nth distributional derivative of h. Also, formula (E.14), in which we consider functions of a single variable, can be extended to functions of several variables,

$$\int_{\mathbb{R}^s} \frac{\partial^n h(x)}{\partial x_i^n} f(x)\, \mathrm{d}x = (-1)^n \int_{\mathbb{R}^s} h(x)\, \frac{\partial^n f(x)}{\partial x_i^n}\, \mathrm{d}x =: \partial_i^n \tilde{h}(f)\,, \qquad \text{(E.15)}$$

where $\partial_i^n \tilde{h}$ stands for the nth distributional derivative of h with respect to x_i, and where $\int_{\mathbb{R}^s}$ stands for s integral signs from $-\infty$ to ∞. These are definitions of distributional derivatives, which show that if f is a C_0^∞ function, then all distributions are infinitely differentiable.

Definition (E.15) is valid in general, even though the derivation illustrated by expression (E.10), which is integration by parts, makes sense only if the distribution is at least a C^1 function. The importance of such a derivation is to show that the distributional definition agrees with the classical one if $h \in C^k$, for $k \geqslant 1$.

As an aside, we note that the above formula has its analogue in the operation of convolutions. The derivative of a convolution of two C^1 functions is obtained by differentiating either one of these functions.

Example E.3. To illustrate expression (E.14), let us consider the nth derivative of the Dirac delta,

$$\partial^n \tilde{\delta}(f) = \int_{-\infty}^{\infty} \delta^{(n)}(x)\, f(x)\, \mathrm{d}x = (-1)^n \int_{-\infty}^{\infty} \delta(x)\, f^{(n)}(x)\, \mathrm{d}x = (-1)^n f^{(n)}(0)\,,$$

where, for the last equality, we use property (E.1).

Again, the above description emphasizes that the Dirac delta can be regarded as a differential operator acting on f, and that it has a meaning only as part of an integrand. To see that distributions allow us to generalize the concept of differentiation of functions, let us consider the following example.

Example E.4. As argued in Example A.2, expression $\nabla^2(1/r)$, including the singularity at the origin, is tantamount to $-4\pi\delta(r)$. To obtain this result in a more rigorous manner, we consider distributions, with $f \in C_0^\infty(\mathbb{R}^3)$: a compactly supported function, as discussed on page 267.

Let us write expression (E.11) as

$$\frac{\partial h}{\partial x_i}(f) := -h\left(\frac{\partial f}{\partial x_i}\right). \tag{E.16}$$

Consequently, the Laplace operator is

$$(\nabla^2 h)(f) = h(\nabla^2 f).$$

Using the last expression, we write the statement to be proven as

$$\iiint\limits_{\mathbb{R}^3} \frac{1}{r}\nabla^2 f \mathrm{d}V = -4\pi f(0), \quad f \in C_0^\infty. \tag{E.17}$$

Let us consider the left-hand side, which we can view as an improper Riemann integral. We write it as

$$\lim_{\substack{R\to\infty \\ \epsilon\to 0^+}} \iiint\limits_{B_R \setminus B_\epsilon} \frac{1}{r}\nabla^2 f \mathrm{d}V,$$

which is the integral over a ball with radius R whose middle of radius ϵ is excluded; since R is a nonnegative quantity, we consider only $R \to +\infty$ and $\epsilon \to 0^+$. Furthermore, since f is compactly supported, the limit as $R \to \infty$ is attained for any R such that the support of f is contained in the interior of B_R; thus, we need to consider only the limit as $\epsilon \to 0^+$,

$$\lim_{\epsilon\to 0^+} \iiint\limits_{B_R \setminus B_\epsilon} \frac{1}{r}\nabla^2 f \mathrm{d}V.$$

To proceed, we invoke identity (A.9), which is valid for scalar fields, to write

$$u\nabla^2 v - v\nabla^2 u = \sum_{i=1}^{3} \frac{\partial}{\partial x_i}\left(u\frac{\partial v}{\partial x_i} - v\frac{\partial u}{\partial x_i}\right).$$

Letting $u = 1/r$ and $v = f$, and using the fact that $\nabla^2(1/r) = 0$, everywhere except at the origin, we obtain

$$\lim_{\epsilon\to 0^+} \iiint\limits_{B_R\setminus B_\epsilon} \sum_{i=1}^{3} \frac{\partial}{\partial x_i}\left(\frac{1}{r}\frac{\partial f}{\partial x_i} - f\frac{\partial}{\partial x_i}\left(\frac{1}{r}\right)\right) \mathrm{d}V\,.$$

Invoking the Divergence Theorem, we write this expression as

$$\lim_{\epsilon\to 0^+} \iint\limits_{S_R\cup S_\epsilon} \sum_{i=1}^{3} N_i\left(\frac{1}{r}\frac{\partial f}{\partial x_i} - f\frac{\partial}{\partial x_i}\left(\frac{1}{r}\right)\right) \mathrm{d}S\,,$$

where N_i are the components of a unit vector normal to the integration surface. For spherical surfaces centred at the origin, we can write

$$\sum_{i=1}^{3} N_i \frac{\partial}{\partial x_i} = \frac{\partial}{\partial r}\,,$$

which transforms the limit of the surface integral to

$$\lim_{\epsilon\to 0^+} \iint\limits_{S_R\cup S_\epsilon} \pm\left(\frac{1}{r}\frac{\partial f}{\partial r} - f\frac{\partial}{\partial r}\left(\frac{1}{r}\right)\right) \mathrm{d}S\,.$$

The choice of the sign depends on the orientation of the surface; for the inner sphere we use the negative sign and for the outer sphere we use the positive sign. Since the contribution of the integral over S_R is zero—by our choice of R—we get

$$-\lim_{\epsilon\to 0^+} \iint\limits_{S_\epsilon} \left(\frac{1}{r}\frac{\partial f}{\partial r} - f\frac{\partial}{\partial r}\left(\frac{1}{r}\right)\right) \mathrm{d}S\,. \tag{E.18}$$

To evaluate such a limit, we interpret the integral using the Mean-value Theorem as

$$-\lim_{\epsilon\to 0^+} \iint\limits_{S_\epsilon} g\,\mathrm{d}S = -\lim_{\epsilon\to 0^+} 4\pi\epsilon^2\overline{g}\,,$$

where the bar denotes the mean value of the integrated function on the sphere with radius ϵ. In the first term of integrand (E.18), $g = \partial_r f/\epsilon$ and the limit becomes

$$-\lim_{\epsilon\to 0^+} 4\pi\epsilon\overline{\frac{\partial f}{\partial r}}\,,$$

which, in view of $f \in C_0^\infty(\mathbb{R}^3)$, tends to zero. In the second term of integrand (E.18), $g = -f/\epsilon^2$ and the limit becomes

$$-\lim_{\epsilon\to 0^+} 4\pi\overline{f}\,,$$

where the mean value tends to $f(0)$ as $\epsilon \to 0^+$. Hence, the left-hand side of equation (E.17) is $-4\pi f(0)$, as required, and as expected in view of Example A.2.

This discussion exemplifies the fact that the theory of distributions generalizes the concept of differentiation of functions. A similar approach can be used for the Fourier transform.

E.2.2 *Fourier transform of distributions*

In a manner similar to the one presented in Appendix E.2.1, let us use the fact that we can perform regular operations on $\int hf\,\mathrm{d}x$, even though h might not be a function. Herein, let us consider the Fourier transform of h.

The Fourier transform of a distribution is tantamount to to the effect of the Fourier transform of a distribution on a test function. Thus, in accordance with expression (E.2), for a single-variable distribution, we write

$$\int\limits_{-\infty}^{\infty} \hat{h}(x)\, f(x)\,\mathrm{d}x = \int\limits_{-\infty}^{\infty} \left(\frac{1}{\sqrt{2\pi}} \int\limits_{-\infty}^{\infty} h(p)\, e^{-\iota x p}\,\mathrm{d}p \right) f(x)\,\mathrm{d}x\,,$$

where the expression in parentheses is the transform discussed in Appendix D.3.1. To proceed, we invoke Fubini's theorem, which allows us to exchange the order of integration,

$$\int\limits_{-\infty}^{\infty} h(p) \left(\frac{1}{\sqrt{2\pi}} \int\limits_{-\infty}^{\infty} f(x)\, e^{-\iota x p}\,\mathrm{d}x \right) \mathrm{d}p = \int\limits_{-\infty}^{\infty} h(p)\, \hat{f}(p)\,\mathrm{d}p\,.$$

Using the arbitrariness of the integration variable, we can write this result as

$$\int\limits_{-\infty}^{\infty} \hat{h}(x)\, f(x)\,\mathrm{d}x = \int\limits_{-\infty}^{\infty} h(x)\, \hat{f}(x)\,\mathrm{d}x\,. \tag{E.19}$$

Concisely, it can be expressed as

$$\left\langle \hat{h}, f \right\rangle = \left\langle h, \hat{f} \right\rangle;$$

this result remains valid for distributions of several variables, $x_1, x_2, \ldots, x_n$, and for the inverse Fourier transform.

Let us exemplify formula (E.19) by considering the Fourier transform of a specific distribution.

Example E.5. For the Dirac delta, we write expression (E.19) as

$$\int\limits_{-\infty}^{\infty} \hat{\delta}(x)\, f(x)\,\mathrm{d}x = \int\limits_{-\infty}^{\infty} \delta(x)\, \hat{f}(x)\,\mathrm{d}x. \tag{E.20}$$

Following property (E.1), we obtain

$$\int_{-\infty}^{\infty} \delta(x) \hat{f}(x) \, dx = \hat{f}(0) . \tag{E.21}$$

To interpret this result, we consider transform (D.16), namely,

$$\hat{f}(p) = \frac{1}{\sqrt{2\pi}} \int_{-\infty}^{\infty} f(x) \exp\{-\iota x p\} \, dx .$$

Setting $p = 0$, we get

$$\hat{f}(0) = \frac{1}{\sqrt{2\pi}} \int_{-\infty}^{\infty} f(x) \, dx = \int_{-\infty}^{\infty} \frac{1}{\sqrt{2\pi}} f(x) \, dx . \tag{E.22}$$

Examining this expressions in view of expressions (E.20) and (E.21), we conclude that $\hat{\delta} = 1/\sqrt{2\pi}$. In other words, the Fourier transform of the Dirac delta is a constant function for the entire domain. Furthermore, since the inverse of an inverse is the original function, $\hat{\hat{\delta}} = \delta$, it follows that $\delta = \widehat{1/\sqrt{2\pi}}$, which allows us to regard δ as the Fourier transform of a constant function. If we endow the transformation variables with a physical meaning of time and frequency, we can make the following interpretation of this result. An impulse, represented by δ, contains all frequencies, which exhibit the same amplitude.

Regarding δ as the Fourier transform of a constant function is consistent with expression (E.5). To see this, we recall expression (E.1),

$$\delta_n(x) = \frac{1}{2\pi} \int_{-n}^{n} e^{\iota x p} dp ,$$

which can be written as

$$\frac{1}{\sqrt{2\pi}} \int_{-\infty}^{\infty} f_n(p) e^{\iota x p} dp ,$$

where

$$f_n(p) = \begin{cases} 0, & p \in (-\infty, -n) \cup (n, \infty) \\ \dfrac{1}{\sqrt{2\pi}}, & p \in (-n, n) \end{cases} . \tag{E.23}$$

In view of the definition of the inverse of the Fourier transform, we see that $f_n(p)$ is the Fourier transform of $\delta_n(x)$. Also, we see that, as expected, expression (E.23) tends to the constant function: $1/\sqrt{2\pi}$, as n goes to infinity.

Let us note that to obtain the distributional Fourier transform, function f need not be compactly supported; however, it must be rapidly decreasing at infinities, together with all its derivatives. We refer to the resulting distributions as tempered distributions.

E.3 Symbol

To study partial differential equations, which we can write formally as

$$F\left(x, \frac{\partial^\beta u}{\partial x^\beta}\right) = 0\,,$$

where $|\beta| \leqslant m$, it is convenient to consider a linear differential operator defined by the following polynomial:

$$P\left(x, \frac{\partial}{\partial x}\right) = \sum_\alpha a_\alpha(x)\frac{\partial^\alpha}{\partial x^\alpha}\,,$$

where

$$a_\alpha(x) = \frac{\partial F}{\partial\left(\dfrac{\partial^\alpha u}{\partial x^\alpha}\right)}\left(x, \frac{\partial^\beta u}{\partial x^\beta}\right).$$

To associate the differential operator with the corresponding algebraic polynomial, we replace

$$\left[\frac{\partial}{\partial x_1}, \ldots, \frac{\partial}{\partial x_n}\right]$$

by

$$[\iota\xi_1, \ldots, \iota\xi_n]$$

to get

$$P(x, \iota\xi) = \sum_\alpha a_\alpha(x)(\iota\xi)^\alpha\,.$$

$P(x, \iota\xi)$ is called the symbol of differential operator $P(x, \partial/\partial x)$. It is a polynomial of degree m in ξ whose coefficients depend on x. Let us also define

$$D_j := -\iota\frac{\partial}{\partial x_j}\,;$$

in other words,

$$D := -\iota \left[\frac{\partial}{\partial x_1}, \ldots, \frac{\partial}{\partial x_n} \right] .$$

This notation implies that the symbol of

$$P(x, D) = \sum_{\alpha} a_{\alpha}(x) D^{\alpha}$$

is

$$P(x, \xi) = \sum_{\alpha} a_{\alpha}(x) \xi^{\alpha} .$$

The symbol is also denoted by $\sigma(F)(\xi)$.

E.4 Principal symbol

Following the definition of the symbol, we state the definition of the principal symbol.

Definition E.2. The principal symbol of

$$P(x, D) = \sum_{|\alpha| \leqslant m} a_{\alpha}(x) D^{\alpha}$$

is

$$P_m(x, \xi) = \sum_{|\alpha| = m} a_{\alpha}(x) \xi^{\alpha} .$$

P_m is a polynomial homogeneous of degree m in ξ.[2] The principal symbol is important for studying oscillatory functions, as implied in the following fundamental theorem of the asymptotic series.

Theorem E.1. *If f is a smooth real-valued function, then as $\omega \to \infty$, we can write*

$$\begin{aligned} P(x, D) &= \exp[-\iota\omega f] P(x, D) \exp[\iota\omega f] \\ &= \omega^m P_m(x, \mathrm{d}f) + O\left(\omega^{m-1}\right) . \end{aligned}$$

The symbol describes how a differential equation acts on functions that have their support contained in a small neighbourhood of point x. If a given function varies rapidly then the highest-order derivatives are dominant and, hence, the principal symbol contains the most important information.

[2]Readers interested in the symbol of a differential expression might refer to Renardy and Rogers (1993, pp. 38-42).

Closing remarks

By introducing the concept of distributions, we enhance the mathematical description of physical phenomena. Historically, a description of the propagation of waves was the motivation to introduce distributions.[3]

The concept of distributions is used explicitly in Chapter 4 and Appendix F.

[3]Readers interested in history of distributions might refer to Schwartz (2001, Chapter VI).

Appendix F

Green's functions

The concept-driven revolutions are the ones that attract the most attention and have the greatest impact on the public awareness of science, but in fact they are comparatively rare. In the last five hundred years we have had six major concept-driven revolutions, [...], *besides the quantum-mechanical revolution that Kuhn took as his model. During the same period there have been about twenty tool-driven revolutions.* [...] *George Green's discovery, the Green's function, is a mathematical tool rather than a physical concept.* [...] *It gave the world a new bag of mathematical tricks, useful in exploring the consequence of theories and for predicting the existence of new phenomena that experimenters could search for. The Green's function was a tool of discovery, like a telescope and the microscope, but aimed at mathematical models and theories instead of being aimed at the sky and the microbe.*

Freeman Dyson (2001, Appendix VIb)

Preliminary remarks

Recognizing patterns is essential for mathematical examinations of physical phenomena. It allows us to avoid unnecessary steps in rediscovering known approaches to such examinations. More importantly, it allows us to gain an insight into the essence of a phenomenon in question and to view it as a coherent part of Nature. This coherency—whether or not it is her intrinsic property—is crucial for our understanding of mathematical models. Green's function, named in honour of George Green (1828), is a tool that allows us to search for patterns and to recognize both the coherency and subtle differences within theories. This insight is achieved due to the fact that a significant part of mathematical physics consists of differential equations and Green's functions are the fundamental solutions of these equations; they represents responses to impulses.

In this appendix, we examine patterns of solutions of the electrostatic, wave and elastodynamic equations that allow us to avoid solving these equations from the beginning each time that expressions for a source or the side conditions are modified. These patterns belong to the theory of Green's functions. Also, we examine intrinsic differences of behaviour among these solutions, which depend on their spatial dimensions.

We begin this appendix with a formulation of Green's function for Coulomb's law. Notably, electromagnetism is a physical context in which George Green (1828) introduced the function that bears his name. Subsequently, we formulate Green's functions for the wave equation in different spatial dimensions. We conclude this appendix by examining Green's functions in the context of the elastodynamic equations, and we comment on relationships between these functions and the characteristics of these equations.

F.1 Electrostatic equation

As stated in the quote at the beginning of this chapter, Green's functions consists of mathematical tricks; their applications are within the realm of differential equations. Let us exemplify a Green's function using an equation discussed in Appendix C and stemming from the Maxwell equations. Recalling Coulomb's law, we invoke expression (C.3),

$$\nabla \cdot E(x) = \frac{\rho}{\epsilon_0}\,, \tag{F.1}$$

where E stands for the electric-field intensity at $x \equiv (x_1, x_2, x_3)$, ρ for the electric-charge density and ϵ_0 is a proportionality constant. Considering an electrostatic case, we write expression (C.18) as

$$E = -\nabla\phi\,, \tag{F.2}$$

where ϕ stands for the electrostatic potential. Using this result in expression (F.1) and invoking the definition of the Laplace operator, we write

$$\nabla^2\phi(x) = -\frac{\rho(x)}{\epsilon_0}\,. \tag{F.3}$$

This is an electrostatic equation, which is an elliptic differential equation; its derivatives are with respect to x only; there is no temporal dependence.

Herein, we define Green's function, $G(x, x_0)$, as the potential due to the point source of unit magnitude at $x_0 \equiv (x_{0_1}, x_{0_2}, x_{0_3})$; in other words,

G is a solution of equation (F.3) for

$$\frac{\rho}{\epsilon_0} = \delta(x - x_0)\,, \tag{F.4}$$

where $\delta(x - x_0)$ is a concise notation for $\delta\big((x_1, x_2, x_3) - (x_{0_1}, x_{0_2}, x_{0_3})\big) \equiv \delta(x_1 - x_{0_1}, x_2 - x_{0_2}, x_3 - x_{0_3})$, which, for the Dirac delta,[1] is equivalent to $\delta(x_1 - x_{0_1})\delta(x_2 - x_{0_2})\delta(x_3 - x_{0_3})$. Hence, using expression (F.4) in equation (F.3), we write

$$\nabla^2 G(x) = -\delta(x - x_0)\,. \tag{F.5}$$

For a three-dimensional space, the units of δ are m^{-3}, since—as can be illustrated by examining expression (E.4)—the units of δ are the reciprocals of the units of its argument, which is a consequence of δ being homogeneous of degree negative one in that argument, and since—as implied by the property of δ stated above expression (F.5)—in n dimensions, δ is homogeneous of degree $-n$;[2] this property of δ follows also from the n-dimensional analogue of expressions (E.7). Thus, considering the units of ∇^2 in equation (F.5), we conclude that G has the units of m^{-1}. The units of ϕ, on the other hand, are volts. Hence, strictly speaking, G is not a potential.

To express ϕ in terms of G, we use the fact that—since they both are scalar-valued functions—we can relate them by Corollary A.1 on page 222 to write

$$\iiint\limits_V \left(\phi\nabla^2 G - G\nabla^2\phi\right) \mathrm{d}V = \iint\limits_S \left(\phi\nabla G - G\nabla\phi\right) \cdot N \mathrm{d}S\,.$$

Considering a large enough volume and assuming that the integrand decreases with distance more than $1/\|x\|^2$, we can neglect the contribution of the surface integral to write

$$\iiint\limits_V \phi\nabla^2 G\,\mathrm{d}V = \iiint\limits_V G\nabla^2\phi\,\mathrm{d}V\,.$$

We can express $\nabla^2\phi$ and $\nabla^2 G$ by invoking, respectively, equation (F.3), and equation (F.5). Thus, we write

$$\iiint\limits_V \phi\,\delta(x - x_0)\,\mathrm{d}V = \frac{1}{\epsilon_0}\iiint\limits_V G\,\rho\,\mathrm{d}V\,.$$

[1] Readers interested in examples and properties of δ might refer to Arnold (2004, Section 9.1).

[2] Readers interested in the homogeneity of δ in n dimensions might refer to Arnold (2004, p. 79).

Using property (E.3) of the Dirac delta, we obtain

$$\phi(x_0) = \frac{1}{\epsilon_0} \iiint_V G(x, x_0)\rho(x)\, \mathrm{d}V\,, \tag{F.6}$$

which is the electrostatic potential expressed in terms of G. To find an explicit expression for G, we recall that, as discussed in Examples A.2 and E.4,

$$\iiint_V \nabla^2 \left(\frac{1}{\|x\|}\right) \mathrm{d}V = \begin{cases} 0 \\ \\ -4\pi \end{cases};$$

-4π corresponds to the volume that includes a singularity at the origin and 0 to the volume that does not. Herein, where the singularity is a point source at x_0, we write

$$\nabla^2 \left(\frac{1}{4\pi\|x - x_0\|}\right) = -\delta(x - x_0)\,.$$

In view of equation (F.5), we see that

$$G(x, x_0) = \frac{1}{4\pi\|x - x_0\|}\,, \tag{F.7}$$

which is an explicit expression for G. This is a solution of equation (F.5) and, hence—by definition—it is the Green's function for equation (F.3) in three dimensions; G corresponds to the Laplace operator and the Dirac delta in (x_1, x_2, x_3).[3]

Also, we refer to expression (F.7) as the fundamental solution of equation (F.3), since we can express its general solution in terms of G. Recalling solution (F.6) and invoking the symmetry of the Green's function, $G(x, x_0) = G(x_0, x)$, which follows from expression (F.7) and implies that the result is the same if we exchange the locations of the source and the observer, we write explicitly the general solution of equation (F.3) as

$$\phi(x) = \frac{1}{\epsilon_0} \iiint_V \frac{\rho(x_0)}{4\pi\|x - x_0\|}\, \mathrm{d}x_0\,; \tag{F.8}$$

G can be viewed as a weighting function that scales—with respect to the distance between x and x_0—the value of the electrostatic potential at x due to charge density at x_0. Solution (F.8) is a continuous superposition

[3] Readers interested in the relation between Green's functions and spatial dimensions might refer to Arfken *et al.* (2013, Section 10.2), to Barton (1989, Sections 4.4.2 and 11.4), and to Bleistein (1984, Sections 4.5, 5.3 and 6.3).

of solution (F.7), which corresponds to the impulse source, and the charge-density source.

Expression (F.8) exemplifies the use of Green's functions for solving differential equations. Formally, we can write equation (F.3) as

$$\mathcal{L}\, f(x) = F(x)\,, \tag{F.9}$$

where $\mathcal{L}$ is the differential operator. If we know G, the general solution of equation (F.9) is

$$f(x) = (F * G)(x) := \int\limits_{\mathbb{R}} \cdots \int\limits_{\mathbb{R}} F(x_0)G(x - x_0)\, \mathrm{d}x_0\,, \tag{F.10}$$

where $*$ is the convolution and x_0 stands for the integration variables.

F.2 Wave equations

F.2.1 *Three spatial dimensions*

To introduce Green's functions for the wave equation, we consider a modification of equation (5.5), which we write as

$$\nabla^2 f(x,t) - \frac{1}{v^2}\frac{\partial^2 f(x,t)}{\partial t^2} = F(x,t)\,, \tag{F.11}$$

where F represents the source. In a manner similar to the one discussed in Appendix F.1, let us consider the case where F is a point source, which herein is a function of both space and time. The corresponding solution can be denoted by G; explicitly, it results from

$$\nabla^2 G(x,t) - \frac{1}{v^2}\frac{\partial^2 G(x,t)}{\partial t^2} = \delta(x - x_0)\delta(t - t_0)\,, \tag{F.12}$$

where $\delta(x - x_0)$ stands for $\delta(x_1 - x_{0_1})\delta(x_2 - x_{0_2})\delta(x_3 - x_{0_3})$. Herein, the units of the right-hand side are $\mathrm{m}^{-3}\mathrm{s}^{-1}$. Hence, the units of G are $\mathrm{m}^{-1}\mathrm{s}^{-1}$.

Equation (F.12) is analogous to equation (F.5). Comparing these equations, we see that they differ by the sign in front of the delta. In equations (F.5), the signs stems from expression (F.2), where—according to its definition—the gradient points to the increase of the scalar field, and—according to a physical convention—the electric force on a positive charge is towards the lower potential. From the mathematical viewpoint, we are free to choose either sign in front of δ; this change in sign results only in the change in sign of G, as we can see by comparing expression (F.7) with expressions (F.19), (F.21) and (F.22), below.

To seek the explicit form of G, we take the Fourier transform of equation (F.12), with respect to t. Using property (D.20) for the temporal derivative and definition (D.16) with property (E.3) for $\delta(t-t_0)$, we obtain

$$\nabla^2\hat{G}(x,\omega)+\frac{\omega^2}{v^2}\hat{G}(x,\omega)=\delta(x-x_0)\frac{1}{\sqrt{2\pi}}e^{-\iota\omega t_0}. \tag{F.13}$$

Note that, according to definition (D.16),

$$\hat{G}(x,\omega)=\frac{1}{\sqrt{2\pi}}\int\limits_{-\infty}^{\infty}G(x,t)\exp(-\iota\omega t)\,\mathrm{d}t\,,$$

which means that $\hat{G}$ contains the factor in front of the integral; according to the same definition and property (E.3),

$$\hat{\delta}(t-t_0)=\frac{1}{\sqrt{2\pi}}\int\limits_{-\infty}^{\infty}\delta(t-t_0)\exp(-\iota\omega t)\,\mathrm{d}t=\frac{1}{\sqrt{2\pi}}\exp(-\iota\omega t_0)\,,$$

which means that the transform integral is evaluated. Equation (F.13) is rotationally symmetric around x_0; we expect the solution to have the same symmetry. Hence, it is convenient to investigate the problem in spherical coordinates centred around x_0. If $\hat{G}$ depends only on $r=\|x-x_0\|$, not on the angles, its Laplace operator spherical coordinates is[4]

$$\nabla^2\hat{G}=\frac{1}{r^2}\frac{\partial}{\partial r}\left(r^2\frac{\partial\hat{G}}{\partial r}\right)=\frac{1}{r}\frac{\partial^2(r\hat{G})}{\partial r^2}\,,$$

which allows us to write equation (F.13) as

$$\frac{\partial^2(r\hat{G})}{\partial r^2}+\frac{\omega^2}{v^2}r\hat{G}=r\delta(r)\frac{1}{\sqrt{2\pi}}e^{-\iota\omega t_0}\,. \tag{F.14}$$

Equation (F.14) can be written as an ordinary differential equation for

$$\hat{u}(r)=r\hat{G}(r,\omega)\,, \tag{F.15}$$

where ω is fixed, namely,

$$\frac{\mathrm{d}^2\hat{u}(r)}{\mathrm{d}r^2}+\frac{\omega^2}{v^2}\hat{u}(r)=0\,,$$

and where we consider $r>0$. The solution of this equation is

$$\hat{u}(r)=A\exp\left(\frac{\iota\omega r}{v}\right)+B\exp\left(-\frac{\iota\omega r}{v}\right)\,,$$

[4] *See also*: Slawinski (2015, Exercise 6.3).

where A and B may depend on ω. Recalling substitution (F.15), we write the solution of equation (F.14) as

$$\hat{G}(r,\omega) = \frac{A(\omega)}{r}\exp\left(\frac{\iota\omega r}{v}\right) + \frac{B(\omega)}{r}\exp\left(-\frac{\iota\omega r}{v}\right). \tag{F.16}$$

To determine coefficients A and B, we write the inverse Fourier transform,

$$G(r,t) = \frac{1}{\sqrt{2\pi}}\int_{\mathbb{R}}\left(\frac{A(\omega)}{r}\exp\left(\frac{\iota\omega r}{v}\right) + \frac{B(\omega)}{r}\exp\left(-\frac{\iota\omega r}{v}\right)\right)\exp(\iota\omega t)\,\mathrm{d}\omega \tag{F.17}$$

$$= \frac{1}{\sqrt{2\pi}}\int_{\mathbb{R}}\left(\frac{A(\omega)}{r}\exp\left(\iota\omega\left(t+\frac{r}{v}\right)\right) + \frac{B(\omega)}{r}\exp\left(\iota\omega\left(t-\frac{r}{v}\right)\right)\right)\mathrm{d}\omega\,. \tag{F.18}$$

This expression describes a superposition of plane waves whose amplitudes diminish with time. The waves corresponding to A propagate towards the source, and those corresponding to B propagate away from it. Since we do not consider waves propagating inwards, we set $A = 0$. Substituting the remaining term of solution (F.16) into equation (F.13), we obtain

$$B(\omega)\left(\nabla^2\frac{\exp\left(-\frac{\iota\omega r}{v}\right)}{r} + \frac{\omega^2}{v^2}\frac{\exp\left(-\frac{\iota\omega r}{v}\right)}{r}\right) = \delta(r)\frac{1}{\sqrt{2\pi}}\exp(-\iota\omega t_0)\,,$$

which can be integrated over a ball, B_R, centred at x_0, and with a fixed radius, R,[5]

$$B(\omega)\iiint\limits_{B_R}\left(\nabla^2\frac{\exp\left(-\frac{\iota\omega r}{v}\right)}{r} + \frac{\omega^2}{v^2}\frac{\exp\left(-\frac{\iota\omega r}{v}\right)}{r}\right)\mathrm{d}V = \frac{1}{\sqrt{2\pi}}\exp(-\iota\omega t_0)\,.$$

To evaluate the integral of the first summand, we use Theorem A.1 and the fact that $\nabla^2 := \nabla\cdot\nabla$; the second integral is evaluated in the spherical coordinates,

$$\iint\limits_{\partial B_R}\frac{\partial}{\partial r}\frac{\exp\left(-\frac{\iota\omega r}{v}\right)}{r}\mathrm{d}S + \int_0^{2\pi}\int_0^{\pi}\int_0^{R}\frac{\omega^2}{v^2}\frac{\exp\left(-\frac{\iota\omega r}{v}\right)}{r}r^2\sin\theta\,\mathrm{d}r\mathrm{d}\theta\mathrm{d}\phi$$
$$= \frac{\exp(-\iota\omega t_0)}{\sqrt{2\pi}}\frac{1}{B(\omega)}\,.$$

[5]We encounter such a domain of integration in Example E.4.

An intermediate step in the integration is

$$
\begin{aligned}
&4\pi \exp\left(-\frac{\iota\omega R}{v}\right)\left(-\frac{\iota\omega R}{v}-1\right) \\
&+4\pi\frac{\omega^2}{v^2}\left(-\frac{vR}{\iota\omega}\exp\left(-\frac{\iota\omega R}{v}\right)+\frac{v^2}{\omega^2}\exp\left(-\frac{\iota\omega R}{v}\right)-\frac{v^2}{\omega^2}\right) \\
&=\frac{\exp(-\iota\omega t_0)}{\sqrt{2\pi}}\frac{1}{B(\omega)},
\end{aligned}
$$

which results in

$$B(\omega)=-\frac{1}{4\pi}\frac{\exp(-\iota\omega t_0)}{\sqrt{2\pi}},$$

which after substituting into equation (F.18) and using expression (E.7) becomes

$$G(x,t,x_0,t_0)=-\frac{\delta\left(t-t_0-\dfrac{\|x-x_0\|}{v}\right)}{4\pi\|x-x_0\|}. \tag{F.19}$$

This is the explicit form of the solution of equation (F.12), which means that it is the fundamental solution of equation (F.11). Herein, G represents the response of an isotropic homogeneous continuum to an impulse, as illustrated in Figure F.1, which can be viewed as the amplitude of a spherical wavefront generated by an impulse point source at location x_0 and time t_0. As expected, in a manner similar to expression (F.7), the amplitude of G is a function of the distance between x and x_0. Moreover, this Green's function—in contrast to expression (F.7) resulting from Coulomb's law, which is time-independent—depends on time.

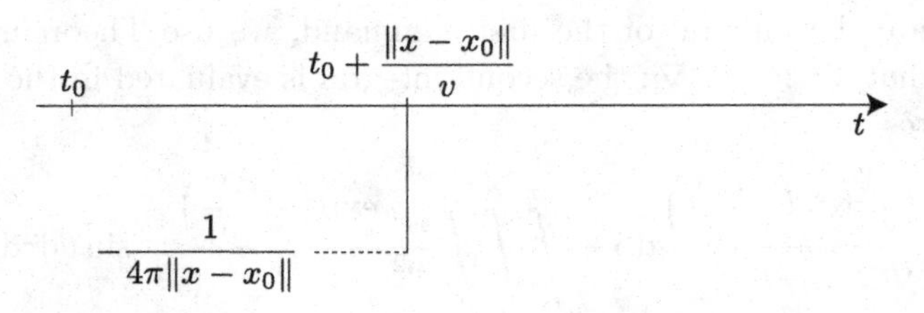

Fig. F.1 Illustration of expression (F.19), which is the Green's function for the wave equation in three spatial dimensions. An impulse generated at time t_0 propagates with speed v. It reaches an observer at time $t_0+\|x-x_0\|/v$, where x stands for (x_1,x_2,x_3), with the amplitude inversely proportional to the distance travelled.

Solution (F.19) can be used to obtain the general solution of equation (F.11), which involves an arbitrary source. This is achieved by a continuous superposition of G and F,

$$f(x,t) = \int\limits_{\mathbb{R}} \iiint\limits_{\mathbb{R}^3} F(x_0,t_0)G(x,t,x_0,t_0)\,\mathrm{d}x_0\,\mathrm{d}t_0\,, \tag{F.20}$$

which is analogous to solution (F.8).

F.2.2 *Two spatial dimensions*

Let us examine expression (F.19) to consider Green's functions in two spatial dimensions. We can use the fact that a line source in three spatial dimensions results in a solution that is invariant along the direction of that line, and thus can be viewed as a function in two spatial dimensions. To obtain this solution, we integrate $G(x,t,x_0,t_0)$ along the source line, say, along the x_{0_3}-axis and set x_3 to zero. Thus, using y as the integration variable, we write

$$G_2(x,t,x_0,t_0) = -\int\limits_{-\infty}^{\infty} \frac{\delta\left(t - t_0 - \dfrac{\sqrt{(x_1-x_{0_1})^2+(x_2-x_{0_2})^2+y^2}}{v}\right)}{4\pi\sqrt{(x_1-x_{0_1})^2+(x_2-x_{0_2})^2+y^2}}\,\mathrm{d}y\,,$$

which is Green's function for the wave equation in two spatial dimensions. Letting

$$s = \frac{\sqrt{(x_1-x_{0_1})^2+(x_2-x_{0_2})^2+y^2}}{v}\,,$$

we write

$$G_2(x,t,x_0,t_0) = -\frac{1}{2\pi}\int\limits_{a}^{\infty} \frac{\delta(t-t_0-s)v}{\sqrt{(vs)^2-(x_1-x_{0_1})^2-(x_2-x_{0_2})^2}}\,\mathrm{d}s\,,$$

where $a = \sqrt{(x_1-x_{0_1})^2+(x_2-x_{0_2})^2}/v$. If $t - t_0 > a$,

$$G_2(x,t,x_0,t_0) = -\frac{1}{2\pi}\frac{v}{\sqrt{(v(t-t_0))^2-(x_1-x_{0_1})^2-(x_2-x_{0_2})^2}}\,,$$

otherwise, $G_2 = 0$. Invoking the Heaviside step, we write concisely,

$$G_2(x,t,x_0,t_0) = -\frac{1}{2\pi}\frac{H\left(t-t_0-\dfrac{\sqrt{(x_1-x_{0_1})^2+(x_2-x_{0_2})^2}}{v}\right)}{\sqrt{(t-t_0)^2-\dfrac{(x_1-x_{0_1})^2}{v^2}-\dfrac{(x_2-x_{0_2})^2}{v^2}}}\,, \tag{F.21}$$

which can be viewed as the amplitude of a circular wavefront generated by an impulse point source at x_0 and t_0.

Examining expressions (F.19) and (F.21) with Figures F.1 and F.2, which correspond to the three- and two-dimensional continua, respectively, and where both are subject to an impulse, we see that the response of the former is also an impulse, since G is expressed in terms of δ, but the response of the latter is not; it persists, since G_2 is expressed in terms of H. Viewed as a wavefront, we can see the response in two spatial dimensions as a spreading disc, not only a circle, as opposed to the response in three spatial dimensions, which is only a spreading sphere, not a ball.

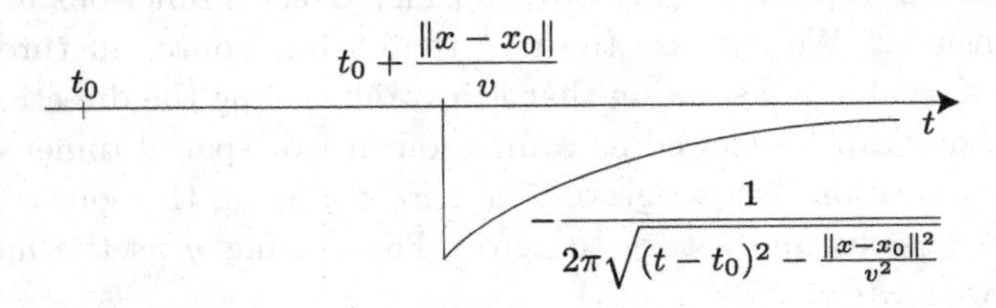

Fig. F.2 Illustration of expression (F.21), which is the Green's function for the wave equation in two spatial dimensions. An impulse generated at time t_0 propagates with speed v. It reaches an observer at time $t_0 + \|x - x_0\|/v$, where x stands for (x_1, x_2), with the amplitude inversely proportional to the square of the distance travelled. There is also a temporal diminishment associated with the fact that—in contrast to Figure F.1, where the effect of an impulse disappears upon its passing—the effect of an impulse persists for $t > t_0 + \|x - x_0\|/v$ with the amplitude inversely proportional to the elapsed time, $t - t_0$.

F.2.3 *One spatial dimension*

To obtain Green's function for the wave equation in one spatial dimension, following the approach analogous to the one used in Appendix F.2.2, we integrate G_2 along the x_{0_2}-axis and set x_2 to zero. Thus,

$$G_1(x, t, x_0, t_0) = -\frac{1}{2\pi} \int_{-\infty}^{\infty} \frac{H\left(t - t_0 - \frac{\sqrt{(x_1 - x_{0_1})^2 + y^2}}{v}\right)}{\sqrt{(t-t_0)^2 - \frac{(x_1 - x_{0_1})^2}{v^2} - \frac{y^2}{v^2}}} \, \mathrm{d}y\,,$$

which can be written as

$$G_1(x, t, x_0, t_0) = -\frac{1}{2\pi} \int_{-a}^{a} \frac{v}{\sqrt{a^2 - y^2}} \, \mathrm{d}y\,,$$

where $a = \sqrt{v^2(t-t_0)^2 - (x-x_0)^2}$, with a positive radicand. Letting $y = a\sin\alpha$, we obtain

$$G_1(x,t,x_0,t_0) = -\frac{v}{2},$$

for $v^2(t-t_0)^2 - (x-x_0)^2 > 0$, and $G_1 = 0$, otherwise. In other words,

$$G_1(x,t,x_0,t_0) = \frac{v}{2}\Big(H\left(\|x-x_0\| - v(t-t_0)\right) - H\left(\|x-x_0\| + v(t-t_0)\right)\Big). \tag{F.22}$$

The magnitude of Green's function stated in expression (F.22) and illustrated in Figure F.3 not only persists, as for the case of a two-dimensional continuum in expression (F.21), but is constant. The geometrical spread-

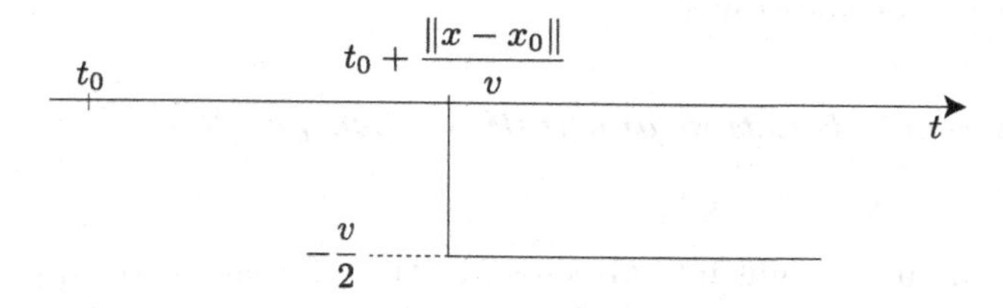

Fig. F.3 Illustration of expression (F.22), which is the Green's function for the wave equation in one spatial dimension. An impulse generated at time t_0 propagates with speed v. It reaches an observer at time $t_0 + \|x-x_0\|/v$; its amplitude remains constant, $-v/2$.

ing, which diminishes the wavefront amplitude in three and two spatial dimensions cannot have any effect in one spatial dimension, since there is no spreading.

The pattern exhibited by Green's functions in expressions (F.19), (F.21) and (F.22), with Figures F.1, F.2 and F.3, respectively, allows us to gain an insight into solutions whose behaviour depends on the spatial dimensions. In all cases discussed, the response is causal; there is no effect prior to the arrival of the signal. However, except for the case of the three spatial dimensions—for which the propagation of the signal can be viewed as an expanding spherical shell with no trace of its passage—there is an "afterglow", which diminishes but persists, in two dimensions, and persists without diminishing, in one dimension.

Such behaviours of the wave-equation solutions could be investigated also in terms of their range of influence and the domain of dependence.[6] Furthermore, these behaviours are illustrations of Huygens's principle.

[6] *See also*: Slawinski (2015, Sections 6.5.4, in particular, Fig. 6.5.1).

Examining the arguments of the Green's functions on the right-hand sides of expressions (F.19), (F.21) and (F.22), we see that they are in the units of time, which are also the units of the horizontal axes in Figures F.1, F.2 and F.3. This property is a consequence of the rotational symmetry of the wave equation with a point source, which allows us to parametrize the entire problem—regardless of its spatial dimension—by a single parameter, in a manner of a sphere or a circle parametrized by the distance or travel-time from its centre.[7,8] Herein, the parameter is time, which allows us to consider not only the amplitude of G upon reaching point x, which is a function of distance, but also its amplitude as a function of time at a given location. The same symmetry is present for the electrostatic Green's function in expression (F.7); therein—since there is no time dependence—the single parameter is distance.

F.2.4 *Green's function and initial value problem*

F.2.4.1 *Impulse response*

To gain a further insight into Green's functions for the wave equations and the relation between the side conditions and the source term, let us examine the relation between the solution of equation (F.12) and the solution of equation (5.5), namely,

$$\nabla^2 f(x,t) - \frac{1}{v^2}\frac{\partial^2 f(x,t)}{\partial t^2} = 0\,, \tag{F.23}$$

with its side conditions, which are the initial conditions given by

$$f(x,0) = f_0(x) \tag{F.24}$$

and

$$\frac{\partial f}{\partial t}(x,0) = f_N(x)\,. \tag{F.25}$$

First, we show that—for the particular initial conditions—we can satisfy the problem stated in equations (F.23)–(F.25), which has no source term, by a solution of equation (F.12), which has an impulse-source term.

To compare a solution of equation (F.12) to a solution of system (F.23)–(F.25), we seek the solution of the latter in a manner similar to the one

[7] Also, the presence of that symmetry allows us to use the d'Alembert approach to solve the wave equation in any spatial dimension, provided that the initial conditions exhibit that symmetry.

[8] *See also*: Slawinski (2015, Section 6.5.1 and Exercise 6.4).

presented for the former. Thus, we take the Fourier transform with respect to x and use property (D.20) to get

$$\frac{\mathrm{d}^2 \hat{f}(k,t)}{\mathrm{d}t^2} + v^2 \|k\|^2 \hat{f}(k,t) = 0\,,$$

with $\hat{f}(k,0) = \hat{f}_0(k)$ and $\mathrm{d}\hat{f}(k,0)/\mathrm{d}t = \hat{f}_N(k)$. A solution satisfying this ordinary differential equation, where k is fixed, is

$$\hat{f}(k,t) = \hat{f}_0(k)\cos(\|k\|vt) + \hat{f}_N(k)\frac{\sin(\|k\|vt)}{\|k\|v}\,, \tag{F.26}$$

which, in view of the relation between the derivatives of the sine and cosine functions, can be written concisely as

$$\hat{f}(k,t) = \hat{f}_0(k)\frac{\partial \hat{g}(k,t)}{\partial t} + \hat{f}_N(k)\hat{g}(k,t)\,, \tag{F.27}$$

where

$$\hat{g}(k,t) := \frac{\sin(\|k\|vt)}{\|k\|v}\,. \tag{F.28}$$

To obtain the solution of system (F.23)–(F.25), we need to find the inverse Fourier transform of expression (F.28), which for three spatial dimensions, is

$$g(x,t) = \frac{1}{(2\pi)^{3/2}} \iiint\limits_{\mathbb{R}^3} \frac{\sin(\|k\|vt)}{\|k\|v} e^{\iota k\cdot x}\,\mathrm{d}k\,. \tag{F.29}$$

To perform this integration, we use the fact that the rotational invariance of the wave equation is inherited by the rotational invariance of g, even if the side conditions are not rotationally symmetric. Since it is easier to integrate rotationally invariant functions in terms of the spherical coordinates, we state expression (F.28) in such coordinates. We note that vector x is fixed; hence, we can use it to rotate the coordinates in such a way that θ is measured from x. In other words, we rotate the coordinates $x \mapsto \tilde{x} = (0,0,\|x\|)$. Then, $\tilde{k}_1 = \|k\|\sin\theta\cos\phi$, $\tilde{k}_2 = \|k\|\sin\theta\sin\phi$ and $\tilde{k}_3 = \|k\|\cos\theta$ with $\mathrm{d}k = \|k\|^2 \sin\theta\,\mathrm{d}\|k\|\,\mathrm{d}\theta\,\mathrm{d}\phi$, and the scalar product, $x\cdot k$, which is preserved under the rotation of the coordinates, becomes $\|k\|\|x\|\cos\theta$. Thus, equation (F.29) can be written as

$$\begin{aligned} g &= \frac{1}{(2\pi)^{3/2}} \int_0^{2\pi}\int_0^{\pi}\int_0^{\infty} \frac{\sin(\|k\|vt)}{\|k\|v} e^{\iota\|k\|\|x\|\cos\theta}\|k\|^2 \sin\theta\,\mathrm{d}\|k\|\,\mathrm{d}\theta\,\mathrm{d}\phi \\ &= \frac{1}{\sqrt{2\pi}\,v} \int_0^{\infty} \|k\|\sin(\|k\|vt)\left(\int_1^{-1} e^{\iota\|k\|\|x\|\tau}\,\mathrm{d}\tau\right)\mathrm{d}\|k\|\,, \end{aligned}$$

where we let $\tau := \cos\theta$. The integral in parentheses is

$$\frac{e^{-\iota\|k\|\|x\|}-e^{\iota\|k\|\|x\|}}{\iota\|k\|\|x\|}.$$

The remaining trigonometric function can be written in terms of exponentials. Together, these steps result in

$$g=\frac{\iota}{\sqrt{2\pi}\,\|x\|v}\int\limits_0^\infty\frac{e^{\iota\|k\|vt}-e^{-\iota\|k\|vt}}{2\iota}\left(e^{\iota\|k\|\|x\|}-e^{-\iota\|k\|\|x\|}\right)\mathrm{d}\|k\|,$$

which can be written as

$$g=\frac{\int\limits_0^\infty\left(e^{\iota\|k\|(vt+\|x\|)}-e^{\iota\|k\|(\|x\|-vt)}+e^{-\iota\|k\|(vt+\|x\|)}-e^{-\iota\|k\|(\|x\|-vt)}\right)\mathrm{d}\|k\|}{2\sqrt{2\pi}\,\|x\|v}.$$

Substituting $\|k\|$ for $-\|k\|$ in the exponentials with $+\iota$, we obtain

$$g=\frac{1}{2\sqrt{2\pi}\,\|x\|v}\int\limits_{-\infty}^\infty\left(e^{-\iota\|k\|(vt+\|x\|)}-e^{-\iota\|k\|(\|x\|-vt)}\right)\mathrm{d}\|k\|.$$

By recognizing—in view of expression (E.7)—that this integral is the difference between the Dirac delta at $vt+\|x\|$ and $vt-\|x\|$, and using property (E.3), we complete the evaluation of this integral,

$$g=\frac{\sqrt{2\pi}}{2\|x\|v}\left(\delta(\|x\|+vt)-\delta(\|x\|-vt)\right),$$

which is the sought integration of expression (F.29). This is not a function, in a classical sense; it is a distribution. Considering only the causal part and—invoking the homogeneity property of $\delta(ax)=\delta(x)/|a|$, where $|\,|$ stands for the absolute value—expressing $\delta(\|x\|-vt)$ as $\delta(t-\|x\|/v)/v$, we write

$$g(x,t)=-\frac{\sqrt{2\pi}}{2v^2}\frac{\delta\left(t-\dfrac{\|x\|}{v}\right)}{\|x\|}; \tag{F.30}$$

which we can view as a Green's function for the Cauchy problem whose initial conditions are stated in terms of impulses; the units of g are $\mathrm{s\,m^{-3}}$. In other words, expression (F.30) is a solution for a particular form of problem (F.23)–(F.25). Examining solutions (F.19) and (F.30), we express g in terms of G as

$$g(x,t)=\frac{(2\pi)^{3/2}}{v^2}G(x,t;0,0). \tag{F.31}$$

Thus, the two solutions are related by a scalar, whose units, in view of the comments below expressions (F.12) and (F.30), are $(\mathrm{s/m})^2$, which is consistent with the units of speed, v.

We conclude that a particular form of the initial-value problem stated in equations (F.23)–(F.25) is tantamount to equation (F.11) with a particular source term. The essence of this equivalence is the impulse. This result is to be expected, since subjecting a continuum that is undisturbed to an impulse at a given point and at the particular instant, $\delta(x-x_0)\,\delta(t-t_0)$, can be seen as initial conditions.

F.2.4.2 *General response*

To use the impulse response in formulating more general solutions, we return to solution (F.27)—and, since, upon applying the Fourier transform, the multiplication becomes convolution—we write

$$f(x,t)=f_0(x)*\frac{\partial g(x,t)}{\partial t}+f_N(x)*g(x,t)\,, \tag{F.32}$$

where g is given by expression (F.30). Expression (F.32) is the solution of the initial-value problem given by expressions (F.23)–(F.25). We refer to it as a fundamental solution, since knowing conditions f_0 and f_N suffices to write the solution for the Cauchy problem without solving it each time that f_0 and f_N are given.

We wish to express problem (F.23)–(F.25) in terms of the wave equation with a source term. In other words, the role of the initial conditions is to be assumed by the source term.

In the context of condition (F.25), we propose the source term at x_0 and t_0 to be

$$\delta(t_0)h_N(x_0)\,;$$

where the explicit form of h_N is to be such that the solution corresponding to this term be the same as the part of solution (F.32) that corresponds to condition (F.25), namely,

$$(f_N*g)(x,t)\,. \tag{F.33}$$

According to equation (F.20), the solution corresponding to this term is

$$\int\limits_{\mathbb{R}}\iiint\limits_{\mathbb{R}^3}\delta(t_0)h_N(x_0)G(x,t,x_0,t_0)\,\mathrm{d}x_0\,\mathrm{d}t_0\,,$$

which, in view of property (E.3), is

$$\iiint\limits_{\mathbb{R}^3}h_N(x_0)G(x,t,x_0,0)\,\mathrm{d}x_0\,. \tag{F.34}$$

In view of expression (F.19), we see that $G(x,t,x_0,t_0) = G(x-x_0,t,0,t_0)$; using this property in expression (F.34), we write

$$\iiint_{\mathbb{R}^3} h_N(x_0)G(x-x_0,t,0,0)\,\mathrm{d}x_0\,. \tag{F.35}$$

In view of relation (F.31), expression (F.35) is equivalent to

$$v^2(h_N * g)(x,t)\,, \tag{F.36}$$

where we use the definition of convolution in three dimensions,

$$a * b(y) = \frac{1}{(2\pi)^{3/2}}\iiint_{\mathbb{R}^3} a(z)b(y-z)\mathrm{d}z\,. \tag{F.37}$$

Comparing expressions (F.33) and (F.36), we see that

$$h_N(x_0) = \frac{1}{v^2}f_N(x_0)\,, \tag{F.38}$$

which is the explicit form of h_N, as required.

Considering condition (F.24), we propose the source term to be

$$\delta'(t_0)h_0(x_0)\,.$$

According to equation (F.20), the solution corresponding to this term is

$$\int_{\mathbb{R}}\iiint_{\mathbb{R}^3} \delta'(t_0)h_0(x_0)G(x,t,x_0,t_0)\,\mathrm{d}x_0\,\mathrm{d}t_0\,,$$

which—in view of $G(x,t,x_0,t_0) = G(x-x_0,t-t_0,0,0)$ and the derivative of the Dirac delta—is

$$\int_{\mathbb{R}}\iiint_{\mathbb{R}^3} \delta(t_0)h_0(x_0)\frac{\partial}{\partial t_0}G(x-x_0,t-t_0,0,0)\,\mathrm{d}x_0\,\mathrm{d}t_0\,.$$

Using the chain rule, we see that it equals to

$$-\iiint_{\mathbb{R}^3} h_0(x_0)\frac{\partial}{\partial t}G(x-x_0,t,0,0)\,\mathrm{d}x_0\,,$$

which can be written as

$$-\,v^2\left(h_0 * \frac{\partial g}{\partial t}\right)(x,t)\,. \tag{F.39}$$

The part of solution (F.32) that corresponds to condition (F.24) is

$$\left(f_0 * \frac{\partial g}{\partial t}\right)(x,t)\,; \tag{F.40}$$

thus, comparing expressions (F.39) and (F.40), we see that

$$h_0(x_0) = -\frac{1}{v^2} f_0(x_0)\,, \tag{F.41}$$

which is the explicit form of h_0. Using expressions (F.38) and (F.41), we write the source term corresponding to the inital value problem (F.23)–(F.25) as

$$F(x_0, t_0) = -\frac{1}{v^2} f_0(x_0)\, \delta'(t_0) + \frac{1}{v^2} f_N(x_0)\, \delta(t_0)\,, \tag{F.42}$$

or equivalently

$$v^2 \nabla^2 f(x,t) - \frac{\partial^2 f(x,t)}{\partial t^2} = f_N(x_0)\, \delta(t_0) - f_0(x_0)\, \delta'(t_0)\,,$$

which is a wave equation with a source term, $F(x_0, t_0)$. Below, in Example F.1, we illustrate this relation for a one-dimensional continuum.

Example F.1. To relate solution (F.32) and function (F.22) to the d'Alembert solution, which applies to the wave equation in one spatial dimension and is given by expression (2.29), we recall expression (F.32),

$$f(x,t) = f_0(x) * \frac{\partial g(x,t)}{\partial t} + f_N(x) * g(x,t)\,,$$

which is the solution of the initial-value problem given by expressions (F.23)–(F.25), without a source term. Considering relation (F.31) in a single spatial dimension at $x_0 = 0$ and $t_0 = 0$, we obtain

$$g(x,t) = \frac{\sqrt{2\pi}}{v^2} G_1(x,t,0,0) = -\frac{\sqrt{2\pi}}{2v} \left(H(x - vt) - H(x + vt)\right). \tag{F.43}$$

Invoking definition (F.37), for the convolution in one dimension,

$$a * b(y) = \frac{1}{\sqrt{2\pi}} \int\limits_{\mathbb{R}} a(z) b(y - z) \mathrm{d}z\,,$$

and using properties of the Heaviside step and its derivative, shown in Example E.2, we write

$$f(x,t) = -\frac{1}{2v} \int\limits_{\mathbb{R}} \Bigg(f_0(x - z) \frac{\partial}{\partial t} \left(H(z - vt) - H(z + vt)\right) \\ + f_N(x - z) \left(H(z - vt) - H(z + vt)\right) \Bigg) \mathrm{d}z$$

and, hence,

$$f(x,t) = \frac{1}{2}\int_{\mathbb{R}} f_0(x-z)\Big(\delta(z-vt)+\delta(z+vt)\Big)\mathrm{d}z + \frac{1}{2v}\int_{-vt}^{vt} f_N(x-z)\mathrm{d}z\,.$$

Using property (E.3) of the Dirac delta, we write

$$f(x,t) = \frac{1}{2}\Big(f_0(x-vt)+f_0(x+vt)\Big) + \frac{1}{2v}\int_{x-vt}^{x+vt} f_N(\zeta)\mathrm{d}\zeta,$$

which is tantamount to solution (2.29), as desired.

As stated in footnote 7, for the spherically symmetric initial conditions, we could extend this result to the wave equation in three spatial dimensions.

F.3 Elastodynamic equations

In Appendix F.2, we use the rotational symmetry of the wave equations, which stems from the isotropy of continua, together with the intrinsic rotational symmetry of a point source. Let us extend our discussion beyond such symmetries by considering the special case of equation (B.24), which is the elastodynamic equation that corresponds to the anisotropic but homogeneous continua, which means that the elasticity parameters, c_{ijkl}, and the mass density, ρ, do not depend on x. However, let us include a source term, F, which herein—in contrast to equations (F.3) and (F.11)—is a vector. Thus,

$$\rho\frac{\partial^2 u_i(x,t)}{\partial t^2} - \sum_{j=1}^{3}\sum_{k=1}^{3}\sum_{l=1}^{3} c_{ijkl}\frac{\partial^2 u_k(x,t)}{\partial x_l \partial x_j} = F_i\,, \quad i\in\{1,2,3\}\,, \tag{F.44}$$

where u is the displacement vector.

In a manner similar to the approach in Appendices F.1 and F.2, we consider an impulse source and let the corresponding solution be G_{im}, which is a second-rank tensor.[9] Thus, we rewrite equation (F.44) as

$$\rho\frac{\partial^2 G_{im}(x,t)}{\partial t^2} - \sum_{j=1}^{3}\sum_{k=1}^{3}\sum_{l=1}^{3} c_{ijkl}\frac{\partial^2 G_{mk}(x,t)}{\partial x_l \partial x_j} = \delta(x-x_0)\delta(t-t_0)\delta_{im}\,, \tag{F.45}$$

where δ_{im} is the Kronecker delta, and $i, m\in\{1,2,3\}$.

[9] Readers interested in further discussion on Green's tensor for the elastodynamic equations might refer to Chapman (2004, pp. 112–113), Červený (2001, pp. 80–84), Dahlen and Tromp (1998, Section 15.7.1) and Stolt and Weglein (2012, Chapter 6).

Equation (F.45) is analogous to equations (F.5) and (F.12). Examining these equations, we see that, in equation (F.5), G is a scalar-valued quantity whose argument is the position, x; in equation (F.12), it is also a scalar-valued quantity but its arguments are both position and time, t; in equation (F.45), it is a tensor-valued quantity whose arguments are both position and time. The presence or absence of time is a consequence of belonging to an elliptic versus a hyperbolic differential equation. The tensor-valued, as opposed to the scalar-valued, quantities are a consequence of belonging to a vector versus a scalar differential equation. In reference to the sign in front of δ, discussed on page 297, equation (F.45) exhibits the opposite convention to equation (F.12). To compare explicitly, we write equation (F.12) as

$$\sum_{i=1}^{3}(\lambda + 2\mu)\frac{\partial^2 G}{\partial x_i^2} - \rho\frac{\partial^2 G}{\partial t^2} = \delta(x - x_0)\delta(t - t_0)\,, \tag{F.46}$$

where, due to isotropy, the only independent elasticity-tensor components are c_{1111}, which is denoted commonly in terms of the Lamé parameters by $\lambda{+}2\mu$ and corresponds to the P waves, and $c_{1212} =: \mu$, which corresponds to the S waves. For comparison of the impulse sign, we write equation (F.45) as

$$\sum_{j=1}^{3}\sum_{k=1}^{3}\sum_{l=1}^{3} c_{ijkl}\frac{\partial^2 G_{mk}}{\partial x_l \partial x_j} - \rho\frac{\partial^2 G_{im}}{\partial t^2} = -\delta(x-x_0)\delta(t-t_0)\delta_{im}\,, \quad i, m \in \{1,2,3\}\,.$$

Note that, in spite of the similarity of form, expression (F.46) is the wave equation corresponding to a wave in an isotropic continuum and the latter equation is the elastodynamic equation containing three waves in an anisotropic continuum.

Expression (F.45) stands for nine equations; however, due to the symmetry of G_{im}, there are only six independent equations. Physically, this symmetry means that exchanging the source and receiver positions does not affect the solution, provided that both the positions and orientations are exchanged. If the impulse in direction d_i at x_0 results in a displacement in direction d_j at x, the symmetry implies that the impulse at x in direction d_j results in a displacement in direction d_i at x_0, as illustrated, respectively, in the left and right diagrams of Figure F.4. In other words, since $d_j = G(x, x_0)d_i$ and $d_i = G(x_0, x)d_j$, it follows that $G(x, x_0) = G(x_0, x)^T$, which is equivalent to $G_{ij} = G_{ji}$.

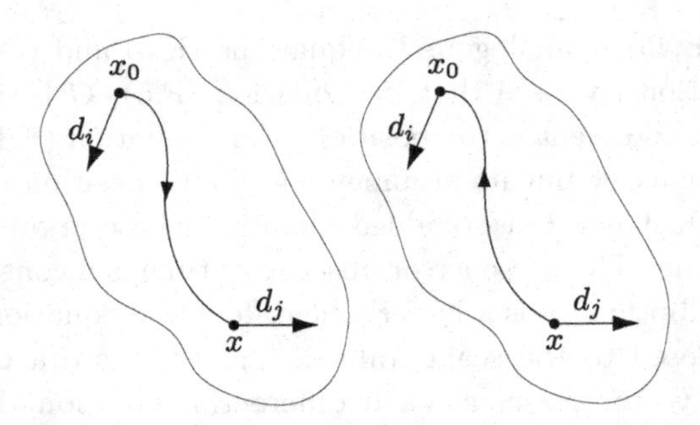

Fig. F.4 The symmetry of Green's tensor implies that there is no effect of exchanging the positions, x_0 and x, of an impulse source and its response, provided it is done together with the change of the corresponding displacement directions, d_i and d_j. Also, the signal trajectory and traveltime remain the same, which means that exchanging sources and receivers has no effect on them.

Applying the Fourier transform in the four variables to equation (F.45), using property (D.20) for the partial derivatives, and invoking definition (D.16) with property (E.3) for $\delta(x-x_0)\delta(t-t_0)$, we obtain

$$-\rho\omega^2\hat{G}_{im}(\kappa,\omega)+\sum_{j=1}^{3}\sum_{k=1}^{3}\sum_{l=1}^{3}c_{ijkl}\kappa_l\kappa_j\hat{G}_{mk}(\kappa,\omega)=\frac{1}{(2\pi)^2}e^{-\iota\kappa\cdot x_0}e^{-\iota\omega t_0}\delta_{im}\,,$$

with $i,m\in\{1,2,3\}$, which can be written as

$$(2\pi)^2e^{\iota\kappa\cdot x_0}e^{\iota\omega t_0}\sum_{k=1}^{3}\left(\sum_{j=1}^{3}\sum_{l=1}^{3}c_{ijkl}n_ln_j-\rho v^2\delta_{ik}\right)\hat{G}_{mk}=\delta_{im}\,,$$

with $i,m\in\{1,2,3\}$, where $n=\kappa/\|\kappa\|$ and $v=\omega/\|\kappa\|$. For the fixed values of κ and ω, the term in parentheses is a 3×3 matrix, and $\hat{G}_{mk}$ is the inverse of

$$(2\pi)^2e^{\iota\kappa\cdot x_0}e^{\iota\omega t_0}\left(\sum_{j=1}^{3}\sum_{l=1}^{3}c_{ijkl}n_ln_j-\rho v^2\delta_{ik}\right),\quad i,k\in\{1,2,3\}\,.\tag{F.47}$$

In contrast to the electrostatic equation and the wave equations with a scalar-valued function, the resulting Green's tensor cannot be expressed in terms of a single parameter only.

The similarity between the term in parentheses of expression (F.47) and expressions (2.66) and (4.67) suggests the existence of relationships between Green's tensor and the eikonal equation, characteristic surfaces, the Christoffel equation and, in general, propagation of discontinuities in hyperbolic equations. Notably the three eigenvalues of this matrix are associated with the three eikonal equations and, hence, with the wavefront velocities of three distinct waves that propagate within an anisotropic Hookean solid, as discussed in Section 3.4.1.

Closing remarks

As illustrated by the electrostatic, wave and elastodynamic equations, Green's functions appear in a variety of contexts as solutions of differential equations associated with impulse sources. The importance of the solutions corresponding to impulse sources is greater than the immediate description of such physical sources. It allows us to describe the response of a system to an arbitrary source; as stated by Arnold (2004, Section 9.2):

> [the superposition] principle is the basis of all so-called linear physics, [even though] it is in essence a simple fact of linear algebra. [...] Physically, if we throw one stone into water, we obtain a certain picture of the waves. If we throw a second stone, we obtain a second picture of the waves. If we throw both stones together, the resulting picture of the waves is the same as if we added the disturbances from throwing the first stone to those from throwing the second. [...] If we want to solve a linear differential equation with an arbitrary function F on the right-hand side, it suffices to learn how to solve this equation with the δ-function on the right, then apply the principle of superposition, except that instead of sums one must write integrals.

As exemplified by integral (F.10), the solution is a continuous linear combination of the impulse-response solution, G, and the source, F. To obtain such a solution, mathematically, we need integrability properties, and, physically, we need a given linear theory to be accurate enough for the problem in question.

In view of the presence of the Dirac delta and the Heaviside step, the methodology introduced by Green (1828) anticipated by over a century the formalism of distributions established by Sobolev and Schwartz in the first half of the twentieth century.[10] Strictly speaking, Green's functions are distributions.

[10]Readers interested in the life of George Green might refer to his biography by Cannell (2001). This book contains several appendices, including Schwinger (2001) and Dyson (2001), where modern applications of Green's ideas are described, particularly in the context of quantum electrodynamics.

As indicated by relationships of Green's functions to solutions of the wave equations and by their relationships to the characteristics of the equations of motions, Green's functions are intimately related to wavefronts within elastic continua. Even though not used explicitly in the main text, implicitly, Green's functions are ubiquitous throughout this book, and many formulations could be recast in terms of these functions.

List of symbols

In one regard, mathematics is like poetry. Every discipline uses its own system of representation [...] that helps to convey the old facts and new speculations in the field. [...] but only mathematics matches poetry in tying innovation to notation.

James Robert Brown (2008, p. 84)

$a, b, c, \ldots$ auxiliary parameters or functions
$e_1, \ldots, e_n$ basis in $\mathbb{R}^n$
f, g, h functions
f' first derivative of function f with respect to the argument
$f^{(n)}$ nth derivative of function f with respect to the argument
$\hat{f}$ Fourier transform of function f
$\check{f}$ inverse Fourier transform of function f
i, j, k, l, ℓ summation indices
$n = (n_1, \ldots, n_i)$ unit tangent vector
r, s parameters for parametrization of curves and surfaces
t time
$p = (p_1, \ldots, p_n)$ point in the dual space to $\mathbb{R}^n$
$x = (x_1, \ldots, x_n)$ point in $\mathbb{R}^n$ represented by coordinates in a given basis
$A, B, C, \ldots, X, Y, Z$ tensors, matrices, vectors and auxiliary constants or functions
α, β multiindices; see page 12
A, ϕ elastodynamic potentials; see expression (B.31) on page 247
$\mathcal{A}, \varphi$ elastodynamic wave functions; see pages 248–249
A, ϕ electromagnetic potentials and wave functions; see pages 260–262
$C^k(\mathbb{R}^n)$ space of k-times continuously differentiable functions on $\mathbb{R}^n$
$C_0(\mathbb{R}^n)$ space of compactly supported functions on $\mathbb{R}^n$; see page 267

$\Im$ imaginary part
$\Re$ real part
$\mathbb{R}$ real numbers
$\mathbb{N}$ natural numbers: $1, 2, 3, \ldots$
$N = (N_1, \ldots, N_n)$ unit normal vector
$X_1, \ldots, X_n$ components of vector X in a given basis
D_X directional derivative in direction of X; see expression (1.65) on page 31
∇ nabla operator: in Cartesian coordinates, $\left(\partial/\partial x_1, \partial/\partial x_2, \ldots, \partial/\partial x_n\right)$; for polar coordinates see Exercise 4.13 on page 177
∇^2 Laplace operator: in Cartesian coordinates, $\partial^2/\partial x_1^2 + \partial^2/\partial x_2^2 + \cdots + \partial^2/\partial x_n^2$; for polar coordinates see Exercise 4.13 on page 177
$\square$ d'Alembert operator, see page 28
$o\left(\cdot\right)$ Landau symbol *little-o*, see page 140
$O\left(\cdot\right)$ Landau symbol *big-O*, see page 140
$:=$ definition
$\sim$ asymptotically equivalent to
$\|\|_p$ p-norm; see page 267
$\|\|$ Euclidean norm: $\|\|_2$
$(\, , \,)$ open interval
$[\, , \,]$ closed interval
$\langle \, , \, \rangle$ scalar product; see expression (D.3)
$\cdot$ Euclidean scalar (dot) product, multiplication
ι imaginary unit: $\sqrt{-1}$
σ stress tensor
ε strain tensor
ρ mass density

Bibliography

Achenbach, J. (2003). *Reciprocity in elastodynamics* (Cambridge University Press).

Achenbach, J. D. (1984). *Wave propagation in elastic solids* (North-Holland).

Aki, K. and Richards, P. G. (2002). *Quantitative seismology*, 2nd edn. (University Science Books).

Aleksandrov, A. D., Kolmogorov, A. and Lavrentev, M. (eds.) (1999). *Mathematics: Its contents, methods and meaning* (Dover).

Appel, W. (2005). *Mathématique pour la physique et les physiciens*, 3rd edn. (H&K).

Arfken, G., Weber, H. and Harris, F. (2013). *Mathematical methods for physicists: A comprehensive guide*, 7th edn. (Elsevier/Academic Press).

Arnold, V. I. (1984). *Ordinary differential equations*, 3rd edn. (Springer-Verlag).

Arnold, V. I. (1989). *Mathematical methods of classical mechanics*, 2nd edn. (Springer-Verlag).

Arnold, V. I. (1990). *Singularities of caustics and wave fronts* (Kluwer).

Arnold, V. I. (1991). *The theory of singularities and its applications.* (Press Syndicate of the University of Cambridge).

Arnold, V. I. (1992). *Catastrophy theory* (Springer-Verlag).

Arnold, V. I. (2004). *Lectures on partial differential equations* (Springer).

Baker, B. and Copson, E. (1939). *The mathematical theory of Huygens' principle* (Oxford, Clarendon Press).

Barton, G. (1989). *Elements of Green's functions and propagation: Potential, diffusion and waves* (Oxford University Press).

Batterman, R. W. (2002). *The devil in the details: Asymptotic reasoning in explanation, reduction, and emergence* (Oxford University Press).

Bleistein, N. (1984). *Mathematical methods for wave phenomena* (Academic Press).

Bleistein, N., Cohen, J. and Stockwell, J. (2001). *Mathematics of multidimensional seismic imaging, migration, and inversion* (Springer-Verlag).

Bleistein, N. and Handelsman, R. A. (1986). *Asymptotic expansions of integrals* (Dover).

Boccara, N. (1997). *Distributions* (Ellipses).

Bos, L. and Slawinski, M. (2010). Elastodynamic equations: Characteristics, wavefronts and rays, *The Quarterly Journal of Mechanics and Applied Mathematics* **63**, 1, pp. 23–37.

Brown, J. R. (1994). *Smoke and mirrors: How science reflects reality* (Routledge).

Brown, J. R. (2008). *Philosophy of mathematics: A contemporary introduction to the world of proofs and pictures*, 2nd edn. (Routledge).

Brown, J. R. and Slawinski, M. A. (2017). *On foundations of seismology: Bringing idealizations down to Earth* (World Scientific).

Bunge, M. (1967). *Foundations of physics* (Springer-Verlag).

Cannell, D. M. (2001). *George Green, mathematician and physicist: The background to his life and work* (SIAM).

Červený, V. (2001). *Seismic ray theory* (Cambridge University Press).

Chapman, C. (2004). *Fundamentals of seismic wave propagation* (Cambridge University Press).

Colley, S. J. (2012). *Vector calculus*, 4th edn. (Pearson).

Copson, E. T. (1965). *Asymptotic expansions* (Cambridge University Press).

Courant, R. and Hilbert, D. (1989). *Methods of mathematical physics* (Wiley).

Dahlen, F. A. and Tromp, J. (1998). *Theoretical global seismology* (Princeton University Press).

De Bruijn, N. G. (1981). *Asymptotic methods in analysis* (Dover).

Duistermaat, J. (1996). *Fourier integral operators*, *Progress in Mathematics*, Vol. 130 (Birkhäuser).

Dyson, F. (2001). Homage to George Green: How physics looked in the nineteen-forties, in *George Green, mathematician and physicist: The background to his life and work* (SIAM).

Epstein, M. (2010). *The geometrical language of continuum mechanics* (Cambridge University Press).

Erdélyi, A. (1956). *Asymptotic expansions* (Dover).

Feynman, R. P. (1967). *The character of physical law* (MIT Press).

Feynman, R. P., Leighton, R. B. and Sands, M. L. (1964). *The Feynman lectures on physics* (Addison-Wesley).

Folland, G. B. (1992). *Fourier analysis and its applications* (Brooks/Cole).

Folland, G. B. (1995). *Introduction to partial differential equations*, 2nd edn. (Princeton University Press).

Garrity, T. A. (2001). *All the mathematics you missed but need to know for graduate school* (Cambridge University Press).

Gel'fand, M., Graev, M. I. and Shapiro, Z. Y. (1969). Differential forms and integral geometry, *Functional Analysis and its Applications* **3**, pp. 101–114.

Gowers, T. (ed.) (2008). *The Princeton companion to mathematics* (Princeton University Press).

Green, G. (1828). An essay on the application of mathematical analysis to the theories of electricity and magnetism, *arXiv:0807.0088* .

Guillemin, V. and Sternberg, S. (1990). *Geometric asymptotics*, revised edn., no. 14 in Mathematical Surveys (AMS).

Gurtin, M. E. (2003). *An introduction to continuum mechanics* (Academic Press).

Hadamard, J. (1903). *Leçons sur la propagation des ondes et les équations de l'hydrodynamique* (Broché).

Hadamard, J. (1932). *Le problème de Cauchy et les équations aux dérivées partielles linéaires hyperboliques* (Broché).

Hamilton, W. R. (1828–1837). *Theory of systems of rays with three supplements* (Royal Irish Academy).

Hörmander, L. (1983). *The analysis of linear partial differential operators I-IV*, Grundlehren der matematischen Wissenschaften (Springer).

Hutchinson, G. E. (1962). Sir George Gabriel Stokes, 1819-1903, in *The enchanted voyage and other studies* (Yale University Press).

Jeffreys, H. (1939). The times of the core waves, *M.N.R.A.S Geophys. Suppl.* **4**, pp. 548–561.

Jeffreys, H. (1962). *Asymptotic approximations* (Oxford University Press).

John, F. (1982). *Partial differential equations*, 4th edn. (Springer).

Kline, M. (1974). *Mathematical thought from ancient to modern times* (Oxford University Press).

Kot, M. (2014). *A First Course in the Calculus of Variations* (American Mathematical Society).

Kravtsov, Y. A. and Orlov, Y. I. (1999). *Caustics, catastrophes and wave fields.* (Springer-Verlag).

Kreyszig, E. (1964). *Differential geometry* (University of Toronto Press).

Lanczos, C. (1949/1986). *The variational principles of mechanics* (Dover).

Landau, L. D. and Lifschitz, E. M. (2006). *The classical theory of fields*, Vol. 2, fourth revised edn. (Elsevier).

Lang, S. (1968). *Linear algebra* (Addison-Wesley).

Love, A. (1944). *A treatise on the mathematical theory of elasticity* (Dover).

Luneberg, R. K. (1944). *Mathematical theory of optics* (Brown University).

Malvern, L. (1969). *Introduction to the mechanics of a continuous medium* (Prentice Hall).

Marsden, J. E. and Tromba, A. (1981). *Vector calculus*, 2nd edn. (W. H. Freeman & Co.).

McOwen, R. C. (2003). *Partial differential equations: methods and applications*, 2nd edn. (Prentice Hall).

Morse, P. M. and Feshbach, H. (1953). *Methods of theoretical physics* (McGraw-Hill).

Moskovits, M. (ed.) (1977). *Science and society* (Anansi).

Nye, J. F. (1999). *Natural focusing and fine structure of light: Caustics and wave dislocations* (Institute of Physics Publishing).

Olver, F. W. J. (1974). *Introduction to asymptotics and special functions* (Academic Press).

Olver, F. W. J. (1997). *Asymptotics and special functions.* (AK Peters).

Poincaré, H. (1886). Sur les intégrales irrégulières, *Acta Mathematica* **8**.

Poincaré, H. (1999). *La valeur de la science* (Flammarion).

Porteous, I. R. (1994). *Geometric differentiation for the intelligence of curves and surfaces* (Cambridge University Press).

Rauch, J. (1991). *Partial differential equations* (Springer-Verlag).

Reed, M. and Simon, B. (1980). *Methods of modern mathematical physics: Functional analysis*, Vol. 1 (Academic Press).

Renardy, M. and Rogers, R. (1993). *An introduction to partial differential equations* (Springer).

Roach, G. (1982). *Green's functions*, 2nd edn. (Cambridge University Press).

Rochester, M. G. (2010). Note on the necessary conditions for P and S wave propagation in a homogeneous isotropic elastic solid, *Journal of Elasticity* **98**, 1, pp. 111–114.

Rogister, Y. and Slawinski, M. A. (2005). Analytic solution of raytracing equations for linearly inhomogeneous and elliptically anisotropic velocity model, *Geophysics* **70**, pp. D37–D41.

Rudin, W. (1991). *Functional analysis*, 2nd edn. (McGraw-Hill).

Schwartz, L. (1997). *Un mathématicien aux prises avec le siècle* (Editions Odile Jacob).

Schwartz, L. (2001). *A mathematician grappling with his century* (Springer).

Schwinger, J. (2001). The Greening of quantum field theory: George and I, in *George Green, mathematician and physicist: The background to his life and work* (SIAM).

Slawinski, M. A. (2010). *Waves and rays in elastic continua* (World Scientific).

Slawinski, M. A. (2015). *Waves and rays in elastic continua*, 3rd edn. (World Scientific).

Slawinski, M. A. (2016). *Waves and rays in seismology: Answers to unasked questions* (World Scientific).

Slawinski, M. A. (2018). *Waves and rays in seismology: Answers to unasked questions*, 2nd edn. (World Scientific).

Snieder, R. (2006). *A guided tour of mathematical methods for the physical sciences*, 2nd edn. (Cambridge University Press).

Spivak, M. (1999). *A comprehensive introduction to differential geometry* (Publish or perish).

Stolt, R. H. and Weglein, A. B. (2012). *Seismic imaging and inversion: Application of linear inverse theory* (Cambridge University Press).

Strauss, W. (2008). *Partial differential equations: An introduction* (John Wiley and Sons).

Synge, J. and Schild, A. (1978). *Tensor calculus* (Dover).

Tao, T. (2008). The Fourier transform, in T. Gowers (ed.), *The Princeton companion to mathematics* (Princeton University Press).

Thornley, M. (2001). The mathematics of George Green, in *George Green, mathematician and physicist: The background to his life and work*, 2nd edn. (SIAM), pp. 183–204.

Thurston, R. N. (1969). Definition of a linear medium for one-dimensional longitudinal motion, *The Journal of the Acoustical Society of America* **45**, pp. 1329–1341.

Trèves, F. (1980). *Introduction to pseudodifferential and Fourier integral operators*, The University Series in Mathematics (Plenum).

Trèves, F. (2006). *Basic linear partial differential equations* (Dover).

Udías, A. (1999). *Principles of seismology* (Cambridge University Press).

Zimmer, R. (1990). *Essential results of functional analysis* (The University of Chicago Press).

Index